Critical Geopolitics of the Polar Regions

Focusing on both Polar Regions, this book provides a comprehensive understanding of political processes related to the rapidly changing Arctic and Antarctic, where the environmental impacts of human activities are extremely visible.

Environmental changes in the Arctic and the Antarctic are increasingly seen as barometers of the global impact of human activities, while newly arising economic opportunities in both Polar Regions prompt predictions that they will be the site of future conflicts. This book maps and analyses the different actors involved in the politics of the Polar Regions to explain why similar patterns of interpretation of such major issues have become dominant in practical, popular and formal geopolitical discourses. Disentangling the politics, the author illustrates how the ordering principles have evolved, explains recent dynamics in political processes and provides the groundwork needed to better forecast future trends. By focusing on the Americas, the only continent that borders both Polar Regions, the author shows how geographic proximity inspires interaction and cooperation among state and non-state actors in very different ways.

This volume will be of interest to scholars and students of political science, political geography, international relations, global governance and cultural studies. It will have an international appeal particularly in the Americas, and other countries with growing interests in the Polar Regions.

Dorothea Wehrmann studied Social Sciences with a major in Sociology, InterAmerican Studies and Political Communication at the University of Osnabrück and Bielefeld University (Germany). She holds a PhD in Political Science, is affiliated with the Center for InterAmerican Studies, the Institute for World Society Studies and the Bielefeld Graduate School in History and Sociology and works as a researcher at the German Development Institute.

InterAmerican Research: Contact, Communication, Conflict
Series Editors:
Olaf Kaltmeier, Josef Raab, Wilfried Raussert, Sebastian Thies

The Americas are shaped by a multitude of dynamics which have extensive, con-flictive and at times contradictory consequences for society, culture, politics and the environment. These processes are embedded within a history of interdepen-dence and mutual observation between North and South which originates in the conquest and simultaneous 'invention' of America by European colonial powers.

The series will challenge the ways we think about the Americas, in particu-lar, and the concept of area studies, in general. Put simply, the series perceives the Americas as transversally related, chronotopically entangled and multiply interconnected. In its critical positioning at the crossroads of area studies and cultural studies the series aims to push further the postcolonial, postnational, and cross-border turns in recent studies of the Americas toward a model of horizontal dialogue between cultures, areas, and disciplines.

The series pursues the goal to 'think the Americas different' and to explore these phenomena from transregional as well as interdisciplinary perspectives.

Entangled Heritages
Postcolonial Perspectives on the Uses of the Past in Latin America
Edited by Olaf Kaltmeier and Mario Rufer

Mobile and Entangled America(s)
Edited by Maryemma Graham and Wilfried Raussert

Practices of Resistance
Narratives, Politics, and Aesthetics across the Caribbean and its Diasporas
Edited by Wiebke Beushausen, Miriam Brandel, Joseph Farquharson, Marius Littschwager, Annika McPherson and Julia Roth

Political Protest and Undocumented Immigrant Youth
(Re-)framing Testimonio
Stefanie Quakernack

Critical Geopolitics of the Polar Regions
An Inter-American Perspective
Dorothea Wehrmann

For more information about this series, please visit https://www.routledge.com/Inter American-Research-Contact-Communication-Conflict/book-series/ASHSER-1426

Critical Geopolitics of the Polar Regions

An Inter-American Perspective

Dorothea Wehrmann

Routledge
Taylor & Francis Group

LONDON AND NEW YORK

First published 2019
by Routledge
2 Park Square, Milton Park, Abingdon, Oxon OX14 4RN

and by Routledge
605 Third Avenue, New York, NY 10017

First issued in paperback 2021

*Routledge is an imprint of the Taylor & Francis Group, an informa
business*

© 2019 Dorothea Wehrmann

Publisher's Note
The publisher has gone to great lengths to ensure the quality of this reprint but
points out that some imperfections in the original copies may be apparent.

British Library Cataloguing-in-Publication Data
A catalogue record for this book is available from the British Library

Library of Congress Cataloging-in-Publication Data
A catalog record has been requested for this book

Typeset in Times New Roman
by Apex CoVantage, LLC

ISBN 13: 978-1-03-209437-3 (pbk)
ISBN 13: 978-1-138-48581-5 (hbk)

In memory of Tante Margret

Contents

Figures

Tables

Boxes

Note

Unless stated otherwise, all translations and figures are provided by the author.

Foreword

Mathias Albert

There can be little doubt that the Arctic and the Antarctic have much in common: For what are the most extensive parts of Earth's cryosphere, few glaciologists would probably doubt that statement. Even beyond that, many – particularly natural science – disciplines also would find that many relevant research subjects within their purview quite naturally extend to both Polar Regions. In the social sciences, and in particular in political science and International Relations, the focus on *polar* issues that extend to both Polar Regions seems much less pronounced, at the expense of more attention drawn to either the Arctic or the Antarctic, and over recent years probably far more to the former than the latter. Of course, from a general social science perspective, there is a marked difference between the two: the Arctic has for a long time been inhabited by humankind, while the Antarctic is the habitat of a more recent, mostly fleeting, and quite specialised human population (although both areas have for a long time been the subject of human phantasies and projections elsewhere).

Needless to say, a focus on one rather than on both of the Polar Regions is also warranted by the fact that both are subject to vastly different political histories and legal regimes. Regardless of many other issues, the Antarctic Treaty, for example, effectively has prevented anything like the "new Cold War" geopolitical discussions that have become prominent in relation to the Arctic over recent years. Nonetheless, it is *both* Polar Regions that have entered the purview of International Relations and other social science scholars through the realisation that due to the effect of climate change (as well as, in turn, their role in that process) they constitute *global regions*. However, it is still probably the case that particularly IR scholars who do not have a wide expertise in studying global climate change (or are of the rare breed that even specialises in the politics of one of the Polar Regions), have little clue as to the general issues, let alone the more intricate details of either Arctic or Antarctic politics – and how wider, then, the lack of understanding and the research gap when it comes to both realms of polar politics, or even *inter*-relations between them.

The present book's prime achievement is that it takes a great leap forward in filling that gap. Probably first among other achievements, Dorothea Wehrmann does not merely provide a comparative analysis, but rather focuses on what she calls polar *entanglements*. While not subduing differences, the emphasis here is on a range of similarities in what is a complex, yet globally connected discursive space of both state and non-state actors. Through meticulous empirical analysis, Dorothea convincingly demonstrates that when it comes to various forms of geopolitical reasoning, many of the geopolitical imaginaries and interpretations regarding the two Polar Regions actually overlap and in this sense are "entangled". While this analysis would provide a range of interesting insights in and for itself, it is further enriched by actually not being global in the sense of seeking to analyse polar discourses everywhere, but by inserting this analysis in a research framework that particularly focuses on *inter-American* entanglements. Showing the insertion of Argentinian, Canadian, Chilean and US discourses in global geopolitical discourses, and how these impact on policy processes and identity constructions is a first in the social science literature on the Polar Regions. It occupies a space between more globally or more locally oriented analyses, while on an analytical level in between those does not lose sight of either. In addition, it is a study that shows how studies on the Polar Regions can fruitfully be related to other thematic fields. It is in this sense that I not only see this as useful study in and for itself, but as hopefully an indicator that the analysis of the Polar Regions in the social sciences is in a process of maturing and moving forward.

Acknowledgements

"There is no planet B" emphasised the former United Nations Secretary General Ban Ki-Moon at the 2016 Arctic Circle Conference in Reykjavík, Iceland, stressing the global significance of the Arctic's rapidly melting sea ice. Not the first person (though one of the more prominent people) to connect the fate of the Arctic with the fate of our planet, Ban Ki-Moon reproduced the growing understanding of the Polar Regions as "barometers", based on the extraordinary visibility of climate change effects in two of the earth's iciest regions – most notably exemplified by the recent collapse of the Larsen C ice shelf in Antarctica and the record low extent of the Arctic sea ice as measured periodically. Over the past decade, however, growing interest in the Polar Regions has been inspired not only by environmental changes in the Arctic and Antarctic, but also by related political processes. Popularising it under the imaginary of a "scramble for polar riches", the international media have hyped the intensifying interest in these areas shown by both polar-rim states and other more distant ones. For a long time too, scholars have mostly focused on the Polar Regions' potential for international conflict and cooperation. As a result, a rather black-and-white picture of the politics of the Polar Regions has often been sketched in analyses that have focused predominantly on the clashing interests of states and neglected the numerous processes that led to the formulation of these interests in the first place.

The writing of this book was particularly motivated by the observation that policy making and action taking in the Polar Regions have effects that are more and more likely to be global in scope. With that in mind, this study takes the global significance ascribed to the changing Polar Regions as its point of departure for an attempt to trace the evolution of views and interests expressed by state and non-state actors with regard to the Arctic and the Antarctic. In so doing, this study "scrambles" and at the same time "(dis)entangles" dominant discourses on the Polar Regions inasmuch as it places the geographically opposite, and in many ways entirely different, Arctic and Antarctic in a relational perspective. By pushing back against the majority of the literature, which focuses on political processes in either the Arctic or the Antarctic, the "entanglement perspective" developed in this study aims to provide a more differentiated understanding of past and recent dynamics in the politics of both.

This book is an edited version of a dissertation accepted by the Faculty of Sociology at Bielefeld University in January 2017. Research for it started with my employment at Bielefeld University in April 2013. Funded by the German Academic Exchange Service and by the Federal Ministry of Education and Research, I was able to conduct 12 weeks of field work in Argentina, Chile, Canada and the U.S. in 2014 and 2015. This study benefitted greatly from the insights provided by 26 interviewees who shared their information and their time with me and to whom I owe very special thanks for their willingness to speak to me, their trust and their general helpfulness. The conceptualisation of this book was greatly inspired by discussions with my colleagues at the Center for InterAmerican Studies, particularly those from the Entangled Americas project. I am extremely grateful for this very supportive interdisciplinary research environment, and for the invigorating exchanges of knowledge with my colleagues and friends, from whom I learned a great deal. Thank you all – Johannes Bohle, Martin Breuer, Clara Buitrago, Joseph Farquharson, Clara Gläve, Cruz Gonzalez, Yaatsil Guevara Gonzalez, Jochen Kemner, Marius Littschwager, Alexander Mosena, Mirko Petersen, Julia Roth, Anne Tittor, Klaus Weinhauer and particularly my always encouraging office mate Paul-Matthias Tyrell. I also owe many thanks to Angelika Epple, Olaf Kaltmeier and Willy Raussert for establishing this research group and to Lukas Rehm for all his support in coordinating our research team.

I owe a deep debt of gratitude to my supervisors, particularly to Mathias Albert, who drew my attention to the Polar Regions in the first place, trusted me to carry on his idea for a project, gave me the freedom to develop my own perspective without laying down any limits to my thought and, at the same time, supported me with prompt and helpful advice whenever I asked for it. Similarly, I wish to thank Christoph Humrich for his continuing interest in my research, for his ever helpful advice and particularly for his encouraging words at different points in time that kept me motivated and helped me more than he probably thought they would to actually finish the dissertation. This book, moreover, benefitted greatly from Mathias and Christoph's close reading of the thesis and their very helpful suggestions for revising the manuscript.

I am also indebted to the many insightful comments made by colleagues in research classes supervised by Lutz Leisering and those affiliated with the Bielefeld Graduate School in History and Sociology and the Institute for World Society Studies, all of whom inspired my thinking with their critical and challenging interventions – many thanks particularly to Kerstin Eppert, Thomas Müller and Katrin Weible for their unfailingly helpful ideas. Many thanks are also due to colleagues who commented on my research at conferences and particularly to my Arctic-research peers Golo Bartsch, Kathrin Stephen and Sebastian Knecht, who supported my studies not only with well-thought-out comments and questions but also with their friendship. I also wish to thank Stephen Curtis whose careful copyediting contributed very much to the readability of this book.

I am in no doubt at all that I could not have written this book without the encouragement provided outside academia by my family and friends, who have not only maintained an interest in my work over the past years but, more importantly, supported me by sharing different steps and challenges in *their own* lives, thus keeping me from getting stuck in what German doctoral students call the "thesis tunnel"! Thank you! And particularly I wish to thank Sebastian: There are no words to express my gratitude for all your encouragement and altruism – thank you for being such a wonderful partner in my life!

Irrespective of all shortcomings and errors that only I can be held responsible for, this book will, hopefully, contribute to informed conversations on how we approach the changing Polar Regions. It is also my hope, more generally, that this book will support a growing awareness of, and interest in, political processes taking place in areas as distant and remote as the Arctic and the Antarctic, areas so different from most others that they are difficult to identify with – but of vital importance, nonetheless, as policy making there will ultimately affect, and likewise be affected by, the lifestyles of those living many thousand kilometres distant from them.

Bielefeld, August 2018

Abbreviations

AAC	Arctic Athabaskan Council
AC	Arctic Council
ACGF	Arctic Coast Guard Forum
ACIA	Arctic Climate Impact Assessment
AEC	Arctic Economic Council
AEPS	Arctic Environmental Protection Strategy
AHDR	Arctic Human Development Report
AHHEG	Arctic Human Health Expert Group
AIA	Aleut International Association
AMAP	Arctic Monitoring and Assessment Programme
ANWR	Arctic National Wildlife Refuge
AORF	Arctic Offshore Regulators Forum
ASOC	Antarctic and Southern Ocean Coalition
ATCMs	Antarctic Treaty Consultative Meetings
ATCPs	Antarctic Treaty Consultative Parties
ATS	Antarctic Treaty System
CAFF	Conservation of Arctic Flora and Fauna
CCAMLR	Commission for the Conservation of Antarctic Marine Living Resources
CCU	Circumpolar Conservation Union
CEP	Committee for Environmental Protection
CEPAL	Comisión Económica para América Latina y el Caribe
CFR	Council on Foreign Relations
CIGI	Centre for International Governance Innovation
CLCS	Commission on the Limits of the Continental Shelf
COLTO	Coalition of Legal Toothfish Operators
COMNAP	Council of Managers of National Antarctic Programs
COPESA	Consorcio Periodístico de Chile Sociedad Anónima
CRAMRA	Convention on the Regulation of Antarctic Mineral Resource Activities
CSIS	Center for Strategic and International Studies
DFAIT	Department of Foreign Affairs and International Trade
DIFROL	Dirección de Fronteras y Límites del Estado

DNA	Dirección Nacional del Antárctico
EEZ	Exclusive Economic Zone
ENGOs	Environmental Non-Governmental Organisations
EPPR	Emergency, Prevention, Preparedness and Response
EU	European Union
GCI	Gwich'in Council International
GEC	Global Environmental Change
IAA	Instituto Antártico Argentino
IAATO	International Association of Antarctic Tour Operators
IACHR	Inter-American Commission on Human Rights
ICC	Inuit Circumpolar Council
ICSU	International Council for Science
IMO	International Maritime Organization
INACH	Instituto Antártico Chileno
IPCC	Intergovernmental Panel on Climate Change
IPY	International Polar Year
IR	International Relations
IUCN	International Union for Conservation of Nature and Natural Resources
IUU	Illegal, Unreported and Unregulated Fishing
Mercosur	Mercado Común del Sur
MSR	Minister's Special Representative
NCM	Nordic Council of Ministers
NEP	Northeast Passage
NGOs	Non-Governmental Organisations
NSPD/HSPD	National Security Presidential Directive and Homeland Security Presidential Directive
NSR	Northern Sea Route
NWP	Northwest Passage
OAS	Organization of American States
OGP	Association of Oil and Gas Producers
PAME	Protection of the Arctic Marine Environment
POLAR	Polar Knowledge Canada
PPs	Permanent Participants
RAIPON	Russian Association of Indigenous Peoples of the North
RAPAL	Reunión de Administradores de Programas Antárticos Latino-americanos
SAOs	Senior Arctic Officials
SAR	Search-and-Rescue
SATCMs	Special Antarctic Treaty Consultative Meetings
SCAR	Scientific Committee on Antarctic Research
SCPAR	Standing Committee of Conference of Parliamentarians of the Arctic Region
SDWG	Sustainable Development Working Group
SORS	Southern Oceanic Rim States

UN	United Nations
UNASUR	Unión de Naciones Suramericanas
UNCLOS	United Nations Convention on the Law of the Sea
WMO	World Meteorological Organisation
WWF	World Wildlife Fund

1 Introduction

Critical Geopolitics of the Polar Regions and an inter-American perspective

1.1 Significance of the changing Polar Regions and of polar politics

"Poles apart" and "antipodes" are popular expressions that derive from, and have often been applied to, the Arctic and its counterpart and opposite, the Antarctic. Both the "ends of the earth" are known for their remote and icy environments and increasingly for their profound impacts on the earth's climate. The multiple disparities between the Arctic and the Antarctic that have to do, amongst other things, with the governance of the regions are often identified as the reason why most social scientists focus on either the one or the other when analysing the politics of the Polar Regions. This book challenges this widespread point of view.

> It is true that the two poles are not identical, but they do have a lot in common and much can be learned by comparing the two.
>
> (Jorge Taiana, on behalf of the Argentine delegation on the occasion of the 32nd Antarctic Treaty Consultative Meeting, 2009)

It builds instead on the perspective promoted by the former Argentine Minister of Foreign Affairs, Jorge Taiana, and argues that to talk about and compare the Arctic and the Antarctic solely from the point of view of their being mutual opposites is to ignore the many entanglements that shape political processes connected with them. Why, for instance, are both the Arctic and the Antarctic most often regarded as regions that attract either political cooperation or political conflict? Why do political actors in the Arctic and Antarctic emphasise their polar identity despite the fact that both regions are predominantly known for being icy cold places and are represented as the world's final "frontiers", whose exploration, exploitation and investigation have always been challenging to humans due to their difficult accessibility? Why too have concerns relating to sovereignty, resource exploration, environmental conservation, international cooperation and sustainable development been so much discussed at different points of time in the politics of the Arctic and Antarctic?

> While probably in terms of sheer numbers, the largest scale human impacts of future climate instabilities will affect the poor in 'the South', it is now

becoming clear that some of the most dramatic climate impacts and environmental consequences of carbon-fueled industrialization are affecting the
polar regions.

(Dalby, 2003, p. 189)

In focusing on both Polar Regions, this book provides a more comprehensive
understanding of political processes that relate to the rapidly changing Arctic and
Antarctic – two regions which, as Simon Dalby observes, exhibit the environmental impacts of human activities to the most extreme degree. As early as the year
2000, Monica Tennberg had already concluded that "the problem of the environment is, above all, a problem of order" (p. 125), which is also why this study is
driven by the aim of exemplifying how the prioritisation of environmental concerns in the Polar Regions is entangled with the geopolitical reasoning that has
shaped their politics. This book, therefore, helps its readers understand how the
ordering principles operate that ultimately affect and guide policy making and
influence collective and individual actions at a time when the effects of climate
change and the related environmental changes in the Polar Regions are increasingly considered as matters of global concern.

More specifically, the aims of this work are threefold. First, it examines systematically how actors represent the changing Arctic and Antarctic in discourses
in order to outline corresponding and conflicting positions and views and trace
their emergence in different contexts over a period of more than 25 years. Second,
by taking a relational perspective,[1] it sketches different patterns of interpretation that have become dominant in the politics of both Polar Regions and shows
how a similar reasoning has been applied to both at different points in time and
how this has influenced negotiations on policies and political actions to manage
the changing Arctic and Antarctic. Third, unlike the majority of studies, which
derive predominantly from a focus on either the Arctic or the Antarctic, this book
applies an inter-American regional perspective and consequently also considers
the space *in between* both Polar Regions. The examples from the Americas show
how political dynamics originating outside the Polar Regions have an impact on
policy making in them and concerning them – one example being the experience of domination during a colonial past. In focusing on the double continent
of the Americas – the only continent that borders both Polar Regions – this book,
moreover, shows how geographic proximity inspires interaction and cooperation
among state and non-state actors in American polar-rim countries very differently.
Overall, and especially by addressing these central gaps in research, the book
identifies entanglements (and disentanglements) among actors and discourses that
help to explain why representations and actor constellations in the politics of the
Polar Regions have changed.

But what exactly characterises the "changes" that are taking place in the Polar
Regions? In terms of climate, the mean average temperature in the Arctic and Antarctic is rising at twice the rate that it is in other regions, which is why both Polar
Regions are increasingly depicted as "climate change barometers" (cf. Hansom
and Gordon, 2014; Shadian, 2014). Although manifest in particular localities, the

warming effects relate to the phenomenon called Global Environmental Change (GEC), by which the Polar Regions are considered to be most vulnerable to and most affected (cf. Intergovernmental Panel on Climate Change, 2014). The record lowest seasonal minimum ice extents observed almost annually in the Arctic are arguably the most widely known effect of the warming climate. The environmental and ecological changes caused by the warming climate are numerous and range from melting sea ice and glaciers and collapsing ice shelves to rising sea levels, coastal erosion, the migration of species and changing ecosystems. In Antarctica, these changes are "comparably dramatic" although they attract "less international attention" (Ali and Pincus, 2015, p. 1). Environmental changes in the Arctic, however, have already led "to more radical social and economic consequences" (Duyck, 2015, p. 27). In this regard, environmental changes in the Polar Regions are, to different degrees, considered a key driver – or a "threat multiplier" (Bratspies, 2015, p. 175) – of a variety of economic, social and political changes as well as of security concerns that have also been addressed in Arctic and Antarctic politics over the past decades.

Economically, the Polar Regions are often depicted as being rich in resources. The often cited U.S. Geological Survey (2008), still considered the most telling contribution of its kind, states that the Arctic is "the largest unexplored prospective area for petroleum remaining on earth" with oil and gas deposits that account for "almost 10 percent of the world's known conventional petroleum resources" (p. 1) of which 84 percent are presumed to be located in offshore areas. The loss of sea ice and consequently the easier access to offshore fossil fuels creates the so-called "Arctic Paradox" (see e.g. Palosaari, 2016) as the use of fossil fuels enhances global warming and thus further contributes to the ice melting. Extensive seismographic examinations are still needed to verify these estimates and offshore exploration is associated with significant environmental risks in the Arctic. However, these predictions and the growing accessibility of resources have already fostered the interest of various actors in conducting explorations. This interest in Arctic oil and gas development has caused political struggles between corporate actors, environmental non-governmental organisations (ENGOs), regulators and Indigenous peoples groups. Oil, gas, iron ore, chromium, copper, gold, nickel, platinum, other minerals and coal are also presumed to exist in the Antarctic (U.S. Geological Survey, 2016). And although the development of mineral resources is prohibited in Antarctica, the exploitation of oil is also a topic of concern in the sub-Polar Regions (cf. the Islas Malvinas/Falkland Islands dispute between Argentina and the United Kingdom). Fishing and whaling are perennially discussed in both Polar Regions, particularly regulation of both the scientific and the traditional use of these living resources, as well as the increasingly observable amount of illegal, unreported and unregulated (IUU) fishing. Both regions have experienced an almost constant increase in tourism in recent decades while, in the Arctic, freight navigation is also expected to grow in the mid term as the Northwest Passage and the Northern Sea Route are becoming more reliably negotiable in the summer season, which will shorten travel distances and be of considerable significance for the transportation of extracted resources.

In both regions too, social changes are being triggered by the transformation and changing use of land and sea. Coastal erosion and the migration of species, for instance, are forcing the about four million people who live in the Arctic – about one tenth of whom are Indigenous peoples – to alter their ways of life: to change their diet because reindeer and fish are more difficult and dangerous to hunt and catch, and because pollutants, released by the melting ice, are contaminating the food chain; to move to other regions because the melting permafrost no longer provides solid ground for their settlements; and to adapt to growing Arctic industries, which are able to develop resources formerly covered by ice, bring new employment opportunities, motivate labourers to migrate to the Arctic regions and encourage investments in local infrastructure. While Antarctica, on the other hand, has no permanent population, social changes are visible in the sub-Antarctic regions and are, for example, related to the observation and scientific exploration of the changing environment. Expeditions to Antarctica mostly depart from Punta Arenas (Chile) and Ushuaia (Argentina), where Antarctica is increasingly outlined as a reference point for cultural identity. The growing numbers of tourists and scientists departing from both cities facilitate the identification of Ushuaia and Punta Arenas as "gateway cities to Antarctica" and have generated investments in the local infrastructure, such as tourist accommodation, university courses, air- and seaports.

In terms of security, journalists, politicians and scholars have often depicted the geopolitical interests associated with the economic potential ascribed to the Polar Regions and also overlapping territorial claims as eventually leading to an escalation of conflict (e.g. Potapov, 2009; Byers, 2009; Beck, 2014). For the Arctic, these predictions boomed particularly in the years 2007 and 2008, when a scientific expedition from Russia planted a flag on the North Pole, when the U.S. Geological Survey was published and the Canadian Navy and Coast Guard resumed their annual Arctic exercise (Operation Nanook). Against this backdrop, the remilitarisation of the Arctic has often been associated with the emergence of a new Cold War. Newspaper articles published in the 2000s and early 2010s in American polar-rim states, for instance, carried headlines such as "An Ice-Cold War", "El Ártico se derrite y desata otra Guerra Fría entre varios países", "Race Is On as Ice Melt Reveals Arctic Treasures", "El Ártico, en peligro", or "As Polar Ice Turns to Water, Dreams of Treasure Abound".

These "Cold War" and "Gold Rush" narratives, as the researchers Le Mière and Mazo (2013, p. 9) label them – Palosaari (2012) uses the term "Arctic 'boom' perspective" instead – also played a significant role in public political and scientific discourses in regard to the Antarctic more than 20 years earlier. The Antarctic Treaty (a "product of cold war rivalry", Press, 2014, p. x) had already provided one answer to the question of "Who owns Antarctica?" in 1959, as its signatories accepted that all territorial claims in Antarctica would be treated as "frozen" from that point on. In the 1980s, however, a debate similar to the more recent one on the Arctic evolved around the "Gold Rush for South Pole Wealth" (Beck, 2014). Particularly during the negotiations for the Convention on the Regulation of Antarctic Mineral Resource Activities (CRAMRA), the consultative parties to the Antarctic

Treaty faced much criticism for their interest in developing resources there. Ultimately, the ratification of CRAMRA failed due to the resistance of two of the consultative parties, Australia and France, whose disapproval was later traced to the major international "World Park" campaigns conducted by environmental groups such as the World Wildlife Fun (WWF) and Greenpeace (British Antarctic Survey, 2015). Antigua and Barbuda together with Malaysia raised the "Question of Antarctica" at the United Nations in 1983 on the basis of their opposition to mineral resource development. They demanded "a more democratic, accountable, and transparent management regime" for Antarctica than that defined in the Antarctic Treaty System. As a consequence of this debate, between 1983 and 2005 the governance of Antarctica was placed on the General Assembly's agenda on a regular basis (cf. Beck, 2004, p. 210). After CRAMRA had failed in 1991, the Protocol on Environmental Protection to the Antarctic Treaty (the so-called Environmental/ Madrid Protocol) was signed, under the terms of which the exploitation of non-living resources is prohibited in Antarctica until at least 2048. Since then, only a minority has been still of the opinion that conflicting geopolitical interests will inevitably provoke a future war in or over the Antarctic (cf. Abdel-Motaal, 2016; *The Economist*, 2015). In the sub-Polar Regions, however, the disputed ownership of the islands Picton, Lennox and Nueva in the Beagle Channel nearly led to war between Argentina and Chile in the late 1970s. The conflict was resolved in 1984. The Falkland Islands/Islas Malvinas conflict between the United Kingdom and Argentina, though, escalated in 1982 and resulted in a war that lasted 74 days. Since then, the islands have remained under British control, which Argentina and its regional partners regard as "a case of colonialism" (Argentine Republic, 2015). They have constantly addressed this matter in all the international forums of which both Argentina and the United Kingdom are members. More recently, applications handed in to the Commission on the Limits of the Continental Shelf (CLCS) by various polar-rim states that aim to extend their exclusive economic zones (EEZs) link the dominating public debates on the governance of resource extraction and territorial sovereignty in the Polar Regions.[2]

As the different processes mentioned earlier demonstrate, the melting "polar caps" not only have an impact on the Earth's climate but are also associated with social, economic and political interests that have been formulated by different actor groups located remote from the Arctic and Antarctic (as illustrated, for instance, by the numerous policy and strategy papers published by non-polar-rim states, such as Germany, France, China and the Netherlands). Moreover, these interests are to different degrees entangled with and shaped by processes, visions and events that relate to and occur in places far away from the Polar Regions. In this regard, Bratspies (2015, p. 171) introduced the term "feedback loops" to describe how, in the case of the changing Arctic, the region is connected to "events unfolding in other parts of the world" as "decisions made elsewhere increasingly influence changes driven by the twin pressures of climate change and globalization". In a similar vein, Stokke (2015, p. 334) argues that "[m]any of the governance challenges drawing the attention of more and more Arctic or global institutions derive from processes and developments outside the region".

Although both Polar Regions have long been described as regional and international arenas, it is particularly against this backdrop that they are increasingly being examined in "global perspectives" and as being "globally embedded" (cf. Keil and Knecht, 2017; Triggs and Riddell, 2007). Nevertheless, political processes in the Arctic and Antarctic are still mostly analysed separately. Although the term "Arctic [not Polar!] Change" (see Stepien et al., 2015; Tennberg, 2015), is, for instance, related to "a combination of complex, interrelated environmental, social, cultural, economic and political transformations and efforts to tackle them and adapt to them" (Tennberg, 2015, p. 408), its focus is mostly limited to the Arctic region even though the Arctic's "internal networks" are acknowledged to be "connected to global markets" (ibid., p. 411).

This book does not challenge the perception that some processes apply solely in the Arctic or the Antarctic, nor does it neglect the various disparities between the two Polar Regions or the fact that both are shaped by "differing currents of change" (Hemmings, 2015, p. 66). It argues, however, that, in order to understand the political dynamics in the Polar Regions that shape the above-mentioned transformations, it is also necessary to speak about and investigate *polar entanglements*. Against this backdrop, this book provides evidence that the social changes, economic changes and security concerns are not solely inspired by actors and discourses in and beyond the Polar Regions, but also by a similar and recurring geopolitical reasoning. In this respect, and following the approach of Critical Geopolitics, this book does not regard "geopolitics" as being solely about power over territory but as also covering the geographical understandings and reasonings that influence representations of environmental changes in the Polar Regions. The more rapidly changing environment is increasingly addressed in positions and priorities negotiated in the politics of the Polar Regions in and beyond the Arctic and the Antarctic and is understood as triggering all the social and economic changes and security concerns touched upon before. The changing environment thus constitutes a common denominator in the politics of the Polar Regions, in the shape of recognition of a "shared responsibility" (cf. Bastmeijer, 2015; Murray and Nuttall, 2014; Tanaka, 2014). As is shown throughout this book, this common denominator has strengthened cooperation among state and non-state actors even in times of crisis. This cooperation has been formalised in the regional and international governance structures that focused particularly on the protection of the environment and were formed in increasing numbers after 1989 (Palosaari and Tynkkynen, 2015; Stokke, 2015).

To put this into context. The end of the Cold War is widely regarded as having laid the foundation for improved dialogue and collaboration among Arctic states. President Mikhail Gorbachev's famous Murmansk speech of 1987 is often considered to have been a turning point in international relations and a "catalytic event" for institutional dynamics in the Arctic (Dodds and Nuttall, 2016, p. 109). The year 1989, however, marked a significant shift for both Polar Regions, because, as outlined earlier, the end of the Cold War and the failure of CRAMRA are generally regarded as having had a significant impact on how the politics of the Polar Regions have been shaped since. And, although environmental issues had previously served as a "'soft', non-ideological, almost non-political theme which started new cooperation between East and West" (Palosaari and Tynkkynen, 2015, p. 88), the *Bahía Paraíso* and *Exxon Valdez* spills in January and March 1989, together with the sinking of the Soviet nuclear submarine *Komsomolets* in the Barents Sea in April 1989 stimulated greater awareness of the fragility of polar ecosystems (Graczyk and Koivurova, 2015). Due to the public attention they received, these accidents are even regarded as the events that made the U.S. become a member of the Arctic Council that it was at first reluctant to join (Young, 1998). At the same time, growing attention to environmental concerns was also noticeable globally. In 1992, for instance, the United Nations Conference on Environment and Development took place, at which the *Climate Change Convention* was agreed upon, the precursor of the *Kyoto Protocol* (1997). But, even "before the Cold War was over, states worldwide had begun to recognize the seriousness of regional, continental and global scale environmental problems" (Castree, 2003, p. 429), as exemplified by the United Nation's (UN) Stockholm Conference on Environment and Development of 1972, which created the UN Environmental Programme, the *Convention on International Trade in Endangered Species* (1973), and the *UN Convention on the Law of the Sea* (1982). Obviously, these processes, which have led to intensified international cooperation in the Arctic and the Antarctic in recent decades, complicate the still prominent perspective promoted in popular science and by numerous politicians and journalists, who describe the "Arctic as a battleground" (Emmerson, 2010), speak of "resource wars" (Howard, 2009), and as a place where geopolitical interests clash between East and West (Bittner, 2016).

This book, though, does not argue that geopolitical interests in the Polar Regions have diminished, but that they have changed. It illustrates how geopolitical positions are constructed and negotiated under an additional frame, as they are no longer seen as solely serving the interests of any actor involved in the politics of the Polar Regions but as having an influence on how the Polar Regions are changing and how the globe is changing too. In the words of the geographer Noel Castree (2003, p. 427):

> [T]here is a geopolitics to how environmental problems are represented. [. . .] While these problems are undoubtedly real, there is no objective perspective on their nature, causes, and solutions. Instead, we have an array of actors – such as states, NGOs, quasi-governmental bodies, and environmental scientists – all claiming to know the 'truth' about these problems (to the extent that what is defined by some actors as an environmental 'problem' is not seen as one by others).

Consequently, and with regard to both Polar Regions, this book argues that representations of environmental changes need to be perceived as being constructed differently in order to impact how priorities are set in the politics of the Polar Regions.

1.1.1 Conceptual matters

Because significant events and processes in policy making have taken place at different times in both regions, this book pays particular attention to the politics of the Polar Regions since 1989. Moreover, it concentrates especially on state and non-state actors who have contributed to the formulation of policy and strategy papers in the main regional "polar" forums (the Arctic Council and the Antarctic Treaty System) and in American polar-rim states (Argentina, Canada, Chile and the U.S.). Although the politics of the Polar Regions are shaped by various actors, institutions, processes and events beyond their boundaries, the double continent of the Americas is directly affected by changes in both Polar Regions. As Dodds (1997, p. 213) pointed out, geographic proximity has been the "key justification for considerable interest in the southern polar region's environmental and political management" in the case of the Southern Oceanic Rim States (SORS) too, and scholars focusing on transnationalisation processes have found that geographic proximity is a particular factor encouraging interactions and cooperation among different actors (cf. Mann, 2006). Despite their differing experiences of and political dealings with the changing Arctic and Antarctic, this book outlines similar tendencies that are observable in Canada, the U.S., Argentina and Chile in national policy formulation processes. Further, growing inter-American political and economic links make it possible to examine whether or not entanglements between the politics of the Arctic and Antarctic transcend national boundaries and constitute the formation of a new political sphere in the Americas.

Prominent examples of already existing inter-American entanglements are the *Inter-American Treaty of Reciprocal Assistance* (1947), the formation of the Organization of American States (1948), the establishment of the Inter-American Court of Human Rights (1979), the Reunión de Administradores de Programas Antárticos Latinoamericanos (RAPAL, 1990), the establishment of the Mercado Común del Sur (Mercosur, 1991) and of the Unión de Naciones Suramericanas (UNASUR, 2008). Whether or not the changing Polar Regions have also contributed to the formation of a new inter-American sphere is a question regarded as being of particular significance since, obviously, the representation of changing local and national positions can have implications for the interests of other neighbouring states and the international community at large and is strongly influenced by the experience of regional environmental changes (as is shown throughout this book in the Polar Regions).

This book is accordingly based on the assumption that, in order to understand why changes take place in representations in discourses and in the constellation of actors, it is also necessary to consider how political positions and agendas transcend the domestic and regional spheres. Different from the majority of studies that particularly derive from a focus on the Arctic or the Antarctic, this book applies an inter-American regional perspective to also consider the space in-between both Polar Regions. This novelty is of particular significance to illustrate how political dynamics originating outside the Polar Regions have an impact on policy making in and on the Polar Regions, such as the experience of domination in a colonial past.

1.2 Terminology

Three central subjects are under analysis in this book:

1 the Polar Regions,
2 polar entanglements, and
3 geopolitical discourses.

The meanings ascribed to these terms can be multiple and they are often applied differently in interpretations. Likewise, their meanings may depend on which of the various actors is using them and which context they relate to.

1.2.1 The Polar Regions

In this book, the Polar Regions are understood first and foremost as political regions in which "a limited number of states [are] linked together by a geographic relationship and by a degree of mutual interdependence" (Nye, 1968, p. vii) along with related non-state actors (Clavin, 2005). Moreover, as the political scientists Olav Schram Stokke and Geir Hønneland (2006, p. 21) pointed out, a shared identity is what particularly characterises a political region, which "must to a considerable extent be reflected in flows of interaction" among state representatives, groups and individuals. The Arctic has been understood as a political region especially since the end of the Cold War. Since then, political relations have been increasingly conducted on a circumpolar basis and are no longer predominantly determined by the Eastern and Western blocs.

The Antarctic, on the other hand, has mostly been perceived as a political region since the preparations for the landmark International Geophysical Year (1957–1958), which laid the foundations for the adoption of the Antarctic Treaty (1959) that "linked together" the 12 original signatories and has subsequently been acceded to by 41 other countries (ATS, 2015). By virtue of being political regions, both the Arctic and Antarctic have been strengthened in the past decades. Various institutions have been formed in which state representatives, non-state actors and individuals collaborate and interact in accordance with common interests. Even outside these institutions, the Arctic and the Antarctic are increasingly regarded as points of reference to circumpolar states.

In a way that is typical of "regionality" (Tennberg, 2000), specific issues of concern are also assigned, framed and addressed as problems (or solutions) for "the Arctic" and/or "the Antarctic". Both regions are sometimes used as referent objects for securitisation (Dodds, 2014), for instance when "the vulnerable Arctic environment" is constructed as a security referent (Palosaari and Tynkkynen, 2015, p. 89f.) or when the use of military force is threatened to secure particular claims to territory. The meanings and significance ascribed to the Arctic and Antarctic regions are nevertheless distinct and are expressed differently by the various actors involved in the politics of the Polar Regions. This book analyses the distinct ascriptions as well as the strategic use of prevalent imaginaries applied to both regions in geopolitical discourses. Despite the growing interdependencies observable in both and their shared sense of regionality, neither is understood as a closed system or container. Instead, the book argues that political developments in the Arctic and Antarctic need to be understood as being entangled with local, national, regional and global processes.

Among others, the political scientist Carina Keskitalo (2015) pointed out that the different political interests of the actors under analysis are also the reason for the differing official definitions provided for the geographical areas that the Arctic (and Antarctic) encompass. While, generally speaking, the Arctic and the Antarctic Circles are regarded as the geographical boundaries of the Polar Regions, no consistent political definition exists for either. Instead, states adjacent to both Polar Regions include different geographic areas in their understanding of what constitutes the Arctic and the Antarctic (cf. Annex). Given the "contingent conglomeration of bounded territories (state sovereignty), local sites of politics and governance, transnational Arctic politics, international institutions, and global discourse" Shadian and Olsen (2016, p. 236) understand the Arctic as a "region of regions". Consequently, and unless a different definition is specifically given, all the chapters in this book refer to the respective official definitions of the Arctic and Antarctic provided by the four countries under analysis, the Arctic Council and the Antarctic Treaty System in order to relate the findings of this analysis to the contexts to which they were meant to refer by the actors themselves.

One reason for the variety of definitions that exist is the general acknowledgement that environmental changes and "environmental problems do not respect the borders of states" (Tennberg, 2000, p. 125) but have an impact on areas that need to be defined according to the problem under analysis. Another reason is that definitions, which include the Arctic or Antarctic territorial and maritime claims of individual states, remain particularly contested in cases in which territorial claims overlap. This is why the several international agreements that apply to the Arctic and the Antarctic need also to be seen as being shaped by the distinct national perspectives of their signatories. The different definitions of places and spaces related to the Polar Regions and subsumed under the terms "Arctic" and "Antarctic" further indicate that the meanings and imaginaries ascribed to both regions are multiple or, to put it differently, that there are many "Arctics" (Berger, 2015, p. 65) – and also many "Antarctics".

1.2.2 Polar entanglements

In this book the term "polar entanglements" addresses two dimensions: first, actors who (re)produce a specific (geopolitical) reasoning and, second, discourses, in which this kind of geopolitical reasoning is introduced. In the context of the politics of the Polar Regions, the entanglement perspective is thus regarded as a conceptual tool that allows a better understanding of the changing political positions and power relations and stands in stark contrast to more materialist and realist accounts of polar politics. More specifically, it allows consideration of the complex relationships among and between spaces and places, actors, themes addressed in discourses, discourses themselves and events. In contrast to terms such as interconnectedness and interrelationships, the term "entanglement" inherently implies various dynamics at different levels, such as domination, dependencies, increasing and decreasing complexities between subjects and objects that are not static and can change over time. Further, "entanglement" is a versatile term in view of its range of application: it can apply to subjects and objects alike (e.g. entangled actors, entangled policies, entangled thematic areas, entangled imaginaries), does not privilege any level of observation (micro, macro, meso/local, domestic, regional or international) or actor group (for instance, it addresses individuals as much as societies, states and international organisations). Processes of entanglement and the related transformations are difficult to reverse as entanglements have an impact on the entities involved (such as individual and collective actors, structures, ideas, discourses and events).

Although other approaches, such as those focusing on transnational relations, also pay attention to relationships among actors, these approaches only consider relations to be transnational if actor relationships involve state actors and at least one non-state actor and only if these interactions are dense and cross borders (Pries, 2010, p. 10), but are not global in scope (Albert et al., 2009, p. 7). By contrast, the entanglement perspective is not limited in scope, but, like approaches focusing on transnational relations, also allows for analysis of network formations or the formation of political, social, economic, etc. spheres. Further, from a methodological point of view, the entanglement perspective binds all results to the material considered and thus allows an explicit reflection on the limitations and actual say of results. However, the relational entanglement perspective is not based on a clear-cut theoretical concept. Instead, it needs to be applied in combination with other concepts, such as the approach of Critical Geopolitics, to determine the subjects that are related to another. Otherwise, by bringing together different dimensions on an arbitrary basis, the entanglement perspective would rather cause fuzziness.

1.2.3 Geopolitical discourses

As outlined earlier, this book is interested in providing a better understanding of how state and non-state actors construct and constitute dominating interpretations and imaginaries in geopolitical discourses that relate to the changing Polar Regions. Discourses here are understood in much the same way as they are in the work of post-structuralist scholars (such as Michel Foucault, Norman Fairclough, Ruth Wodak or Ernesto Laclau and Chantal Mouffe). Similar to Foucault's understanding, for instance, discourses are regarded as statements that are connected with rules according to which statements can or cannot be made (cf. Foucault, 2015), as "relatively stable uses of language serving the organization and structuring of social life" (Wodak and Meyer, 2009, p. 6), which aim at fixing meaning (Laclau and Mouffe, 2001, p. 96). Discursive events, which relate to the question of how "one specific statement appeared rather than another" (Foucault, 2015, p. 21) are then understood as "instances of sociocultural practice" which pertain to social and institutional relations and (dominating) practices (Fairclough, 1992, p. 4). These ascriptions to discourses and to discursive events, to put it very briefly, all stress the significance of the contexts to which they relate, and of the settings and times in which they are/were conducted or occurred. As was already emphasised before, these contexts are taken into special account in this book, as references to information are generally regarded as never being neutral but as being embedded in context and always political (Keskitalo, 2015).

Although much has already been written on discourses and their significance for environmental politics and the politics of the Polar Regions, the analysis of geopolitical discourses in this book contributes to the removal of a blind spot still unaddressed in research on the politics of the Polar Regions: the evolution and (re)production of geopolitical reasoning put forward by different actors in discourses that concern the Arctic *and* the Antarctic and determine one another. But what is meant by geopolitical discourses and by the "slippery" term geopolitics?

Geopolitics is a so-called "floating signifier" (a word that relates not to one stable concept but to numerous concepts), which can generally be understood as "the struggle over hegemony in places and spaces" (Petersen and Wehrmann, 2015). A distinction is often made between a "classical" and a "critical" understanding of geopolitics. In "classical" geopolitics, geographical location and control over territory are regarded as being significant for the power of states. Accordingly, the world is divided into particular geographical zones (e.g. Eastern Europe, a "Pivot Area" in Halford Mackinder's Heartland Theory). Critical Geopolitics, on the other hand, does not understand geographical space as a natural given, but as a concept of order that is based on political logics of territorial representation (Reuber et al., 2012). In other words, Critical Geopolitics scrutinises how political power over territory is constructed through language. Scholars of Critical Geopolitics thus shed light on narratives, metaphors and visual images that are deployed for the purpose of labelling global space (e.g. "Iron Curtain") and that "generate a simple model of the world, which can be used to advise and inform foreign and security policy making" (Dodds, 2014, p. 5).

Hence, although geopolitics is first and foremost "about practice" (Ó Tuathail and Agnew, 2006, p. 94), for instance about the intention to annex territory or to defend sovereignty, it is also about the meaning ascribed to territory – the geographical understanding and reasoning applied in discourses, which influence how political leaders act. Meaning is conveyed via language, which is one tool through which "we understand and constitute our social world" (ibid.) and via discourses as particular settings for language. A discourse can also be regarded as "an ensemble of rules by which readers/listeners and speakers/audiences are able to take what they hear and read and construct it into an organized, meaningful whole" (Ó Tuathail and Agnew, 2006, p. 96).

Accordingly, scholars of Critical Geopolitics – like the post-structuralist scholars mentioned before – pay particular attention to the context in which (geopolitical) reasoning is introduced, or to the "frameworks" applied in geopolitics, as the political geographer Gearóid Ó Tuathail calls them, which need to be revealed in order to understand why events are interpreted in a specific manner (Ó Tuathail, 2006, p. 2):

> Many geopolitical narratives are enframed by essentialized oppositions between 'us' and 'them', the 'civilized' versus the 'fanatical.' Whole regions of the world are divided into oppositional zones, a frameworking we can call 'earth labeling.' For example, geopoliticians use grand spatial abstractions like the Islamic World, the Non-Integrating Gap, the Global South or the Civilized World. Other spatial metaphors like heartlands, faultlines, and axes are popular. All these expressions draw rhetorical force from their ability to reduce the complexity of world politics to a simplified framework.

These frameworks, however, are not only created by politicians. They are also shaped by "institutions such as the national media and education systems" (Dodds, 2014, p. 77) and negotiated in connected discourses. This is why scholars of Critical Geopolitics differentiate between practical geopolitics, popular geopolitics and formal geopolitics – though these are perceived as also being shaped by interconnected discourses (Dodds, 2014) – in order to understand how geopolitical representations are constructed and presented to particular audiences. As Ó Tuathail (2006, p. 7) emphasises, these discourses often emerge from a "historical cannon [*sic*] of narratives about state formation and identity".

Following the assumptions explained previously, in this book geopolitical discourses are understood as representational practices by which meaning ascribed to territory is produced (and fixed) via language. These discourses are regarded as a means of informing particular audiences and of propagating a specific geographical understanding and reasoning with the aim of enforcing political power over territory. Discourses, consequently, follow their own rules as to which statements can or cannot be made, rules that depend, in turn, on the (institutional and sociocultural) contexts to which they relate and in which they are embedded.

The structure of this book follows the tripartite analysis of formal, practical and popular geopolitical discourses promoted by Critical Geopolitics. It also expands to cover the different national and regional levels under analysis, which, in accordance with the relational entanglement perspective presented earlier, are crucial to an understanding of how the dominating patterns of interpretation in the context of the changing Polar Regions are produced.

1.3 Structure of the book

This book focuses on the politics of the Polar Regions to outline how ordering principles evolve that ultimately affect and guide policy making. Its intended contribution is to provide a more sophisticated understanding of changes in the politics of the Polar Regions since 1989, which most social and political sciences studies hitherto have examined by dealing with the Arctic and the Antarctic separately. In expanding on these studies, this book is concerned with the (changing) constellations and geopolitical reasoning of state and non-state actors in policy formulation processes that relate to both Polar Regions. From this relational perspective, which is underpinned by the consideration of multiple positions and contexts, this book offers an explanation as to why similar patterns of interpretations have dominated in the politics of the Polar Regions and why a number of similar and related issues have been treated as topics of concern while others have been given less priority. Obviously, any such elucidation needs to acknowledge that the perspectives of individuals and institutions that prioritise some issues over others usually differ, change over time and depend on various factors such as their degree of concern, strategic considerations and the general significance they ascribe to particular issues, and, moreover, that actors do not always follow a utilitarian principle in policy making. This book, however, provides evidence to substantiate the claim that similar issues have turned into topics of concern in the politics of the Polar Regions and have been given higher priority because of the settings in which issues relating to the changing Polar Regions are negotiated. Accordingly, particular attention is given to the two systematically assessable factors mentioned before that have an impact on the prioritisation of issues: (1) the constellation of actors involved in the politics of the Polar Regions in light of the different national and regional settings, and (2) the reasoning expressed in (practical, popular and formal) geopolitical discourses.

The book's first two chapters lay the groundwork for this endeavour. This chapter has already introduced the subject matter of the study, outlined the political significance of the changing Polar Regions from which its main research questions derive, and discussed the terms that are central to its analysis. The second chapter discusses how the concepts, approaches and ideas of scholars and studies focusing on Transnational Relations, Entanglements, Critical Geopolitics and Imaginaries help to understand entanglements in and between both Polar Regions, which have often been neglected in research on their politics. It argues that the combination of all four theoretical perspectives helps to understand different layers in the evolution of the politics of the Polar Regions, which – as the empirical chapters later present in detail – complement the perspective taken, for instance, in inter-American studies.

In order to examine whether or not entanglements among actors and discourses in the context of the changing Polar Regions also enable one to speak of a newly consolidating political space, particular attention is given to the American hemisphere, as it borders on both Polar Regions. While the impacts of the changing Polar Regions are clearly global in scope, the negotiation of problems ascribed to them often differs in global, international, transnational and regional contexts. To better understand the geopolitical reasoning of state and non-state actors introduced into policy formulation processes and how their views and positions in practical, formal and popular geopolitical discourses move between the national and regional spheres, Chapters 3, 4 and 5 shed light on different perceptions of the changing Polar Regions in discourses at the regional and national levels. Accordingly, each is composed of two parts, which focus first on the role of the actors who mainly contribute to the discourses under analysis, and second on the representation of distinct views and positions in geopolitical discourses that relate to the changing Polar Regions. The guiding research questions are:

1 Who contributes to the discourses under analysis and how?
2 How do actors relate to the changing Polar Regions, what positions and views do they introduce into the discourses under analysis?

The "double" chapter, Chapter 3, addresses polar policy making at both the regional and national levels. First, it pays particular attention to the main regional governance settings that determine policy formulation processes in the Polar Regions and compares different institutional contexts that are important for the Regions' politics. Specifically, it discusses the significance of different international and regional institutions and frameworks to qualify that of the Arctic Council (AC), the Antarctic Treaty System (ATS), the United Nations Convention on the Law of the Sea (UNCLOS) and the Organization of American States (OAS). In so doing, the chapter also investigates the principal actors that shape these settings and collaborate in them. In shedding light on the main intergovernmental forums for policy making in the Polar Regions, the chapter details how state and non-state actors from the American polar-rim states participate and collaborate in the Arctic Council and the Antarctic Treaty System. It examines the positions introduced by representatives of state and non-state actors from Canada, the U.S., Chile and Argentina in the selected regional settings and identifies entanglements that affect the representation of priorities and interpretations in the politics of the Polar Regions. Second, Chapter 3 investigates the evolution of positions and interests at the national level. After a classification and contextualisation of Arctic and Antarctic policies introduced by Canada, the U.S., Argentina and Chile, the chapter scrutinises the extent to which their positions are entangled or disentangled and whether or not they correspond with or challenge views expressed in regional polar policy making.

Geopolitical discourses are also shaped by public opinion, which is why Chapter 4 examines how the changing Polar Regions are represented in national newspaper reporting in the American polar-rim states. Assessing again the two different factors mentioned before (constellation and significance of actors as well as the representation of positions in discourses), Chapter 4 first provides an overview of the newspapers under analysis, notes their political orientation and sets them in the context of news reporting in the four American polar-rim states. In a second step, it illustrates how the different newspapers under analysis have represented the processes and dynamics that have concerned the Arctic and Antarctic at different points in time. To what extent do newspapers published in the polar-rim states represent the politics of the changing Polar Regions in a way that underpins, corresponds with or challenges the geopolitical reasoning (and imaginaries) that predominate in practical geopolitics as identified before? This is the main question to which Chapter 4 provides answers.

Chapter 5 assesses the significance and views of non-governmental theorists and strategists from the American polar-rim states. As explained in more detail in

Chapter 2, Critical Geopolitics scholars have pointed out that the discourses and positions of actors examined in Chapters 3 and 4 are also influenced by strategic considerations put forward by think tanks and academia. Consequently, this chapter investigates the so-called formal geopolitical reasoning that non-governmental theorists and strategists recommend should be followed in view of the changes taking place in the Polar Regions. The analysis provided in this chapter differs from that of the other empirical chapters as its purpose is not to trace how different perspectives evolved. Instead, it shows how the positions identified in the previous chapters are reproduced and challenged by scholars from the American polar-rim states and outlines entanglements involving practical, popular and formal geopolitics. It compares different views and positions in an illustrative rather than a systematic manner, because political practitioners often do not make transparent which scientific publications they take into account. This chapter illustrates the heterogeneity of studies published by scholars that political practitioners can select from to give greater legitimacy to their positions. Like other Critical Geopolitics studies, therefore, the analysis provided in Chapter 5 builds on the understanding that science is not politically neutral and that scientific practices also contribute to the dominance of some patterns of interpretation over others.

Lastly, Chapter 6 discusses and classifies the findings of the preceding chapters to answer the guiding research question of this book: why have similar issues been treated and constructed as topics of concern in the politics of both Polar Regions? Like the empirical chapters, it first evaluates the degree to which the significance of actors and actor constellations has changed in the politics of the Polar Regions. Second, taking into account the different geopolitical discourses under analysis, it assesses how far it is possible to speak of entangled geopolitical discourses influencing the regions' politics. Third, on the basis of the different positions and views regarding the changing Polar Regions introduced by and within the American polar-rim states, it evaluates the extent to which the politics of the Polar Regions is strengthening the evolution of a new inter-American political space. Lastly, the significance and limitations of the findings are discussed and the gaps that remain to be addressed in further research, for instance in the fields of political ecology and democratisation studies, are identified.

Notes

1 In this book, the meaning ascribed to the term "relational" is shaped by the sociologist Mustafa Emirbayer (1997, p. 287), who argued "the very terms or units involved in a transaction derive their meaning, significance, and identity from the (changing) functional roles they play within that transaction. The latter, seen as a dynamic, unfolding process, becomes the primary unit of an analysis rather than the constituent elements themselves". Applied to the research context at hand, no actor, no discourse, no event is considered as operating, dominating or occurring in isolation but as being determined by contexts.

2 The CLCS is the subsidiary body of the United Nations that provides recommendations on the affiliation of disputed territories. According to article 76 of the United Nations Convention on the Law of the Sea, states bordering the Arctic and Southern Oceans have sovereign rights over areas extending 12 miles off their coasts (within the Territorial

Sea). The Exclusive Economic Zone beyond this limit gives them sovereign rights over exploring, exploiting, conserving and managing living and non-living resources in the water, seabed and subsoil and extends to a total of 200 sea miles. This area can be expanded, if the bordering state provides evidence that the continental margin extends for more than 200 nautical miles by establishing the foot of the continental slope.

2 Investigating the politics of the Polar Regions

Research gaps and a new perspective

> There will be considerably more work to be undertaken as we and other scholars demand more critical forms of polar geopolitics (Powell and Dodds, 2014). In other words, we need to 'scramble' discourses, while resisting the temptation to exaggerate, to demand more attention to historical connections and context including colonialism and imperialism, which ensured that the Arctic and Antarctic were intimately tied to colonial-era science, commerce and geopolitics. These regions were not 'poles apart'.
>
> (Dodds and Nuttall, 2016, p. 188)

When focusing on the changing Polar Regions, most studies so far have explored political processes in the Arctic and Antarctic separately and often emphasised either a local, domestic or a global perspective. This book argues that these studies are inadequate inasmuch as they do not consider and identify entanglements that help to clarify why similar patterns of interpretation have become dominant in the politics of both regions. Here I posit that a better understanding of these processes is necessary to "'scramble' discourses" (as Dodds and Nuttall put it) in order to critically engage with the dominant understandings that guide collective and individual actions at a time when humans are increasingly perceived as constituting the most important factor of influence with regard to biological, geological and atmospheric processes on Earth (see, for instance, discourses on the Anthropocene). By identifying this gap, however, I do not mean to generally undermine the contributions of studies focusing on either the Arctic or the Antarctic. These studies have certainly provided a better understanding of the complex political processes and developments taking place in one region or the other, and to which this section proposes to provide a guide. What follows, therefore, outlines the findings that this book draws upon, challenges and aims to expand on.

2.1 The Polar Regions in political science literature

In political science and the neighbouring fields of cultural studies, international law and sociology, the amount of literature on the changing Polar Regions (particularly on the Arctic) has significantly increased in the past decades.

Despite the augmented attention paid to the politics of the Polar Regions, however, and recognition of findings in other research contexts, in political science and, more specifically, International Relations (IR) literature, the changing Polar Regions are still treated as rather marginal. Likewise, political science approaches remain in the minority in the very interdisciplinary field of polar research. Nevertheless, contributions that, generally speaking, either concentrate on policy making or are explicitly theory-driven have explored a variety of aspects that fall within the categories of geopolitics (economic, social and political changes) and security concerns that have a particular connection to the changing Polar Regions.

As I have illustrated elsewhere (Albert and Wehrmann, 2015), *theory-driven contributions* often draw on prominent assumptions from the Critical Geopolitics approach or IR literature and relate particularly to realist, institutionalist and constructivist paradigms (cf. Hønneland, 2013). These analyses focus on the representation of geographical spaces, political participation and region-building and mostly adopt the perspective of regime theory to examine forms of cooperation or rivalries and conflicts. Previous studies have paid attention to institutions, questions of sovereignty, (the strategic use of) dominating narratives, imagination and the power of imaginaries in the politics of the Polar Regions (Christensen et al., 2013; Dodds and Nuttall, 2016; Haase Ligget, 2009; Steinberg et al., 2015; Tennberg, 2000; Young and Osherenko, 1993). With regard to the Arctic, some studies have also considered the strategic use of imaginaries by scientists and their representation in newspaper reporting. A systematic analysis of entanglements between the related discourses conducted by politicians, non-governmental theorists and dominant framings of the media, however, is still missing. Furthermore, when examining the formation and functionality of institutions as well as the evolution of geopolitical reasoning, most studies have focused either on the Arctic or the Antarctic.

This book expands on these studies with respect to three dimensions: *actors, geographical scope* and *time*. Instead of concentrating particularly on actors *or* discourses, it considers both as being determined by each other and illustrates the respective entanglements. It expands their geographical scope by systematically studying the interrelationships between political processes and discourses at regional and national levels in the Arctic and the Antarctic. Further, as the period of investigation starts back in 1989, it is possible to observe how institutions, region-related identities and geopolitical reasoning have changed over time, particularly in view of the growing recognition of environmental changes in the Polar Regions. Being determined by these conceptual decisions, this book traces why similar priorities (such as the development and conservation of the environment) have been set at different times in the politics of the two Polar Regions.

The theory-driven literature represents, however, only a fraction of political science literature that mostly focuses on *policy making* in the politics of the Polar Regions. Especially over the last decade, a great number of the analyses provided by political scientists have paid attention to the different interests, positions and strategies of actors operating there (i.e. Beck, 1990; Brady, 2012; Dodds, 1997;

Dodds and Hemmings, 2009; Infante Caffi, 2006; Kelly and Child; 1988; Parodi, 2007). Growing international interests have arguably inspired this focus in the literature, which, at the same time, is seen as one reason for the recent (re)formulation of Arctic policies by Arctic-rim states.

In regard to the Antarctic too, political scientists have particularly addressed national priorities, efforts at international cooperation and conflicts. But, while policy analyses of the Arctic often discuss particular aspects of policy making, studies on policy making in the Antarctic usually relate their findings to broader international relations. Further, and as I show elsewhere (Albert and Wehrmann, 2015), most policy studies on the Antarctic do not predominantly revolve around the question of whether conflicts or patterns of cooperation have determined political processes there. It is striking, however, that even studies that do not focus on territorial claims and questions of sovereignty in Antarctica still relate to the discovery of geographical spaces and their marking out through the establishment of research bases.

Rather surprisingly, the explicit role ascribed to changing polar environments in policy making remains largely unaddressed in the political science literature. And although studies concentrating on the regional level compare the governance structures of the Arctic Council and the Antarctic Treaty System by relating them to global challenges and future operational possibilities, domestic policy making with respect to the changing Arctic and Antarctic is mostly not considered to be entangled (an exception being Beck's 1995 analysis of Canada's Arctic and Antarctic policy). As, for instance, Manicom (2013) shows in his examination of Canadian Arctic policy, however, intra-state negotiations certainly have an impact on international policy making and vice versa. Moreover, as Bergh's (2012) findings reveal, bilateral cooperation between Canada and the U.S. has also had an impact on their ability to pursue domestic interests in the Arctic region. Furthermore, some studies (e.g. Dodds, 1997; Keskitalo, 2012) have found that the media influence policy making relating to the changing Arctic and have also considered the geopolitical representation of the Arctic in policy papers. None of them, however, has explored the interlinkages between discourses shaped by politicians, scientists and the media, which would help to explain why – as May et al. (2005) found – particular issues dominate domestic (and regional) polar policy making at different times.

When it comes to security and geopolitics, on the other hand, climate change is often put forward as a trigger in both Polar Regions. Clausen and Clausen (2013), for instance, classify climate change as a "threat multiplier", arguing that new economic possibilities (and risks) cause territorial disputes and power politics to be charged up – which is also why questions of security and geopolitics are increasingly prioritised in the Arctic and Antarctic. Murray (2012) takes a different and less widespread stance, arguing that the progressive transformation from a unipolar (U.S.-dominated) to a multipolar world order is enhancing the geopolitical significance of the Polar Regions.

Overall, investigations that concentrate on the security-related and geopolitical consequences of these processes of change in the Polar Regions (such as the "race for resources" or the "quest for territory") mostly refer to questions of the sovereignty of states, regional shifts in power and the representation of these aspects in the Arctic. Since the mid-2000s, such studies have more often differentiated between processes on the global and the regional level (i.e. Heininen et al., 2014). According to Heininen (2004), if one is focusing on geopolitical processes in the Arctic, this differentiation is also necessary in order to understand the contemporary dynamics within the Arctic regime, because new international actors are increasingly influencing what happens in the Arctic and thus contributing to an intensification of international cooperation and to changes in future circumpolar geopolitics. In 1989 Osherenko and Young already pointed out that the Arctic needed to be examined both as a regional and an international arena and recommended considering different actor groups (representing Indigenous peoples, environmental organisations and multinational corporations) to grasp the contrary positions that could likely lead to conflict. Most studies on geopolitics and security in the Arctic and Antarctic, however, still relate primarily to state actors while non-state actor groups that influence security-driven and geopolitical actions in the Polar Regions often remain unaddressed.

Three gaps in the political science literature on the Polar Regions are of particular interest for the conceptualisation of this study, given the research questions that it sets out to analyse:

1   There has been no systematic analysis of the interrelationships between discourses shaped by politicians, non-governmental theorists and the representation of the changing Polar Regions in newspaper reporting. Such an analysis is necessary to better explain why similar imaginaries and patterns of interpretation are prioritised and circulated in discourses on the Polar Regions.

2   Most studies still either focus on the national or regional levels when examining policy making with regard to the Polar Regions, or – as has been the case more often with respect to the Arctic – consider the influence of national policy making on the regional level and vice versa. What is lacking is a perspective that considers discourses on both Polar Regions over a longer period of time and thus includes an interregional dimension. This would expand our knowledge as to whether or not, despite the numerous differences between the two regions, a similar prioritisation (based on a particular representation of both the Arctic and Antarctic) has steered policy making at times of change.

3   The vast majority of political science scholars have focused on the role and interests of states in investigations that relate to the changing Polar Regions. Contrary positions among non-state actors and their varying significance for states and in the politics of the Arctic and Antarctic have often been overlooked. This has also contributed to the formation of dominant patterns of interpretation with respect to the changing Polar Regions.

2.2 A new perspective: entangled actors and discourses in the politics of the Polar Regions

> Because we can know the world only through the conceptual schemas provided by
> our culture and languages, we cannot ever assume that the world is independent of
> the representation conventions we use to describe it.
>
> (Dalby et al., 2006, p. 6)

To address these gaps this book builds on theoretical premises found in the concepts of *Transnational Relations, Entanglements, Critical Geopolitics* and *Imaginaries*. The literature on transnational relations and entanglements discusses the formation of actor networks with respect to space and the intensity of the interrelationships involved. The Critical Geopolitics approach and the central role it ascribes to imaginaries add to the proposed perspective by sensitising for the functional use of geopolitical representations in political communication. Consideration of these concepts – as Dalby indicates earlier – thus also influences the different shades of the perspective developed in this study.

2.2.1 Transnational relations

In political science research on the Polar Regions, the consideration of transnational relations is nothing new. In 1989 researchers already observed a complex network of transnational ties that had been formed in the Arctic for economic purposes by foreign corporations and governments (Osherenko and Young, 1989). Nuttall (1998) later argued that these early findings helped to understand the significance of the Arctic in the global system, although that remained state-centred as those findings implied "the involvement of agencies and other parties that act on behalf of the state" (p. 42). The wider scholarly debate on transnational relations at that time also continued to perceive states as the dominant actors in international relations (see e.g. Risse-Kappen, 1995). In the meantime, research on transnational relations advanced and scholars widely agreed that in the Arctic and Antarctic transnational relations had multiplied and were not limited to economic interdependencies but also included political and cultural connections that impacted policy making. Particularly in the 2000s, research on transnational relations in the Polar Regions expanded significantly across various social science disciplines, yielding studies assessing, for instance, the significance of transnational Indigenous activism in the Arctic (Plaut, 2012; Koivurova, 2010; Shadian and Wirpsa, 2006), exploring the "transnational history" of the Polar Regions and transnational processes linked to UNCLOS (Doel et al., 2014; Naidu, 2008) or to globalisation in the Arctic (Keskitalo and Nuttall, 2015) or considering the strength of transnational issue arrangements (Bruun and Medby, 2014; Kharlampyeva, 2013; Young, 2005). All of these and similar studies provide examples to highlight the idea that the world is not structured by container-like nation states and to emphasise instead cooperation and inter-relationships among societies and individuals.

Like the analyses of transnational relations in the Polar Regions, the more general scholarly debate on transnational relations can be thought of as occurring in two "waves". A first scholarly debate on the subject was encouraged by the political scientists Karl Kaiser (1969), and Joseph Nye and Robert Keohane (1971) who highlighted "reciprocal effects between transnational relations and the interstate system as centrally important to the understanding of contemporary world politics" (Nye and Keohane, 1971, p. 331). Scholars involved in this early debate (e.g. Yale H. Ferguson, Harold Jacobson, Richard W. Mansbach, Edward L. Morse, James N. Rosenau) contributed to a growing awareness of non-state entities' influence on inter-societal interactions and thereby challenged the – at that time predominant – state-centric view of world politics particularly in the discipline of IR. However, as Risse-Kappen pointed out in the 1990s, this first debate failed to provide a differentiated understanding either of how transnational relations concretely influence state policies and international relations or of how interstate and transnational relations interact. Instead, this first debate evolved around predicting whether states or transnational actors would be, or were, more important in world politics. This latter question, in fact, continues to be discussed controversially, leading, for instance, to predictions of the "end of the nation-state". This prediction was later challenged by Pries who pointed out that "among other types of actors nation states remain an important *Bezugseinheit* because they continue to coin spatial references and also societies in nation states continue to shape social processes, social relationships, cultural life and political debate" (Pries, 2010, p. 16, original emphasis).

Since the 1990s, numerous empirical studies have attempted to answer a myriad open questions and have, for instance, expanded knowledge of the conditions under which transnational actors matter (i.e. in migration studies, Faist and Özveren, 2004). Regional disparities (Risse-Kappen, 1995) and the creation of transnational networks have been more often investigated as well (i.e. Keck and Sikkink, 1999; Clavin, 2005). Amongst other things, this second debate has also drawn attention to the existence of transnational political spaces, "sphere[s] in which common representations and identifications are negotiated" (Albert et al., 2009, p. 7). In these spheres, as Nye and Keohane (1971, p. 338) had pointed out earlier, transnational organisations in particular are capable of changing attitudes through the creation of new "myths, symbols, and norms to provide legitimacy for their activities [. . .] elsewhere in the world".

Building on both debates in the context of the politics of the Polar Regions, central questions explored in this book are how transnational actors (e.g. environmental networks such as the Antarctic Southern Ocean Coalition) contribute to geopolitical discourses in the American polar-rim states; how they shape transregional entanglements and thus contribute to the formation/intensification of separate or shared transnational political spaces. I illustrate what this perspective entails in more detail by discussing the following questions:

1   What is meant by the term "transnational relations"?
2   How do transnational relations evolve?

3   What is the difference between transnationalisation and transnationalism and what are their possible effects?

In their "pioneering" work (Kaiser et al., 2010, p. 10), Nye and Keohane defined transnational relations as "contacts, coalitions, and interactions across state boundaries that are not controlled by the central foreign policy organs of governments" (1971, p. 331). Risse-Kappen (1995, p. 3) added that these interactions must occur on a regular basis and Pries (2007, p. 16) later summarised the condition that "transnational relations may encompass state actors, but they necessarily include non-state actors". To specifically demarcate the term transnational, Faist and Özveren (2004, p. 2) suggested that one should speak of trans-state relations "when describing ties that criss-cross the borders of sovereign states".

Transnationalisation processes can be observed if interactions (or ties, Faist and Özveren, 2004) among actors from different actor groups, which may include state actors but must contain at least one non-state actor, are dense and cross borders (Pries, 2010), but are not global in scope (Albert et al., 2009). Further, transnational interactions can be regularised or institutionalised when cross-border networks are formed that exemplify the process of transnationalisation. From a transnational perspective, then, institutions such as the Arctic Council (classified as an intergovernmental forum), for instance, can be understood as "sets of [informal and formalized] procedures and norms that regulate social activities" (Faist and Özveren, 2004, p. 1). Transnationalism, on the other hand, "comprises the 'semantic construction' of common representations and identifications among larger groups of state and non-state actors beyond the nation-state" (Kaelble et al., 2002, p. 10, in Albert et al., 2009, p. 8) and thereby considers the level of communication, "imagery and symbolic practices as well as [. . .] the role which the media play in processes of common identity-building" (ibid.). But how do transnationalisation processes and transnationalisation result in the formation of new networks and entanglements among geographical spaces?

Nye and Keohane (1971, p. 339) already referred to the rapid growth of non-governmental organisations that – if transnational in scope – might link national organisations with common interests in policy making. They regarded non-governmental organisations as stimuli for "the creation of new national affiliates" contributing "to the internationalization of domestic politics" (ibid.). Scholars focusing on transnational relations found that growing interactions among non-state actors could lead to the formation of functionally defined networks. Like transnational communities and organisations based on "dense and continuous sets of social and symbolic ties, characterized by a high degree of intimacy, emotional depth, moral obligation and sometimes even social cohesion" (Faist and Özveren, 2004, p. 9f.), such networks may constitute a significant polity in politics (Ferguson and Jones, 2002). For the existence of neither networks nor transnational communities, networks or organisations, however, is geographical proximity necessary.

To put it concisely, transnational networks have a specific function, are characterised by interactions among non-state actors and may include state actors. Actors, however, are not necessarily located geographically close to one another and may instead be located in very different geographical regions. Nevertheless, as "most influences which transcend national borders emanate not from the globe but the neighborhood" (Mann, 2006, p. 28, in Albert et al., 2009, p. 12), networks may benefit from and, at the same time, provide evidence for regional interconnectedness. Such networks are based on shared interests and values and are not necessarily structured hierarchically. Networks, however, are not equally durable, which is why Faist and Özveren (2004) draw attention to the extensity and intensity of ties that differ in transnational issue networks and in epistemic communities.[1] Earlier, Risse-Kappen (1995) had already assumed that differences in domestic structures as well as the varying degrees of international institutionalisation would have an impact on the influence of transnational actors on state policies. He argued

> in order to affect policies transnational actors have to [. . .] gain access to the political system of their 'target state.' Second, they must generate and/or contribute to 'winning' policy coalitions in order to change decisions in the desired direction. Their ability to influence policy changes then depends on the domestic coalition-building processes in the policy networks and on the degree to which stable coalitions form sharing the transnational actors' causes.
>
> (Risse-Kappen, 1995, p. 25)

These considerations highlight two important factors that determine transnational actors' influence: first, their internal structure, the intensity and extensity of their connecting ties; second, the means they have to participate in domestic, regional or international policy-making processes, which do not solely depend on transnational actors themselves but also on the circumstances and conditions under which they succeed in being, or fail to be, included in such processes, which are also determined by other (state) actors. Against this background, the leverage transnational actors obtain in world politics varies and may also increase inequalities among other political entities involved in policy-making processes. Such inequalities are particularly enforced by dependencies (Nye and Keohane, 1971) and, as global governance studies, for instance, have found, are of special importance at present, as "[a]n increasing number of global problems [. . .] cannot be managed by states alone, but require cooperation between different state and non-state actors" (Koch, 2011, p. 197).

The notion of political spaces, which is drawn upon by political geographers as well as by political scientists, "refers to the ways in which identities and loyalties among adherents to various polities are distributed and related" (Ferguson and Mansbach, 2004, p. 67, in Albert et al., 2009, p. 18). Political spaces directly link to the aforementioned variables determining the policy impact of transnational relations (e.g. the degree of institutionalisation and impact, networks, political authority, practices). In this regard, constructivists stress that "ideas" structure political space (Ferguson and Jones, 2002, p. 11) and argue that

> transnational political spaces are not given and cannot be reduced to an abstract and self-regulating development of globalization or de-nationalization, but

are always established and formed by identifiable actors in their discourses and practices. Communication as a performative act plays a central part in the making of the political and is most often associated with imagery and symbolic practices.

(Albert et al., 2009, p. 19)

Thus, political spaces are contested spheres shaped by interaction and communication among state and non-state actors, and are important to consider if one wishes to understand "the extent to which important aspects of global politics and governance transcend territory or are effectively deterritorialized" (Ferguson and Jones, 2002, p. 7).

As far as the changing Polar Regions are concerned, this interaction and communication also shape understanding of the problems and opportunities ascribed to the changing environment as well as their "ordering"/prioritisation. As mentioned before, both regions are shaped by numerous environmental changes that "do not respect the borders of states" and have a variety of effects. As Tennberg (2000, p. 125) points out, in this regard "making order of problems is not only in the hands of the politicians, diplomats and government officials in international negotiations but these problems are a concern for many other actors as well" since they also impact relations between actors at the domestic and international levels. Vice versa, these relations determine the dominant understanding of the changing Polar Regions, since the ways in which views are represented in communication and the actors who are involved in it shape the meanings ascribed to those regions in discourses. Following Young (1989, p. 83, in Tennberg, 2000, p. 125) who understood regimes as "'human artifacts' which have no existence in meaning apart from individuals or groups of human beings", Tennberg concludes that "human beings are enmeshed within webs of environmental relations and those relations are intersubjectively constituted. Therefore, the environment is a social construction" (ibid.). Meaning ascribed to the environment is thus considered as always being imposed and depending on interpretative practices (Shapiro, 1989).

Against that backdrop, the transnational perspective taken in this book is based on the following assumptions:

1   Politics are not solely reduced to the affairs of states and supra-state organisa-
    tions; instead transnational networks and epistemic communities constitute
    significant polities in the politics of the Polar Regions,
2   Norms and identities contested in the politics of the Polar Regions are shaped
    by discourses influenced by actors and (strategic) discourses beyond the bor-
    ders of states,
3   These discourses shape understandings of the changing Polar Regions as they
    order the problems and opportunities ascribed to them.

A conceptual tool that better grasps these entanglements between numerous, heterogeneous actors, discourses and domestic and regional policy making is introduced next.

2.2.2 Entanglements

Analyses of transnational relations focus on interconnections among actor groups that are composed of at least one non-state actor and consider, for instance, how transnational relations impact discourses. In studies investigating entanglements, on the other hand, the centre of interest lies in the significance of specific and related events, of shared historical experiences or geographical interconnections, which contribute to the formation of transnational relations and transnational political spheres. Before I outline the gaps in and potentials of the entanglement and transnational relations approaches when they are brought into play, the following paragraphs aim to demarcate these two concepts from each other and thus clarify:

1   What is meant by the term "entanglement"?
2   How do entanglements evolve?
3   What are the possible effects of entanglements?

The term "entanglements" has been used by scholars from disciplines such as history, literature studies and political geography in studies examining multidimensional connections – "the myriad entanglements" (Sharp et al., 2000, p. 1) – among spaces and historical and/or political processes (e.g. Conrad and Randeria, 2002; Hecht, 2011; Sheppard, 2012; Kaltmeier, 2014). Entanglements, however, are much less addressed by scholars than, for instance, transnational relations are. One reason for this might be that no specific concept or theory is related to the term entanglement. Instead, it is rather fuzzy and mostly used as a metaphor, suggesting, for instance, "an image of knotted threads" (Sharp et al., 2000, p. 1). It has been applied most often by Postcolonial Studies scholars, as well as those focusing on shared, connected or entangled histories as well as Histoire Croisée (cf. Conrad and Randeria, 2002; Subrahmanyam, 1997; Werner and Zimmermann, 2002). They use the term entanglement to promote a relational perspective, allowing developments at the micro and macro levels to be related to one other, and exchange and interaction with and among smaller entities such as ethnic groups, tribes and localities to be taken into consideration (Conrad and Randeria, 2002). Like the work of transnational relations scholars, studies investigating entanglements argue against the perception of areas as given entities, instead understanding areas as the results of relational processes – as also proposed in political geography (cf. Agnew, 1994; Reuber, 2012) and in relational history (Epple, 2014) – emphasising that:

> The constructedness of areas and their relations to others is highlighted by focusing on mutual observation, comparison, competence, interdependence and interplay. [. . .] Areas are thus imagined spaces of interaction which are both addressed and influenced by the geopolitical strategies of institutional actors, economic interests, media, social movements and daily life experiences.
>
> (Kaltmeier, 2014, p. 178)

From this point of view, within predefined geographic areas entanglements relate to different issue areas (e.g. in the fields of economics, politics and culture) and, as an expansion of the metaphor of entanglements already implies, may lead to the formation of "knottings" or "interfaces" (cf. Kaltmeier, 2014, p. 179) of networks and the emergence of centres of power (Sharp et al., 2000; Dodds, 2014). As in transnational relations, a growing occurrence of entanglements is often related to the "new conjuncture in globalization driven by a liberalization of trade, an expansion of the financial markets, and innovation in information technologies" (Kaltmeier, 2014, p. 172) as well as to the end of the Cold War and the related "bi-polar macro-geopolitical world order" (ibid.).

In accordance with this view, it can also be argued that "entanglements are a precondition for the appearance of power" and that they, at the same time, obtain power themselves (Sharp et al., 2000, p. 24f.). In this sense, like power, entanglements can be understood as positive and negative. The term itself is, therefore, less loaded than, for instance, "interdependencies". Entanglements can result from processes actively contributed to or passively undergone (Tittor, 2014). Against this background, Sérgio Costa (2011) and Conrad and Randeria (2002) also drew attention to asymmetries. Roth too (2014, p. 147) argued that "the pure existence or marcation of entanglement does not imply reciprocity of relations". Following the sociologist Norbert Elias, social scientist Anne Tittor (2014) proposed to consider a possible order of entanglements that generates both intended and unintended dynamics. When it sheds light on the entanglements of different discourses, such a view can be helpful as it considers how entangled discourses produce dynamics that enforce or challenge recurring understandings.

Inter-American Studies, for instance, draws on the concept of Entangled Spaces as it sheds light on how regions and spaces are connected within the Americas and considers differences in the forms and significance of entanglements. Here migration, cultural production (e.g. media flows), diasporic cultures and language diversity are taken into account as well as "dynamics in political economy, such as competing regional integration processes" (Kaltmeier, 2014, p. 172). Through the combination of regional and national foci with transnational approaches, Inter-American Studies aims to raise awareness of cross-connections that otherwise remain hidden. Scholars have found that "concrete cultural, political and economic dynamics, tensions, and processes within the Americas have increasingly created inter-American webs and networks that manifest mutual entanglements between locations, regions, and nations beyond a North – South divide" (Raussert, 2014, p. 62).

In the field of political science, it is less the entanglement of spaces than the entanglement of actors on which scholars concentrate. Jacobson et al. (1986), for instance, examined the entanglement of states and international governmental organisations. Specifically, they shed light on entanglements in the context of a growing web of intergovernmental organisations. Voeten (2014) also evaluated increased entanglements between member states and international organisations in order to answer the question of whether or not participation in international organisations increases cooperation. Both contributions relate to the earlier works of Rosenau, who investigated "linkage politics" (1969), and also to Putnam, who analysed "how diplomacy and domestic politics can become entangled" (1988) by addressing the questions of when and how domestic politics and international relations determine each other. In the field of political science, Putnam was probably the first to develop a conceptual framework to make sense of entanglements. He proposed using the perspective of a "two-level game" to understand how diplomacy and domestic politics interact. In this two-level game, policy formulation processes are understood as comprising two stages, a negotiation and a ratification phase. According to Putnam, the negotiation phase could be characterised as "bargaining between negotiators, leading to a tentative agreement" while in the ratification phase "separate discussions [take place] within each group of constituents about whether to ratify the agreement" (Putnam, 1988, p. 436). Scholars focusing on the politics of the European Union expanded this perspective by developing the concept of multi-level governance (e.g. Hooghe and Marks, 2001; Piattoni, 2010). They outlined the significance of entangled political levels and mutual dependencies among different levels of action (e.g. the supranational EU, the member state, the region and the municipality), which – as is also shown in this book – determine the prioritisation of issues in policy making.

Studies focusing on the politics of the Polar Regions also address multi-level governance, for instance in discussions on the inclusion of new observers at the Arctic Council and with regard to environmental regulations in the Arctic and Antarctic (e.g. Stokke, 2011; Koivurova, 2013). Yet, the term entanglement is rather uncommon and generally not much addressed in research on the politics of the Arctic and the Antarctic. The potentials of a relational perspective on the politics of the Polar Regions, however, have been outlined by Hall, who argued in 1986, in line with Putnam, that "Antarctic affairs have been an integral part of world politics". In his study he challenged the then dominant picture of the signing of the Antarctic Treaty ("the triumph of science over politics") as "superficial and misleading" (Hall, 1986, p. vii) since it only considered the series of events during the International Geophysical Year. Instead, he demanded that it should also be taken into account

> that (i) political factors have played a significant part in Antarctic affairs throughout the twentieth century, and (ii) structural changes in world politics have impacted upon Antarctic affairs throughout the same period.
>
> (ibid., p. viii)

Although he does not use the term "entanglement", he points to the same interlinkages as entanglement scholars do, namely connections among events and actors and an interplay of political and strategic considerations, which also constitute the research interest in this study.

In a nutshell, across the disciplines entanglements are, like transnational relations, primarily perceived as linkages connecting actors and/or events that may contribute to the formation of networks as well as leading to dependencies, the consolidation of hierarchies and domination. Interrelationships, interdependencies and interconnectedness are often used as synonyms for entanglements and are understood as having an impact on involved entities, on events and on the perception of spaces and places. Understood in relational terms, these entities "are not fixed nor stable" (Epple, 2014), since entanglements influence their practices and discourses and also determine knowledge circulation (see also Mignolo, 2002, who speaks about the geopolitics of knowledge in this regard).

Modelled after definitions of transnational relations but less actor-specific, an entanglement can also be defined as an interconnectedness of two or more parts that has grown over time. Entanglements, however, do not have to be permanent. It is also possible for entanglements to dissolve or become reversible. Devolution efforts promoted by politicians in Alaska or Tierra del Fuego, for instance, can be regarded as attempts at disentanglement, while renewed coalitions among actors, such as the former members of the Antarctic and Southern Ocean Coalition collaborating in the newly formed Antarctic Ocean Alliance, can be perceived as forms of re-entanglement.

Although research on disentanglements and re-entanglements relates at least indirectly to the notion of agency and intention, entanglement approaches have so far not placed power relations and actors at the centre of their research. This is why, just as Entangled History scholars faced criticism (e.g. from Conrad and Randeria, 2002), other scholars focusing on entanglements have been criticised for disregarding "mechanisms of stratification, exclusion and structures of power more generally" when studying circulation, exchange, mobility and influence (Bauck and Maier, 2015). Entanglement studies are often perceived as neglecting to recognise that higher levels of interdependency are not caused only by intensifications in the communication and transportation of technologies. The combination of the points of view promoted in transnational relation studies and those focusing on entanglements might help quell this criticism. Both open the "container space" typically addressed in traditional area studies when they regard space not as a given but as a constructed entity. Both, consequently, allow one to observe the formation and impact of actor relations (transnational relations) as well as the occurrence and effects of events and mergers at local, national and regional levels (entanglements). If both perspectives are taken into account, the assumption is that the dynamics caused by actors and events will be better grasped, and the circulation and interrelationships of knowledge (Ette, 2012, p. 238) as well as the "multiplicity and simultaneity of knowledge production in different areas" (Kaltmeier, 2014, p. 180) will be duly considered.

In the context of the changing Polar Regions, this book argues that a combination of the transnational relations and entanglement perspectives allows for a better understanding of dominant patterns of interpretation in discourses and of power relations in political processes. Climate change, for instance, emerged as a topic of public and political concern in the 1980s alongside the discovery of the "Antarctic Ozone Hole" (see amongst others Hirst, 2014, p. 4). Since then, in Argentina, Chile, Canada and the U.S. as well as in regional forums such as the Antarctic Treaty Consultative Meetings and later at the Arctic Council, climate change and its impacts have been discussed differently and have shaped the local, national and regional political agendas to very different degrees. As shown in the empirical analysis (Chapters 3–5), this has been caused, unsurprisingly, by events that have had a distinct significance for the areas of focus (e.g. the ice loss from the Larsen Shelf) and by different constellations of actors, their interests and how much say they have in politics. The empirical material considered in this book made it possible to identify those who are able (or unable) to represent their positions in discourses and to trace the differences in the construction of positions in the domestic and regional contexts. It also shows how these representations change as a result of mutual dependencies in multi-level governance. In this way it becomes possible to observe entanglements among actors, (strategic) alliance formation in networks and changing logics of argumentation.

2.2.3 Critical Geopolitics

Critical Geopolitics adds another dimension to the analysis presented in this book, as it scrutinises "geographical representation and practices" specifically (Dalby and Ó Tuathail, 1998, p. 2) and sheds light on the development of and relations between geopolitical lines of argument and worldviews (Albert et al., 2014). Various scholars focusing on the politics of the Polar Regions already lend support to the Critical Geopolitics approach. They have, for instance, shed light on the relationships between humans and the environment (e.g. Dalby, 2003), examined the significance of this relationship for the development of a circumpolar identity (e.g. the "Arctic Nation" Steinberg et al., 2015), pointed to dominant geopolitical narratives and imaginations and investigated the influence of different actors in discourses on the political governance of spaces (e.g. Powell and Dodds, 2014; Steinberg et al., 2015). In these studies, unlike those focusing on "classical" geopolitics, both state and non-state actors are considered, as both are perceived as contributing to geopolitical discourses.

Particularly with regard to the Arctic and Antarctic, Critical Geopolitics challenges the view taken in "classical" geopolitical studies that mostly discuss the significance of the environment on the basis of its geographical and strategic meaning in world politics (e.g. Child, 1988), which they, for instance, relate to the ownership of resources (Dalby, 2003). Critical Geopolitics adds another layer to the understanding of how the politics of the Polar Regions have evolved, because it also sheds light on the strategic meaning given to territory and related "geographical and spatial assumptions about people and places" (which are generally drawn upon in "all international politics" (Dodds et al., 2013, p. 6 and p. 94)). Critical Geopolitics thereby aims at deconstructing the geopolitical arguments and narratives used by political actors to legitimise their geopolitical interests (Albert et al., 2014). To further illustrate what this perspective entails, the following paragraphs first introduce the main assumptions underlying the Critical Geopolitics approach, then elaborate on the methods used in empirical studies by Critical Geopolitics scholars and lastly discuss the gaps and potentials in Critical Geopolitics that affect research on the politics of the Polar Regions.

The political geographers Ó Tuathail, Dalby and Agnew are regarded as the main representatives of this school of thought, which has mostly developed in the discipline of Political Geography but has also been drawn upon by scholars in the field of International Relations. Accordingly, and as Dodds et al. (2013, p. 6f.) emphasise: "Critical geopolitics has no single theoretical canon or set of methods. It rather advances decidedly diverse critiques of, and alternatives to, conventional analysis of international affairs". Similar to the perspectives on transnational relations and entanglements introduced previously, Critical Geopolitics under-stands geographical space not as a natural given. This means, in effect, that when considering political constellations and discourses, Critical Geopolitics scholars draw attention to geographical representations and practices in order to better understand how political power over territory is constructed. They examine how particular conceptions of place (e.g. the "Third World") are produced by political actors (practical geopolitics), strategic institutions (formal geopolitics) and mass media (popular geopolitics) to understand how these become manifest in specific perceptions of world order (Dodds, 2014; Ó Tuathail, 2006), which may then lead to the production of "political realities" (Reuber et al., 2012, p. 16).

Language constructions used in political rhetoric and cartographic depictions of space are the research objects at the centre of interest in Critical Geopolitics. Critical Geopolitics scholars argue that "rather than objective entities, geopolitical construc-tions, including their cartographic representations, are always 'regionalizations' constructed from a one-sided perspective and disseminated to political ends" (Albert et al., 2014, p. 327). Such regionalisations, for instance the geopolitical images of "East" and "West", are critically reflected in Critical Geopolitics as representations within the binary scheme of the "self" and the "other". These geopolitical images and the functionalised language that accompanies them are thus perceived as instruments for constructing and (re)producing a "simple model of the world, which can then be used to advise and inform foreign and security policy making" (Dodds, 2014, p. 5) and which, at the same time, enforces friend–foe modes of thought that contribute to cooperation and conflict. It is argued that this mode of thought – demarcation from the other – contributes to the "making of identities" (cf. Ó Tuathail, 1996).

Thus, in line with constructivist approaches and post-structuralist scholars, in studies applying Critical Geopolitics identity is not perceived as pre-given but as something that is constantly (re)negotiated in discourses (Mose and Reuber, 2011). National territories in particular "have functioned as seemingly stable platforms for the manufacturing and reproduction of national identities" carried out by institu-tions such as the national media or education system (Dodds, 2014, p. 77). Unlike realist perceptions of geopolitics, which are often criticised for overemphasising conflict and promoting a one-dimensional view of global politics, Critical Geopoli-tics does not focus directly on changing geopolitical power relationships, but on the simplification of social complexity in spatial concepts. Accordingly, it tackles the question of why power relationships change by shedding light on the evolution of constructions/imaginaries that are relevant for geopolitics (Dalby and Ó Tuathail, 1998; Reuber and Wolkersdorfer, 2001). To put it in the words of Ó Tuathail:

> The struggle over geography is also a conflict between competing images and imaginings, a contest of power and resistance that involves not only struggles

to represent the materiality of physical geographic objects and boundaries but also the equally powerful and, in a different manner, the equally material force of discursive borders between an idealized Self and a demonized Other, between "us" and "them".

(O'Tuathail, 1996, p. 14f.)

Hence, from the point of view promoted by Critical Geopolitics scholars, geopolitics is perceived "as a form of discourse, able to produce and circulate spatial representations of global politics" (Dodds, 2014, p. 40). Therefore, the perception of spaces is also seen as resulting from dominant representations in discourses. A sense of belonging is created through identity narratives, which "operate at a variety of geographical scales from the subnational to the pan-regional and finally to the global" that are implicated with one another (ibid., pp. 71 and 88) and allow domination over territory. Similar to research on transnational relations, Critical Geopolitics pays attention to supranational actors, to networks of transnational NGOs and international companies in order to examine the impact of their lobbying on these spatial representations (Reuber and Wolkersdorfer, 2001; Dodds, 2014).

Empirical studies driven by theoretical considerations related to Critical Geopolitics often apply discourse analysis or frame analysis to examine the power of a communication text and of the speaker (e.g. Truedsson, 2013). At the centre of such research is the investigation of textual and visual geopolitical representations and practices propagated in different functional and regional discourses. Differentiation between these various geopolitical discourses allows one to understand how practical, formal and popular geopolitical reasoning are interconnected (and entangled) while at the same time making it possible to consider how geopolitical ordering evolves within these different spheres, for instance, how geopolitical imaginaries are produced and circulate within international organisations. Thus, Critical Geopolitics scholars often investigate the knowledge/power nexus asserted by Foucault and aim to illuminate "the precise way in which influence over a human consciousness is exerted by the transfer (or communication) of information from one location – such as speech, utterance, news report, or novel – to that consciousness" (Entman, 1993, p. 51f.).

In epistemological terms the concept of Critical Geopolitics has, however, also provoked criticism. It is, for instance, argued that the heterogeneity of the approach, which draws on different elements when combining different paradigms, theories and methods (e.g. constructivism, post-structuralism, discourse analysis and frame analysis), leads to theoretical and empirical inconsistencies, which are not discussed and fed back to key assumptions (Reuber, 2012). Further, Ó Tuathail (1992) emphasised even during the emergence of the Critical Geopolitics approach that a purely discourse-centred critical geopolitics might underspecify the institutional functioning of geopolitics within political and civil society and disregard its material side. Critical Geopolitics scholars also often neglect "histories of geopolitics within non-western geopolitical imaginations and politics" (Dodds, 2001, p. 471), exclude postcolonial debates on the unequal production and exchange of knowledge, and pay little attention to military affairs and strategy (ibid.). Another issue that has been raised is the rather essentialist concept of actors on which Critical Geopolitics grounds itself when categorising actors into "political elites" or "intellectuals of statecraft" using geopolitical representations in a strategic manner (Reuber, 2012, p. 169f.). Dodds' proposal to include arguments made by feminist scholars such as Dona Haraway can be regarded as a way to overcome this problem:

> First, we need to explore how geopolitics is made and represented to particular audiences. If we want to understand global politics, we have to understand that it is imbued with social and cultural meaning. The current global political system is not natural and inevitable and the stories we tell about international politics are just that – stories. Some narratives are clearly more important than others and some individuals, such as the President of the United States and the President of Russia, are particularly vociferous and emphatic in determining how the world is interpreted.
>
> (Dodds, 2014, p. 39)

Accordingly, and in drawing also on the ideas promoted by scholars focusing on entanglements, actors need to be considered as behaving "relationally" in contexts, as being determined by and determining "particular audiences" and discourses that cross national boundaries. Furthermore, critics stress that deconstructions of geopolitical interpretations lead to reconstructions, to new geopolitical narratives, which are often not reflected and disclosed as such (cf. Albert et al., 2014). Ó Tuathail (2006, p. 6) himself addressed this criticism when demanding that critical thinkers acknowledge that they are never detached observers, meaning that they can "know the world only through the conceptual schemas provided by our culture and languages".

Despite all the difficulties with and limitations of the Critical Geopolitics' approach, its potential lies in providing knowledge of how regional identities and reciprocal differences are propagated in the strategic constructions, namely geopolitical imaginaries, that are used to legitimise actions. Albert et al. (2014, p. 329) underpin the relevance of this awareness-raising approach concerned with the "relativity and inevitably international, strategic character" of representations and imaginaries when arguing that the "political and media landscape [. . .] increasingly privileges rapid, polarizing, sweeping and polemical constructions of geopolitical realities, particularly in cases of conflict" (see also Chapters 3 and 4). As Dalby pointed out, however, the focus of Critical Geopolitics in the 1990s and 2000s was on "questions concerning how the earth is written and read, and how policy is made" that followed the assumption that earth, being a given context, needed to be adjusted. Anthropogenic climate change and its impact on the environment, so Dalby argues, change the "stage" for geopolitics, which is why he says that Critical Geopolitics needs to focus more on related events, because "we are altering the circumstances that give rise to geopolitical thinking in the first place" (Dalby, 2012, p. 12). Dalby here directly refers to the Arctic and melting sea ice and stresses that "only focusing on the scramble for access to the resources that become available once the ice melts is to completely fail to tackle the larger implications of what is now in motion" (ibid.). And although the concrete implications are still unknown, geo-engineering is already indicating what they might look like (see, for instance, Corry, 2017).

2.2.4 Imaginaries

As explained previously, scholars investigating the Arctic and the Antarctic in light of Critical Geopolitics often focus particularly on imaginaries – on geopolitical representations in language and images (cf. Battarbee and Fossum, 2014; Koivurova, 2008; McGhee, 2007; Shields, 1991; Steinberg et al., 2015; Young, 2012a). Reasons for the remarkable number of images and imaginaries of the Polar Regions that currently exist (resource frontier, nature reserve and terra nullius to name but a few) relate to the Regions' exceptionality, particularly their harsh environment and remoteness. At the same time, few people draw upon their own experiences in the Arctic and Antarctic. Instead, they reproduce stories that have been told by others (e.g. politicians, advocacy groups, scholars and the media). Medvedev (2015, p. 2), for instance, argues: "The North is more often communicated than experienced, imagined rather than embodied". Also myths told about early explorations conducted by the adventurers James Cook, John Franklin, Ernest Henry Shackleton and Roald Amundsen fortify imaginations of "the far North" and "the far South". Images of oilrigs and supertankers, of melting sea ice, of polar bears and penguins, on the other hand, are often instrumentalised by advocacy groups to increase attention to the pristine and endangered environment, while pictures of icebreakers and soldiers wearing white camouflage uniforms are used in movies and newspapers to emphasise potential for conflict (as from the scrambles for the Arctic and for Antarctica). But what are imaginaries, how do they evolve and what is their significance in the politics of the Polar Regions?

As I argue elsewhere (cf. Petersen et al., 2016) the terms "image" and "imaginary" are closely related. Maps provide a good example. In their material form, maps can be classified as images that depict territory. They often lead, however, to the perception of space as a natural given. Geography expressed via cartography seems to ground itself on facts, as something that can be measured. The previous paragraphs on Critical Geopolitics have outlined how cartographic representations are always constructed because their production is based on selectivity (what is/is not depicted? what differences and similarities are/are not considered? etc.). Furthermore, cartography does not portray space as shaped by historical processes and influenced by unequal power relations, as something that has been constructed to enforce political and social order, so that maps become an imaginary that is regarded as a powerful instrument for shaping people's views.

The term "imaginary" has been applied by scholars of different disciplines ranging from History and Literature to Philosophy, Political Science, Sociology and Social Anthropology, but, like "entanglement", it has not so far been related to a specific theory or concept. The philosopher Cornelius Castoriadis is often regarded as a pioneer in research on imaginaries. In his 1975 book "The

Imaginary Institution of Society" (originally in French), he understands imaginaries as "something invented" (1998, p. 127) with the capacity to shape societies and their perception of the world, a concept drawn upon later by the historian Benedict Anderson in his book "Imagined Communities" (2006[1983]). The anthropologist Arjun Appadurai, on the other hand, argues that people live not only in Anderson's imagined communities but in "imagined worlds", "that is, the multiple worlds that are constituted by the historically situated imaginations of persons and groups spread around the globe" (2003, p. 33; also in Petersen et al., 2016, p. 4). In his book of the same name (2004) the philosopher Charles Taylor focuses particularly on "modern social imaginaries" that he understands, for instance, as "that common understanding that makes possible common practices and a widely shared sense of legitimacy" (Taylor, 2004, p. 106). In their understanding of political imaginaries, the political scientists Kathrin Keil and Sebastian Knecht follow Taylor's understanding and identify political imaginaries "as organizing principles and cognitive structures for a widely shared sense of a legitimate governance order in which different actors interact and cooperate towards common goals" (2017, p. 6), which can profoundly shape policy outcomes. And, with regard to the Arctic, Steinberg et al. (2015, p. xii) define imaginaries as "foundational myths that provide a framework and reference for everyday life and for future ambitions".

Against this background, like maps, imaginaries can be understood as lenses, "which serve to describe and interpret the world" (Petersen et al., 2016, p. 2), which are used unconsciously and as strategic framings of interpretation. Imaginaries are therefore always complexity-reducing operators, that "are constructed, characterized and embedded into views of the world" (ibid., p. 7), and are often the internalised perceptions of a collective. They may, however, circulate and expand, they may change and evolve over time, they may be entangled and intersect with other imaginaries, they may form part of different narratives on places and spaces and as such they may disappear and reappear in discourses. Not surprisingly, imaginaries, like different views, are contested and some are more dominant than others. Although in this book imaginaries are primarily understood in functional terms, as steering instruments purposely used by actors, they can also gain momentum on their own, when actors are determined by imaginary selves.

In the politics of the Polar Regions particularly, imaginaries are of great significance. The imaginaries of "Arctic states" in the North and "Polar States" in the South are often applied in domestic politics, for instance, to legitimise expensive research and military activities, depicted as means to defend sovereignty (see Chapters 3 and 5). Keil and Knecht (2017, p. 6) further stress that imaginaries order the world for those governing the Arctic, inasmuch as "the modes of contemporary Arctic governance are left to the interests and political ideas of its residents, government and other stakeholders possibly located far away from what is usually seen as 'Arctic'". Additionally, as Steinberg et al. (2015) have outlined, the high number of imaginaries circulating that relate to the Arctic is what has caused the future of the Arctic (and, to lesser degree, that of the Antarctic) to be disputed.

In regard to the questions under analysis in this study, the four different concepts and approaches (Transnational relations, Entanglements, Critical Geopolitics and Imaginaries) introduced earlier can – once brought into play – help to understand different layers in the production of the politics of the Polar Regions. The macro analysis of transnational relations, for instance, draws upon negotiations of common representations and identifications in transnational political spaces (such as the Arctic and Antarctic) and thereby raises the question of how policy-making processes conducted in transnational spheres transfer to the domestic and local levels. Spatial representations and identifications of the kind addressed by Critical Geopolitics provide transnational relations research with the means to analyse on a meso-level why transnational relations change through shedding light on specific perceptions of the world order (imaginaries) that are (re)produced by practical, formal and popular geopolitical reasoning in discourses. Transnational relations and Critical Geopolitics, consequently, intersect. While it has been argued in transnational relations literature that political spheres are shaped by interaction and communication and structured by ideas, through focusing on discourses shaped by politicians, the media and science, research on Critical Geopolitics provides the means to trace how ideas based on imaginaries (on specific perceptions of world order) evolve and help to construct power over territory. The focus on entanglements adds, amongst other things, the variable of time to analyses of transnational relations and Critical Geopolitics since it connects discourses and actors not only to the occurrence of events, that is, historical experiences, but also to the interplay of political and strategic considerations that change depending on the different contexts. Thus, a focus on entanglements allows one to feed back the creation of discourses and relations among actors to the changing constellations of networks (or "centres of power") and to the transformation of formally shared perceptions of the world, which then influences practices and discourses on entities. At a micro-level, the analysis of imaginaries allows for the deconstruction of functionalised language and images, which helps to understand why entanglements dissolve, why spatial representations and transnational relations change.

2.3   Identifying (dis)entangled actors and discourses

> The task of critical Arctic research is to challenge existing discourses claiming truthfulness; to make visible relations between the economic and political spheres; and to recognize that even knowledge by social scientists is relational, often in relation to action or intention.
>
> (Tennberg, 2015, p. 418)

As Tennberg says, it is important to acknowledge that the findings of the analysis presented in this book are shaped (and limited) not solely by the theoretical considerations set forth previously, but also by the research decisions that determined what was examined in the empirical analysis, by the accessibility of material and by my own knowledge. The different theoretical approaches and assumptions, for example, impinged on the selection of the material to be considered. Following the

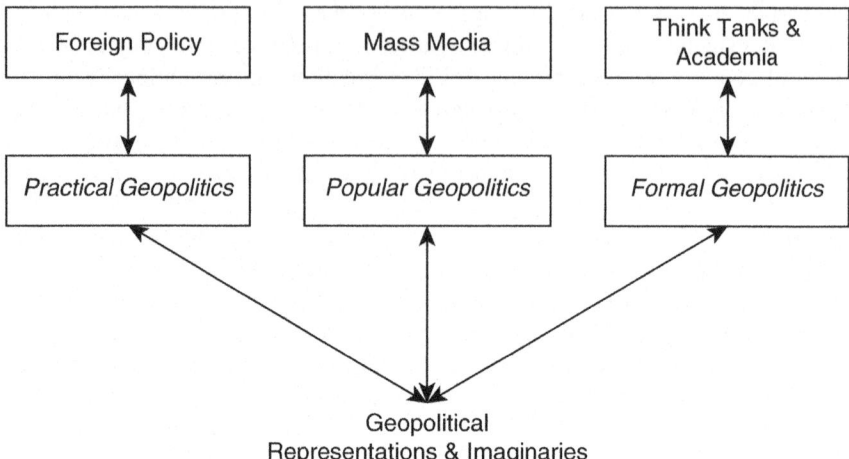

Figure 2.1 Critical Geopolitics discourse typology
(based on Dodds, 2014, p. 42)

critical discourse typology introduced by scholars focusing on Critical Geopolitics (see Figure 2.1), the text corpus analysed includes:

1   policies, statements, strategy and meeting reports released by state and non-state actors in the American Arctic- and Antarctic-rim states and by subsidiary bodies to the Arctic Council or Antarctic Treaty System (Practical Geopolitics),
2   newspaper articles published in American Arctic- and Antarctic-rim states (Popular Geopolitics),
3   think tank papers and academic works (Formal Geopolitics).

With the help of these documents it is possible to outline how entangled national and regional discourses have shaped the politics of the Arctic and the Antarctic since 1989 (cf. Methods, Annex). Information obtained from 26 expert interviews (see Box 7.5, Annex), moreover, provides additional perspectives to complement and challenge the findings of the document analysis.[2]

2.3.1 Policies, statements and strategy reports

In order to identify actors and their concepts that set in train political processes related to the changing Polar Regions, the selection of national policies, statements and strategy reports was based on two superordinate criteria:

1   they should focus explicitly on the Arctic or the Antarctic and
2   they should relate either to the national or the regional level.

These criteria were chosen primarily in order to keep the text corpus manageable. Documents that relate in a minor way to the Polar Regions, such as the U.S. National Security Strategy (2010) that refers to U.S. Arctic interests in only one sentence, are not included. The analysis also only includes non-state actors' participation in the regional bodies. The Antarctic Treaty Secretariat, for instance, provides full, continuous documentation of the statements made by non-state actors in its final meeting reports. The Arctic Council Secretariat likewise allows the participation of non-state actors to be assessed on the basis of participant lists and meeting minutes. The material accessible from the Arctic Council subsidiary bodies, however, differs considerably regarding the text style and the documentation of deliberations at the various ministerial and Senior Arctic Official meetings. Neither the Arctic Council Secretariat nor the Antarctic Treaty Secretariat provides access to internally distributed position papers and working papers, on which statements made by non-state actors in the regional forums are based, as a matter of course. These papers are thus not considered systematically in the analysis; however, they are addressed indirectly whenever interviewees mentioned them.

All reports, government documents and evaluations listed in the Annex (Boxes 7.1 and 7.2) fall into the category of grey literature, which means that these documents do not necessarily reveal how they were produced or why certain pieces of information were included while others were left out. Furthermore, the leading authors and their institutional affiliations are often left unidentified and it is not possible to trace who wrote which sections if more than one author produced the document.

2.3.2 Newspaper articles

In the logics of Popular Geopolitics newspapers are regarded as a form of media with steering power that may influence political agenda settings and the reporting of other media (see e.g. Bottici, 2014; Ó Tuathail, 2006). Newspaper reporting also potentially shapes public opinion and the people's view of events and of political processes. Similarly, this book shows how newspaper reporting (re-)produced and empowered (the circulation of dominant) interpretations and imaginaries in the politics of the Polar Regions in the American polar-rim states. The newspaper analysis considers a total of 29,399 articles[3] published by different newspapers in the American polar-rim states that were selected on the basis of the following criteria:

1   highest average circulation at the national level,
2   significance at the international level, and
3   daily, nationwide appearance to illustrate relevance,
4   production by a different publishing organisation (from other newspapers chosen), and
5   distinct political orientation in order to cover a broad spectrum of opinions.

The selection of newspapers was also discussed with experts whom I interviewed, who confirmed that the following newspapers are those most likely to be consulted by policy makers from state and non-state actor groups.

Table 2.1 Newspapers examined

Canada	U.S.	Chile	Argentina
The Globe and Mail	The New York Times	La Tercera	Clarín
Toronto Star	USA Today	El Mercurio	La Nación
	The Washington Post		Página 12

Although with the rise of new media channels, and particularly of social media, newspapers can no longer nowadays be considered the main medium that shapes public opinion, this development is quite recent. In the 1990s and early 2000s, newspapers were still regarded as the main instrument for influence and as reflecting the public's perception of political processes. Given the period of investigation in this book, therefore, newspaper reporting can properly be regarded as the most telling object for a media analysis.

2.3.3 Think tank and academic works

In the logics of Formal Geopolitics, academics and think tanks are regarded as formal foreign policy actors who "self-consciously invoke an intellectual tradition associated with geopolitics" (Dodds, 2014, p. 41) and thereby contribute to the diffusion of a particular understanding of geopolitical issues, which may become dominant and may even be perceived as common sense. Following this logic, academic literature and assessments by non-governmental strategists in particular are of significant value to understanding priorities in national policy formulation processes as policy makers rely on the policy prescriptions of scholars and consultants who produce knowledge and give orientation. As with politicians and newspapers, however, it is important to keep in mind that, although the "academic field [. . .] still has a certain degree of autonomy from economics and politics, and [. . .] a high potential of self-reflection, [. . .] the production of knowledge – and its funding – is highly political" (Kaltmeier, 2014, p. 176). Think tanks and scholars may thus promote a geopolitical reasoning that underpins their own interests. With that in mind, for the analysis all available and accessible academic literature and think tank reports published in the four American polar-rim states that somehow relate to politics of the Polar Regions in the period of investigation were considered.[4]

The investigation of the material is presented in the empirical chapters (3–5) that follow, which – in light of the previous studies that focused on the politics of the Polar Regions and the theoretical considerations discussed above – concentrate their analysis on Practical Geopolitics, Popular Geopolitics, and Formal Geopolitics particularly on the evolution of

1   transnational relations between state and non-state actors,
2   entanglements between discourses and representations,

3   an inter-American space that can be characterised as a new transnational political sphere in which common representations and identifications that concern the politics of the Polar Regions are negotiated.

Notes

1 Haas (1990) defines epistemic communities as "networks of professionals with an authoritative claim to policy relevant knowledge", whose members are sharing "an episteme with another that they do not necessarily share with other groups or individuals. Such a community shares four common features: a shared set of normative and principled beliefs, which provide a value-based rationale for the social action of community members; shared causal beliefs; shared notions of validity; and a common policy enterprise" (Haas, in Clavin, 2005, p. 427). The Arctic Council can be perceived as a transnational epistemic community, in which politicians and experts form part of a network in which knowledge is co-produced and in which interaction occurs with regularity over time (cf. Risse-Kappen, 1995).
2 The interviewees provided personal insights and examples from their own experience as representatives of state and non-state actor groups contributing to domestic or regional policy making on the Polar Regions. The information they provided was helpful in that it reflected the findings of the document analysis. All interviews were conducted between March 2014 and August 2015 either via email, Skype or at personal meetings in Anchorage, Buenos Aires, Ottawa, Punta Arenas, Santiago, Toronto, Ushuaia and Washington D.C. Following the aim of speaking to as many representatives of different actor groups included in policy formulation processes relating to the changing Polar Regions as possible, a total of 90 persons considered to be relevant were contacted via email beforehand in the American polar-rim states. The word-for-word transcripts and selective protocols of the 26 interviews add up to a total of 226 pages. There is a bias, however, due to the fact that certain actor groups are more represented than others, which does not allow a systematic comparative evaluation of the interviews. Yet, as has been stated before, the information obtained is related to the findings of the document analysis whenever it confirms or challenges them. To comply with research ethics guidelines, the identities of all individuals have been anonymised and all interviewees were informed that their identities would not be revealed.
3 The selected Canadian and U.S. newspapers were accessed and downloaded at the University of Toronto via ProQuest databases and at the Library of Congress, which holds them in electronic form. Within the period of investigation, the five Canadian and U.S. newspapers published a total of 26,969 articles that related in one way or another to the Arctic or the Antarctic. The analysis considers articles in all journalistic genres ranging from features and editorials to readers' letters. After a manual revision of all articles, duplicates and items referring to the Arctic or Antarctic in a different context from the politics of the Polar Regions were removed. A total of 4,899 newspaper articles remained for the analysis. The main national libraries in Chile (la Biblioteca Nacional de Chile) and Argentina (la Biblioteca Nacional Mariano Moreno) did not possess the selected Chilean and Argentinian newspapers in electronic form. A total of 865 non-digitalised articles relating to the Arctic or the Antarctic were accessed in printed form and 1,565 digital articles were accessed via the websites of the selected newspapers for the period from 1 January 1989 to 31 March 2014. Owing to the large volume of newspaper articles published each day by the five selected newspapers and the limited time available to conduct research at the archives in Chile and Argentina, crucial events (listed in Boxes 7.3 and 7.4, Annex) were predefined for each year and selected if they were considered to be of particular significance in local, national or regional contexts or were explicitly addressed in the politics of the Polar Regions. Four weeks before and after the crucial event took place, all articles published by the five selected newspapers were examined.

Thus, in sum, within the period of investigation (1989–2014) 2,430 articles relating to political processes in the Polar Regions and published by the five South American newspapers were retrieved (768 newspaper articles published in Chile and 1,662 newspaper articles published by the selected Argentinian newspapers). As these figures illustrate, the number of newspaper articles considered differs among the countries under analysis, which created a bias that was a factor in the findings presented in Chapter 4.

4 The academic literature and think tank reports considered were retrieved at the following institutions: Library of the University of Alaska Anchorage, University of Toronto Libraries, Departments for Aboriginal Affairs and Northern Development Canada, Library and Archives Canada, Library of Congress (U.S.), Biblioteca Nacional Mariano Moreno (Argentina), Biblioteca del Congreso (Argentina), Biblioteca Antártica de la Dirección Nacional del Antártico (Argentina), Archivo del Ministerio de Relaciones Exteriores y Culto (Argentina), Biblioteca de la Universidad Nacional de Tierra del Fuego, Biblioteca Publica de Punta Arenas, Biblioteca Nacional de Chile, Archivo del Ministerio de Relaciones Exteriores (Chile).

3 Practical geopolitics

Representing the changing Polar Regions in regional and domestic politics

The effects of the changing Arctic and Antarctic environments are experienced, discussed and dealt with very differently in and beyond the Polar Regions. Yet the ways in which state and non-state actors from the American polar-rim states have demanded a say in the politics of the Polar Regions and the framing of their positions have often been alike in the past decades. Why is it, one may wonder, that, despite all the differences, similar patterns of interpretation have become dominant in the politics of the Polar Regions? To explain discursive entanglements of this kind, this "double" chapter pays particular attention to the views and representations introduced in foreign policy (practical geopolitics) at the regional and national levels. It provides the first part of a tripartite analysis continued in the two subsequent chapters. In answer to the question posed earlier and driving this study further, it offers two main explanations:

1 Although the Arctic Council (AC) and the Antarctic Treaty System (ATS) were formed to address different purposes at different times and in different global settings, both diversified since their establishment to deal with the complexity ascribed to the changing Polar Regions and expanded the number of actors involved at times when global interest in the regions grew, when the legitimacy of both polities was challenged, and when resource development and environmental conservation were increasingly under discussion in the Arctic and the Antarctic. In both the AC and the ATS all policy making, however, has remained under the guidance of states (member states to the Arctic Council and Antarctic Treaty Consultative Parties), although the general say of non-state actors has grown over the past decades.

2 At the regional and domestic levels, a similar geopolitical reasoning is most often represented in the Arctic and Antarctic policies and strategies introduced in the four American polar-rim states. In Chile and Argentina, for example, national policies are closely related to developments in the ATS. Both countries have prioritised international cooperation and the strengthening of the ATS as means not only to defend their sovereignty and to keep a say in the politics of Antarctica but also as an opportunity to enhance (particularly inter-American) collaboration with other countries and to gain an international reputation especially with respect to science. In regard to the Arctic and

in contrast to the global significance ascribed to climate change by the AC, however, Canada and the State of Alaska have emphasised the "development of the North" and the use of resource potentials in their respective policies. Moreover, contrary to the positions they have adopted at a regional level, Canada, the State of Alaska and the U.S. have all outlined leadership interests in the Arctic in domestic policies.

But how has the geopolitical reasoning been constructed, reproduced and changed to strategically underpin the specific views of state and non-state actors from the American polar-rim states? And how have similar imaginaries become dominant in the regional and domestic politics of the Polar Regions? Given that:

> How we respond to places ultimately depends on how, where and why we construct places in the first instance and the kind of geographical imaginaries they help to generate and circulate.
>
> (Dodds and Nuttall, 2016, p. 57)

The main answers to these questions that I put forward in this chapter are:

1   The changing Polar Regions have been represented differently in positions adopted by state and non-state actors from the American polar-rim states and in many cases these differing representations illustrate conflicting geopolitical views. However, in both the AC and the ATS an understanding of shared responsibility for dealing with the changing environment in the Polar Regions has grown. Geopolitical positioning has also become more inclusive, particularly in the Arctic where the growing significance ascribed to a global perspective challenges the interests and views of those who fear new modes of dominance that may encourage further processes of disentanglement in the politics of the Polar Regions.

2   Both the AC and the ATS have addressed different topics at different points in time and thus highlighted different management tasks ascribed to the governance of the Polar Regions. "International cooperation" and "the changing environment", however, have been recurring and prioritised topics in both settings. In both, the positions introduced in these regards have shown that "environmental changes" in the Arctic and Antarctic not only have many different impacts but that their management is closely bound up with geopolitical worldviews shaped by past experiences (of domination), by emerging responsibilities (with respect to the future of the planet and the growing populations inhabiting the planet) and by special relationships with particular geographical spaces. The recurring agenda item "international cooperation", on the other hand, has often been addressed as the only way to deal with the depth and complexity of the challenges facing the changing Arctic and Antarctic. In this context, non-state actors were often perceived as contributing to knowledge protection and monitoring, even though the final say remained firmly in the hands of states.

3.1 Regional polities: institutional settings and frameworks

The AC is widely perceived as the main intergovernmental forum for policy making in the Arctic while the ATS is the international forum that administers and manages the Antarctic region. Both polities inherently relate to the Polar Regions, yet, policy making in both is also determined by other international agreements and other polities – such as the United Nations Convention on the Law of the Sea (UNCLOS) and the Organization of American States (OAS). Disagreements over territory in the Polar Regions, for example, are not dealt with among state actors in policy making at the AC and ATS, a fact which underpins the continuing relevance of UNCLOS. Before I focus explicitly on policy making in those two bodies, therefore, the following paragraphs provide an exemplary analysis of how international agreements such as UNCLOS and deliberations conducted in other polities without an explicit polar focus, such as the OAS, impinge upon it.

3.1.1 The United Nations Convention on the Law of the Sea

The 1982 UN Convention on the Law of the Sea provides a significant international framework for the politics of the Polar Regions particularly with regard to matters of sovereignty and (future) resource development. States that wish to request an extension of their control of the continental shelf in the Polar Regions need to pass through the procedures of the Commission on the Limits of the Continental Shelf (CLCS), which scrutinises all applications for sovereignty extensions.[1] Also, as a consequence of the limited ten-year period allowed for the making of CLCS submissions after the ratification of UNCLOS, over the past decade, many circumpolar countries in the Arctic and Antarctic regions have made extensive (and often collaborative) scientific efforts to document the extension of continental shelves.

In the Arctic, all circumpolar states have ratified and support the legal framework of UNCLOS apart from the U.S. The U.S. did, however, sign the agreement and has expressed its commitment to UNCLOS on various occasions (see e.g. the *Ilulissat Declaration*). Russia and Norway submitted their proposals to the CLCS in 2001 and 2006. Norway's claim to an extension of its exclusive economic zone (EEZ) in the Barents Sea was accepted in 2009. In the case of Russia, the Commission proposed additional research (United Nations, 2013). Canada submitted a partial submission in 2013. Concerning its outer limits in the Arctic Ocean, however, Canada filed preliminary information and continued to collect and analyse continental shelf data for another submission planned for 2018 (Global Affairs Canada, 2016). In respect of the northern continental shelf of Greenland, Denmark likewise submitted its proposal to the Commission in 2014 (UNCLOS, 2016).

Newspaper reports in particular have often argued that disputes over ownership will likely lead to interstate conflicts in the Arctic, and some events have also inspired politicians to apply a "conflict rhetoric" to the Arctic – for instance, after Russia's Arktika expedition and its flag planting on the North Pole, which caused extensive and "dismissive" global reactions, "a worldwide media frenzy" (Smieszek and Kankaanpää, 2015, p. 11). As McDorman and Schofield (2015,

p. 220) and others point out, however, such an escalation of interstate territorial disagreements is very unlikely: first, because of the Arctic coastal states' commitment to UNCLOS, second, because of "the long-standing practice" of cooperative problem solving among Arctic states (Stokke, 2015, p. 344) and, third, because even in the areas under dispute disagreeing parties have collaborated bilaterally.[2] Moreover, most territorial disagreements in the Arctic have already been solved:

> the only maritime dispute involving overlapping 200 nm zone claims in the Arctic region is between Canada and the United States in the Beaufort Sea. All the other 200 nm zone overlaps have been resolved – notably, the long-standing [42 years/DW] Norwegian – Russian dispute in the Barents Sea and the potentially more politically fraught Russian – US dispute in the Chukchi Sea.
>
> (McDorman and Schofield, 2015, p. 220)

Likewise, disagreements over the status of the Northwest Passage (NWP) and the Northeast Passage (NEP)/Northern Sea Route (NSR) and their understanding as internal or external shipping straits have not impacted the collaboration between Canada and Russia in the Arctic Council. The Canadian government declared the NWP to be an internal, national waterway in 1973. Since then, Canada has enacted environmental laws with regard to the NWP and demands the expansion of the protection zones from 100 to 200 nautical miles (cf. Jenisch, 2011). Various states (among them the U.S. and the member states of the European Union), however, consider the NWP to be an external shipping strait, fearing, amongst other things, that ships going through the NWP will be regulated and taxed by the Canadian government. To strengthen its own controversial position, Russia supports Canada's claim as it similarly regards the NEP and NSR as internal straits that pass straight through the Russian ocean zones along the country's northern coast. Although the diminishing summer sea ice extent in the Arctic has encouraged numerous predictions regarding the navigation of these shipping straits that offer up to a 40% reduction in travel distance compared to the Suez and Panama Canals, it is unlikely that the still minimal use of these straits will lead to international conflict in the near future.

In the Antarctic region, circumpolar countries such as Chile and Argentina recognise the territory claimed in Antarctica as part of their national territories. Although Argentina's (1943) and Chile's (1940) claims, like those of Australia (1933), France (1924), New Zealand (1923), Norway (1939) and the United Kingdom (1908) are considered as "frozen claims" in the Antarctic Treaty, the seven claimant states are also coastal states and made submissions to the CLCS. With the exception of Australia (2004), Argentina (2009) and the United Kingdom (2006), the submitting states, however, did not include territorial claims in the (sub-) Antarctic region in their applications. In the (unlikely) event of the ATS collapsing in the future, however, New Zealand, Norway and the United Kingdom have reserved the right to make a future outer continental shelf submissions regarding their Antarctic territories.

In 2008 the CLCS made a decision for the first time on territory located in the sub-Antarctic region. It approved most of the Australian submission and defined the extended continental shelves of Australia's sub-Antarctic islands. Argentina's submission is the only one that included "frozen claims" to the continental shelf areas in the Antarctic Territory. Despite international protests, in March 2016 the CLCS accepted Argentina's claims to a wider Atlantic territory but "decided that it was not in a position to consider and qualify those parts of the submission that were subject to dispute" (Fitch, 2016), and thus particularly excluded recommendations regarding the sovereignty question of the sub-Antarctic Islas Malvinas/ Falkland Islands, South Georgia and the so-called Argentine Antarctic Territory. The United Kingdom correspondingly included the Falkland Islands/Islas Malvinas, South Georgia and the South Sandwich Islands in its submission. The CLCS is expected to come to a similar conclusion as in the Argentine case.

As with the territorial disagreements in the Arctic, it is often argued that the disputed sovereignty of the Falkland Islands/Islas Malvinas has affected bilateral relations between Argentina and the United Kingdom and also "their level of cooperation within the ATS" (Liggett, 2015, p. 68). Although the long-lasting dispute over the location of the Antarctic Treaty Secretariat underpins this perception, it should be noted that both countries continued cooperating in the ATS and even conducted bilateral inspections in Antarctica, where inspections are usually undertaken unilaterally (Interview, January 2015), thereby demonstrating that their political relationship is determined by many other factors besides the dispute over the Islas Malvinas/Falkland Islands. On the other hand, during their dictatorships the disputes between Chile and Argentina over the control of the Drake Passage and the Beagle Channel complicated their cooperation in the Antarctic significantly (Interview, January 2015), and earlier, during the International Geophysical Year, Chile and Argentina were even perceived as "reluctant collaborators" (Howkins, 2008). In recent decades, however, Argentina and Chile have been working as partners. Both countries collaborate bilaterally in the Antarctic and also in various alliances such as in Mercosur, UNASUR and OAS.

Overall, however, both the territorial disagreements in the Polar Regions and the CLCS's procedures in approving applications for sovereignty extensions are questioned by those who consider this practice to be a continuation of colonising processes and consider the Arctic and Antarctic as a common heritage (Dodds, 2011, p. 241; Kinossian, 2016; Koivurova, 2009). In this regard, some also argue against the CLCS procedure because it excludes non-polar-rim states and non-state actors such as Indigenous peoples (The Arctic Institute, 2016). Particularly the latter's widely considered "legitimate historic claims to the waters of the North" (ibid.) have only been dealt with at the national and subnational levels. Other international instruments such as the ILO Convention 169 (1989) and the UN Declaration on the Rights of Indigenous Peoples (UNDRIP, 2007), however, promote and protect the rights of Indigenous peoples more generally and are also referred to by those who perceive Indigenous peoples not solely as stakeholders but also as rightsholders.[3]

It is noteworthy that in the AC and ATS territorial disputes *between states* are not addressed but dealt with in accordance with CLCS procedures. Yet, the experience of a colonial past and the criticism predominantly expressed by *non-state actors* with respect to international agreements such as UNCLOS are essential to the reasoning introduced by non-state actors in these regional settings.

3.1.2 The Organization of American States

The formation of positions by and in the four American polar-rim states (Argentina, Chile, Canada and the U.S.) is also influenced by deliberations that take place in other international settings besides the AC and the ATS, notably the Organization of American States (OAS), of which all 35 independent states of the Americas – including the polar-rim states – are members. Furthermore, the OAS is the only regional entity that represents states bordering both Polar Regions and more than one-third (13) of its member countries also form part of the ATS. The OAS has also been addressed by non-state actors from the American polar-rim states: The Inter-American Commission on Human Rights (IACHR), an autonomous body of the OAS, was, for instance, approached by the Arctic Athabaskan Peoples from Canada and the U.S. in 2013. They filed a petition claiming violations of their human rights resulting from rapid Arctic warming and melting. As is shown in the following, however, although climate change, environmental protection (in the American hemisphere) and the issue of the Islas Malvinas/Falkland Islands are regularly addressed in meetings conducted under the OAS's auspices, the Arctic and Antarctica are not generally depicted as being of relevance to the Organization and also seem not to be included in the hemispheric understanding of its 35 member states.

The founding members formed the OAS "to achieve an order of peace and justice, to promote their solidarity, to strengthen their collaboration, and to defend their sovereignty, their territorial integrity, and their independence" (Charter of the Organization of American States, 1948, Art. 1). Although environmental concerns are not prioritised, the OAS also claims to "support [. . .] member states in the design and implementation of policies, and projects to integrate environmental priorities into poverty alleviation and socio-economic development goals" (OAS, 2016). Consequently, the Polar Regions are likely also to be of significance for the hemispheric environmental concerns addressed in the OAS's General Assembly and subsidiary bodies.

While one might assume that the OAS has thus contributed to the formation of a joint inter-American perspective on the changing Polar Regions, in its resolutions and declarations up to 2015 neither the changing Arctic nor Antarctic appears as an explicit reference point. In other words, despite a general consideration of environmental concerns and of climate change, in none of the declarations and resolutions of the General Assembly Proceedings or the accessible Secretary General Reports (2009–2015) are the changing Polar Regions regarded as being of specific significance for the American hemisphere – the only exception being the disputed Falkland Island/Islas Malvinas. This lack of attention to the changing

Polar Regions is notable, as, in contrast to the OAS, other regional organisations have shown a much greater interest in the politics of these regions (for example, the European Union (EU) that even introduced an EU Arctic Policy in 2016). Moreover, as the hemispheric perspective taken by OAS has included neither the Arctic nor the Antarctic, the understanding of the American hemisphere promoted by OAS can thus be perceived as undermining that of American polar-rim states, who stress a hemispheric understanding that includes the Polar Regions. However, the shared positions of OAS member states regarding (1) the protection of the environment, (2) climate change, and (3) the disputed sovereignty over the Islas Malvinas/Falkland Islands certainly provide options to connect to discourses conducted in the AC and ATS in the future.

3.1.2.1 The protection of the environment

Although in the past decades, environmental issues and the significance ascribed to the environment have remained subordinate topics in the OAS General Assembly, environmental concerns have been embedded in a reasoning similar to that which predominated in the AC and ATS. In the 1990s, for example, the OAS often addressed environmental concerns in the light of sustainable development and stressed its significance for "future generations" (cf. e.g. the *Declaration of Asunción*) while in the 2000s environmental protection was particularly related to human rights and security. Although the OAS General Assembly had discussed the matter of nature conservation before, it was in 1989 that the OAS Permanent Council received an extended mandate to assess the creation of an inter-American system for nature conservation (*OAS Resolution No. 1016*, 1989; *OAS Resolution No. 1050*, 1990). The protection of the environment was thus already identified as an issue pertaining to the entire American hemisphere and requiring collaborative measures on the part of the American states. In the early 1990s, however, the protection of the environment was primarily related to the development of resources and linked to economic aims and not directed against global warming and climate change (cf. e.g. *Resolution No. 1114* from 1991). This changed in 1994 with the *Declaration of Belem do Para*, in which the OAS member states explicitly referred to climate change and declared:

> bearing in mind, among other things, the importance of the political will of all states to adopt and implement appropriate measures in environmental management, with respect for biodiversity, climate changes and the elimination of pollution caused by toxic waste.

Although the understanding of environmental protection had already been broadened by that time, the OAS member states also stressed their aim "to strengthen cooperation among member states with a view to contributing to environmental protection and promoting sustainable development in the region" (*OAS Resolution No. 1286*, 1994). Environmental conservation thus continued to be primarily understood as being linked to sustainable development and the "prudent use of natural resources" (cf. the 1995 *Resolution on the Inter-American Program of Action for Environmental Protection*).

It was in the 2000s that environmental issues were increasingly linked to human rights and security: With the adoption of *Resolution No. 1896* in 2002, for instance, the OAS member states resolved to "remain seized of the issue human rights and the environment" and to "encourage institutional cooperation in the area of human rights and the environment". In the subsequent 2003 *Resolution No. 1926*, it was more explicitly stated that the member states acknowledged "a growing awareness of the need to manage the environment in a sustainable manner to promote human dignity and well-being".

Particularly in the regional and domestic politics of the Polar Regions, environmental concerns are often related to security concerns. In its declarations and resolutions, however, the OAS did not allude to the Polar Regions in that connection. Only *Resolution No. 1185* of 1992 referred indirectly to the Polar Regions in the context of overfishing in the high seas, when the member states declared that they were "[a]larmed at the environmental degradation caused by overfishing on the high seas, particularly in the waters adjacent to areas subject to national jurisdiction of American states".

3.1.2.2 Climate change

Apart from the aforementioned reference to climate change(s) in 1994, hardly any other OAS resolution or declaration referred to climate change until 1999. In 1999 for the first time a resolution was adopted that was entitled *Climate Change in the Americas* (No.1674). Like all subsequent annual resolutions it emphasised:

> [t]he urgent need for all member states to begin the process of planning for adaptation to global climate change and to put in place measures to mitigate the possible adverse effects of climate change in the Americas.

These resolutions further demanded that member states consult within the OAS "on the ways and means of addressing climate change in the Americas".

Though the cautious formulation "possible adverse effects of climate change" hints at an understanding of climate change that is still imbued with much uncertainty and is not directly related to natural catastrophes, in 1999 the OAS member states did not doubt that climate change was taking place and referred to it as a "challenge". The 2008 Resolution (No. 2429) further builds on the wording of the 1999 resolution and outlines that the OAS member states take into account the findings of the Intergovernmental Panel on Climate Change (IPCC) "on the adverse effects of climate change, in connection, inter alia, with flooding risks and the dangers of sea-level rise". Moreover, in this and in subsequent resolutions the General Assembly emphasised "[t]hat climate change is a shared concern of all humankind, and that its effects have an impact on sustainable development and could have consequences for the full enjoyment of human rights".

Under consideration of the Fourth Assessment Report of the IPCC, the 2010 *Resolution No. 2588 on Climate Change in the Countries of the Hemisphere* uses a stronger wording when referring to the "responsibility" that

> the Organization of American States (OAS) member states and the international community share [. . .] of finding effective and equitable solutions to climate change in accordance with the principle of common but differentiated responsibilities and their respective capabilities.

In line with the more general understandings of environmental issues analysed before, the OAS also connected climate change to the question of vulnerability. In the 2012 Declaration of Cochabamba, for instance, it acknowledged

> that there are environmental threats and diverse problems in the region to do with food production, access, and consumption, which are exacerbated by extreme weather conditions, water shortage, and climate change that can or do adversely affect agriculture production and the poorest populations.

Meeting in Cochabamba, a city located in the plurinational state of Bolivia, one-third of which is covered by part of the Andean mountain range and which is known for its multi-ethnic population, the General Assembly subscribed for the first time to the pan-Andean concept of "Mother Earth" (also known as "Pachamama") in a declaration in which the member states recognised that

> Mother Earth is a common expression for the planet earth in a number of countries and regions, which reflects the interdependence that exists among human beings, other living species, and the planet that we all inhabit.

Thereby, the General Assembly no longer objectified nature and the environment as things needing to be controlled, dominated, managed and preserved by humans but changed its dominant understanding of nature, acknowledging also the agency ascribed to the planet and environment, for instance by groups of Indigenous peoples. In 2014, however, the new significance given to climate change and the environment in the western hemisphere was not included in the renewed Strategic Vision of the Organization of American States (*Resolution No. 2814*). The vision statement adopted did not refer to environmental concerns or to the "Mother Earth" concept and subsequent resolutions once again placed climate change in the context of sustainable development.

Overall and as in other more general agreements on environmental concerns, even when dealing with climate change, no reference was made to the Polar Regions by the OAS General Assembly in the period of investigation. However, the issues mentioned in these contexts – such as rising sea levels – did relate directly to the changing polar environment and could have provided entry points for an exchange of knowledge among the AC, the ATS and the OAS. Moreover, a similar reasoning is applied when referring to "uncertainty" and "shared responsibility" in the context of climate changes as in discourses in the AC, the ATS and in the American polar-rim states.

3.1.2.3 The disputed sovereignty over the Islas Malvinas/Falkland Islands

"The Question of the Malvinas Islands" is the only topic addressed by the OAS General Assembly that is of direct relevance to the politics of the Polar Regions – although, strictly speaking, the Islas Malvinas/Falkland Islands are located in the sub-Antarctic region and thus do not directly pertain to policy making in regional polar polities. The question of sovereignty over the Malvinas/Falklands is addressed on an annual basis, and the related resolutions and declarations are often even placed at the beginning of the summary of proceedings. Although the United Kingdom is a permanent observer at the OAS, the declarations and resolutions are always based on presentations by the head of delegation of the Argentine Republic as an OAS member state. The content of all declarations only differs slightly but they all refer to the islands under dispute as the Malvinas Islands (reproducing the Argentine understanding and not the British name "Falkland Islands").

In every declaration and resolution adopted during the period of investigation, the member states declare that "the question of the Malvinas is one of enduring interest in the hemisphere". Both parties – Argentina and the United Kingdom – are usually either called on "to move ahead in their efforts to reach a definite solution to all their differences as soon as possible" (e.g. OAS Resolution No. 984, 1989; OAS Resolution No. 1049, 1990) or to "resume the negotiations in order to find, as soon as possible, a peaceful solution to the sovereignty dispute" (OAS Resolution No. 1100, 1991; OAS Declaration No. 2, 1992). In other words, the OAS member states have been promoting peaceful negotiations on the disputed islands and have not stated their support for any planned military intervention by Argentina. Indeed, OAS member states place the will of the island's inhabitants above any geological justification for territorial belonging and emphasise the general interest of the OAS in maintaining and intensifying its relationships with the United Kingdom (see also *Declaration No. 48* from 2006). In the 2007 *Malvinas Declaration* (No. 53), the OAS members, however, explicitly referred to the disputed territory (very likely in reaction to the 2006 CLCS submission made by the United Kingdom) and added to the before mentioned

> that despite those ties and shared values, it has not yet been possible to resume the negotiations between the two countries with a view to solving the sovereignty dispute over the Malvinas Islands, South Georgia and the South Sandwich Islands and the surrounding maritime areas.

Thus, even though Argentina has a greater say than the United Kingdom at the OAS (particularly because of its membership) and has used the OAS to regularly express its position on the Islas Malvinas/Falkland Islands, the OAS has not strengthened Argentina's position in its resolutions and declarations. Quite the opposite. By prioritising the will of the inhabitants and emphasising the "shared values" and ties with the United Kingdom, the OAS has actually weakened Argentina's claim.

In sum, although the OAS has not directly referred to the Polar Regions in the declarations, resolutions and Secretary General's reports that I have examined, many of the positions it has adopted also apply to questions of concern in the politics of the Polar Regions (the conservation of nature, climate change, the understanding of sustainable development, and also the disputed Islas Malvinas/Falkland Islands). Except on the latter issue, however, as the preceding elaborations illustrate, the OAS has not contributed to the formation of a common inter-American perspective on the Polar Regions in the past decades. When the positions introduced by the American polar-rim states in the AC and ATS are considered later in this chapter, it will become clear, nonetheless, that there is a great deal of potential for such a perspective.

3.1.3 Institutional settings provided by the Arctic Council and by the Antarctic Treaty System

To assess the extent to which different state and non-state actors from the American polar-rim countries are able to introduce their different views into the politics of the Polar Regions, the main regional governance settings – the Arctic Council and the Antarctic Treaty System – would seem to be of particular significance. But how important actually are the AC and the ATS as far as policy making in the Polar Regions is concerned? To identify structural dynamics and power relations that have an impact on whose understandings are listened to, circulated and reproduced, it is also important to ask what the formal settings provided by these regional governance institutions are and who participates in policy formulation processes there?

3.1.3.1 The significance and structure of the Arctic Council

The AC was formed following the historic Rovaniemi Meeting in 1989, which resulted in the negotiation of the Arctic Environmental Protection Strategy (AEPS) and its adoption in 1991. This meeting, which led to institutionalised circumpolar cooperation, was attended by the eight circumpolar countries – the Arctic 8 or "Arctic states" as they are often called: Canada, Denmark, Greenland (a self-governing territory under the Danish crown), Finland, Iceland, Norway, Russia, Sweden and the U.S. – and by three Indigenous peoples' organisations, the Saami Council, the Inuit Circumpolar Conference (later renamed the Inuit Circumpolar Council or ICC), the Association of Indigenous Minorities of the North, Siberia and the Far East of the Russian Federation (later renamed the Russian Association of Indigenous Peoples of the North, RAIPON). In 1996, the state parties and Indigenous peoples' organisations with observer status to the AEPS established the AC as an intergovernmental forum. Its aim was to enhance "cooperation, coordination and interaction among the Arctic States, with the involvement of the Arctic indigenous communities and other Arctic inhabitants on common Arctic issues" (Ottawa

Declaration, 1996, Art. 1a). While Arctic military security was explicitly excluded from the AC's mandate, as "pushed by the USA" (Graczyk and Koivurova, 2015, p. 305), sustainable development and environmental protection in particular were addressed as envisioned fields of cooperation and interaction (ibid.). As examined in more detail in the next section, however, thematic priorities have shifted in the past 20 years and military security and geo-strategic issues are no longer entirely left out of the AC's deliberations. This shift has been triggered by changing global settings or, as Graczyk and Koivurova (2015, p. 311) point out, by "the simple fact that the council was conceived and established in a different political reality, when the impacts of climate change were little recognized or explored".

The AC was established as a soft-law forum for non-binding cooperation and discussion, which means that it negotiates on recommendations and concludes multilateral agreements that commit states only politically, not legally and are reached by consensus of the Arctic 8. This soft-law characteristic and focus on programme activities "such as knowledge building and capacity enhancement rather than international regulation" has been "an important feature" of various Arctic-specific institutions (Stokke, 2015, p. 332). In the recent past, and contrary to the common practice of other circumpolar institutions, however, the AC member states have also negotiated three legally binding agreements "pertaining to the most critical issues related to economic activities in the region" (Graczyk and Koivurova, 2015, p. 318): the Agreement on Cooperation on Aeronautical and Maritime Search and Rescue in the Arctic (2011), the Agreement on Cooperation on Marine Oil Pollution, Preparedness and Response in the Arctic (2013), and the Agreement on Enhancing International Arctic Scientific Cooperation (2016).

Not solely against this backdrop, but generally, as Kankaanpää and Young (2012, p. 1) emphasise, the AC is perceived as an institution that "has achieved considerable success in identifying emerging issues, framing them for consideration in policy venues and raising their visibility on the policy agenda". The Council is also regarded as the "promoter or voice of the Arctic" (Heininen, 2004, p. 33) and accordingly, members of the AC are perceived as having an important say in the Arctic region. At the 9th Ministerial meeting in Iqaluit, for instance, the Danish Foreign Minister declared the AC to be "the primary forum for policy making in the region" (Arctic Council, 2015c). This view is also shared by Indigenous peoples' organisations that have obtained permanent participant status in the intergovernmental forum and acknowledge the rather exceptional rights granted as a result of this status in the AC. Jim Gamble from the Aleut International Association, for instance, stated that "the Arctic Council remains the preeminent international forum for dialogue and action on Arctic issues, and the only forum to include the voices of the Indigenous Peoples of the Arctic in a meaningful way" (Aleut International Association, 2013, p. 57).

The successes ascribed to the AC are certainly related to its transformation over the past 20 years. As is shown in the following, its structure has changed with the establishment of new subsidiary bodies coordinating its Arctic-issue-related work (see Figure 3.1) and with the growing number of participants involved in its work in general. In addition, the variety and specification of topics addressed by the Council

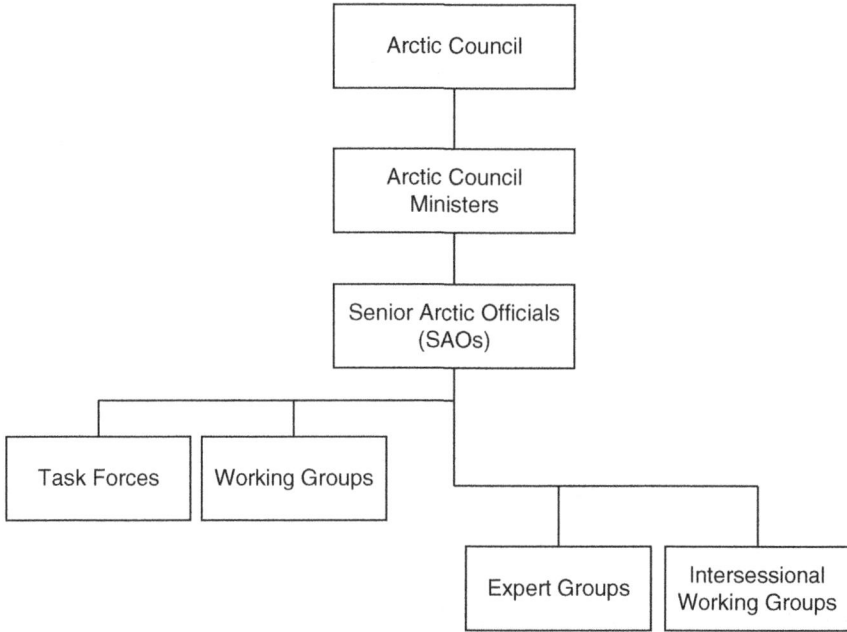

Figure 3.1 Institutional architecture of the Arctic Council

has expanded. Decision-making competencies, however, remain in the hands of the Arctic states, while Indigenous peoples' organisations with permanent participant status – later also given to the Aleut International Association (AIA) in 1998, the Arctic Athabaskan Council (AAC) in 2000 and to the Gwich'in Council International (GCI) in 2000 – have full consultation rights, contribute to agenda-setting processes and to deliberations in all meetings. Although these consultation rights are not linked to their consent, their say is generally regarded as being "close to a de facto power of veto" (Koivurova and Heinämäki in Duyck, 2015, p. 25) in the event that all permanent participants (PPs) reject a particular proposal.

The structure of the AC diversified with the implementation of new working groups and the formation of task forces and expert groups as additional subsidiary bodies. In 1997, the AEPS was incorporated by the Arctic Council as well as the related working groups (the Arctic Monitoring and Assessment Programme, AMAP; Conservation of Arctic Flora and Fauna, CAFF; Protection of the Arctic Marine Environment, PAME and Emergency, Prevention, Preparedness and Response, EPPR). All present working groups operate under a specific mandate according to which they initiate all AC projects and actions, which is why the work of the AC is often regarded as following the bottom-up principle (Graczyk and Koivurova, 2015; Spence, 2015). Before being implemented, however, these projects are discussed at every level of the AC architecture.

With the 2009 Tromsø Declaration, the first task forces were established. Since then, task forces have been mandated to identify issue-related measures, to provide recommendations (e.g. on future needs, mechanisms and actions), to negotiate international instruments, and to report on progress at the next ministerial meeting. Additional expert groups support the agenda-setting and policy formulation processes of working groups. All AC subsidiary bodies – working groups, task forces and expert groups – are implemented by Ministers after the respective agreement has been reached and acknowledged following the biennial ministerial meetings on the declarations in question. Generally, these declarations set the agenda for the ensuing two years and, as I argue elsewhere (Wehrmann, 2016a), these agendas are also shaped by the chairmanship programmes of the "Arctic states". The AC chairmanship as well as the chairmanships of the working groups rotate among the Arctic States. Task forces are usually implemented for the duration of an AC chairmanship and work actively on specific issues delegated to them over the period from the conclusion of one ministerial meeting to the conclusion of the next (cf. Arctic Council, 2013a). They become inactive after they have provided the desired reports to Senior Arctic Official group meetings and ministerial meetings. Senior Arctic Officials (SAOs), like Arctic Council Ministers, are appointed by each of the Arctic states. Working groups and task forces usually work in close cooperation; however, they are not always interrelated. The Task Force to Facilitate the Circumpolar Business Forum, for instance, provided information independently to SAOs and Ministers. Expert groups, on the other hand, are usually connected to working groups. The Arctic Human Health Expert Group (AHHEG), for instance, is a subsidiary body of the Sustainable Development Working Group and also responsible for that Group's human health agenda (Arctic Council, 2015b). All programs and projects carried out by AC subsidiary bodies are "under the guidance and direction of SAOs" (Arctic Council, 2013a, p. 7).

Although the diversification of the AC provides additional settings for deliberations on complex issues where different actor groups can exchange their points of view, these deliberations are particularly shaped by member states (and to a lesser degree by PPs). Interviewees further stated that, for non-state actors, the current large number of AC subsidiary bodies causes difficulties in terms of coordination, financing and especially of efficiency as similar issues are discussed in different settings (cf. Interviews, August 2015). Further, actors with fewer resources than others (e.g. permanent participants) are concerned about the growing human and financial resources needed by those who wish to participate in all subsidiary bodies (Javo, 2015). As elaborated in more detail in the next section which endeavours to shed light on the question of who participates in policy formulation processes in the AC and ATS, although the participation of non-state actors has generally increased since the formation of the AC, the processes and output that it generates remain under the guidance of the Arctic 8.

3.1.3.2 The significance and structure of the Antarctic Treaty System

In Antarctica all activities are regulated by the Antarctic Treaty and the related agreements, which are collectively known as the Antarctic Treaty System (ATS).

Antarctica has no permanent population and is accordingly regarded as terra nullius – nobody's land.[4] The Antarctic Treaty was adopted in 1959 after the landmark International Geophysical Year (1957–1958) as a result of the Washington Conference on Antarctica, "en un momento histórico en que la principal preocupación era la Guerra Fría" [at a historic moment in which the main preoccupation was the Cold War] (Ortúzar, 2010) and has since been acceded to by 53 countries (ATS, 2015). Any disagreement concerning the interpretation and application of the treaty that is not resolved in negotiations is settled under the authority of the International Court of Justice (Antarctic Treaty, Art. XI). The main purpose of the Antarctic Treaty has been to ensure "in the interest of all mankind that Antarctica shall continue for ever to be used exclusively for peaceful purposes and shall not become the scene or object of international discord" (Antarctic Treaty Secretariat, 2016). At the time of its negotiation, the 12 original signatories of the Antarctic Treaty had significant interests in Antarctica: Argentina, Australia, Chile, France, New Zealand, Norway and the United Kingdom, on the one hand, claimed territorial sovereignty over areas in Antarctica, which in the cases of Argentina, Chile and the United Kingdom overlapped. Belgium, Japan, the Soviet Union, South Africa and the U.S., on the other hand, did not claim any territory in Antarctica but conducted exploration activities there. Although neither the U.S. nor the Soviet Union recognised the territorial claims put forward, both reserved the right to assert their own claims and built research stations in any sector as a sign that they did not accept the division of Antarctica into sectors (Interview, January 2015). The negotiation and adoption of the Antarctic Treaty did not provide a solution to the disputes on overlapping territorial claims, but the signatory states adopted Article IV, which states:

> No acts or activities taking place while the present Treaty is in force shall constitute a basis for asserting, supporting or denying a claim to territorial sovereignty in Antarctica or create any rights of sovereignty in Antarctica. No new claim or enlargement of an existing claim to territorial sovereignty in Antarctica shall be asserted while the present Treaty is in force.

Accordingly, all territorial claims that were made before the adoption of the Antarctic Treaty have since been treated as "frozen claims", which has allowed the demilitarisation of the Antarctic continent (in accordance with Art. I.1–2, Antarctic Treaty, 1959) "and provided for its cooperative exploration and future use" (U.S. Department of State, 2016).

Although, particularly in the 1980s and 1990s, the ATS has been critically depicted as a "closed entity", "a vehicle for domination by an exclusive club of largely wealthy countries" (Stokke, 1991, p. 139), or as a "colonial club" (Howkins, 2016, p. 164),[5] it has also been widely regarded as a stable system (Liggett, 2009) and the Antarctic Treaty is often considered one of "the most successful multilateral agreements negotiated in the twentieth century" (Liggett, 2015). This success is often ascribed to the foresight of its drafters. Although in the 1950s the treaty was predominantly negotiated to address geopolitical tensions and political security (its purpose being to demilitarise Antarctica and secure peace there), at a time when environmental concerns did not yet provide an additional impetus, it laid the "foundations of environmental governance in the Antarctic" as Article IX provides the "right of parties to develop new regulatory mechanisms that target 'matters of common interest' such as the 'preservation and conservation of living resources in Antarctica'" (see also Liggett, 2015, p. 63). Furthermore, the aforementioned purpose to ensure that Antarctica "in the interest of all mankind" shall forever be used exclusively for peaceful purposes (as stated in the preamble of the treaty) already refers to the principle of the common heritage of mankind, which was not officially introduced until 1979 (in the Agreement Governing the Activities of States on the Moon and Other Celestial Bodies, Wolfrum, 1983). In the context too of collaboration in view of climate change, the ATS has received global political recognition as an example of an institution "where sovereignty matters were set aside and where peace and science were moved to the fore" (Liggett, 2015, p. 61).

Although the flexibility built into the Antarctic Treaty allowed the expansion of the regulatory framework applying to Antarctica (see Figure 3.2), the adoption of the new agreements and treaties that at present constitute the ATS permitted the regional governance system to manage emerging issues over time. More recently, however, the ATS and particularly the Antarctic Treaty Consultative Parties (ATCPs) have been criticised for "a lack of enthusiasm" and "a lack of administrative capacity to deal with the complexity of human activities and environmental concerns in the Antarctic" (Liggett, 2015, p. 61). This criticism pertains particularly to issues such as growing commercial activities and scientific whaling, the regulation of tourism, biological prospecting and illegal unregulated and unreported (IUU) fishing, concerns regarding continental shelf delimitation and related sovereignty claims, as well as challenges arising from climate change (Chaturvedi, 2011; Liggett, 2015).

In contrast to the Arctic Council, the ATS is based on a set of binding international agreements to which all contracting parties are legally committed. Its primary significance is undisputed in the Antarctic region particularly due to the broad international acceptance expressed through the number of its signatories. Like all other ATS agreements, the Antarctic Treaty, which forms the basis of the System, is "open for accession by any State which is a Member of the United Nations, or by any other State which may be invited to accede to the Treaty with the consent of all the Contracting Parties whose representatives are entitled to participate in the meetings provided for under Article IX of the Treaty" (Antarctic Treaty, Art. XIII). According to Article IX.2, contracting parties need to demonstrate their interests in Antarctica "by conducting substantial scientific research activity there, such as the establishment of a scientific station or the despatch of a scientific expedition" to receive consultative status. While most of the Antarctic Treaty Consultative Parties with voting rights have established a scientific station, this is not obligatory. The Netherlands, for instance, became a Consultative Party in 1990 without maintaining a scientific base in Antarctica. Nevertheless, conducting research in Antarctica is still a very costly undertaking due to the logistics

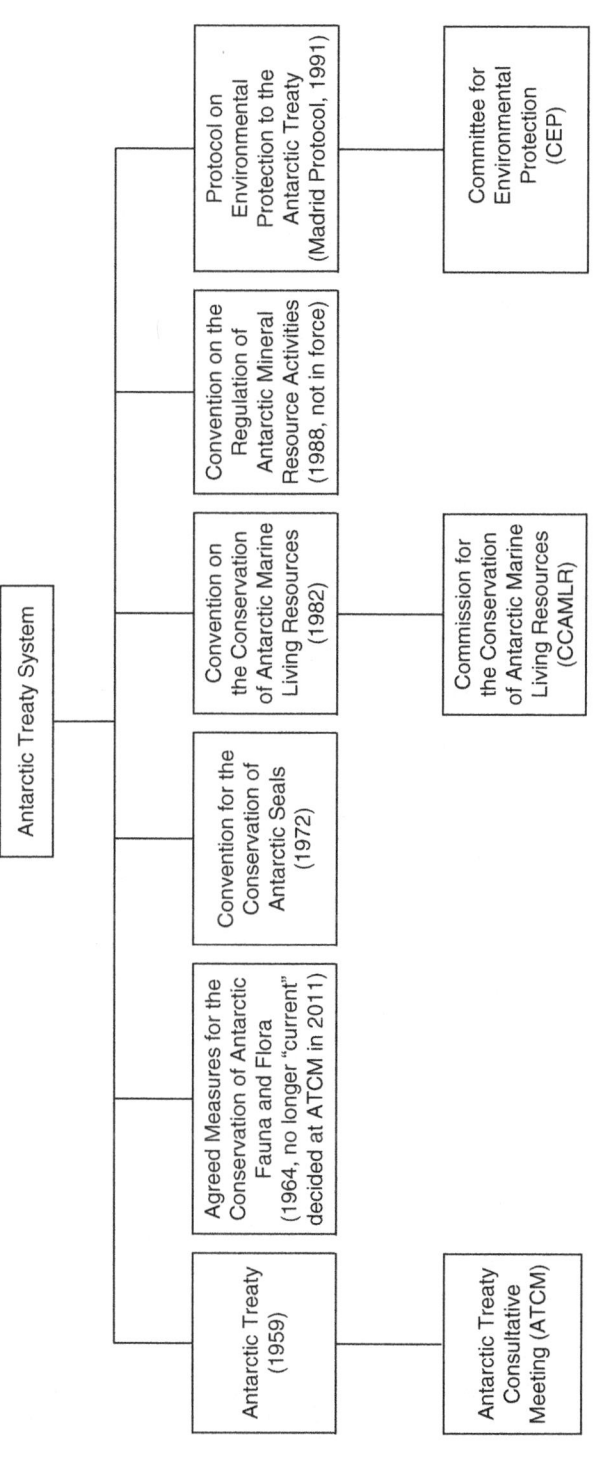

Figure 3.2 Institutional architecture of the Antarctic Treaty System

and special conditions necessitated by the harsh environment. In this regard, particularly so-called emerging countries have had difficulties in justifying Antarctic campaigns domestically (Interview, March 215).

In addition to the 12 original signatories, 18 countries have obtained voting rights in the past. They belong to the group of the 30 ATCPs who decide all matters by consensus. The Consultative Parties also form the Council of Managers of National Antarctic Programs (COMNAP), which connects national Antarctic programs and promotes the international exchange of information and the foundation of international partnerships. The 24 non-Consultative Parties that joined the Antarctic Treaty without meeting the requirements of Article IX.2 are excluded from decision-making. With the adoption of the Canberra Recommendation XII-6 by the ATCPs in 1983 they were, however, invited to attend Antarctic Treaty Consultative Meetings (ATCMs).

The framework of the Antarctic Treaty expanded for the first time at the third ATCM in Brussels in 1964, when the Consultative Parties negotiated the Agreed Measures for the Conservation of Antarctic Fauna and Flora (later superseded by the practical provisions of the Environmental Protocol and its Annexes, cf. Antarctic Treaty, 2011). At the sixth ATCM in Japan in 1970 the participants decided to convene a special conference to negotiate the Convention for the Conservation of Antarctic Seals in Antarctica (Sollie, 1983). This special conference was the first opening of Antarctic Treaty meetings to intergovernmental and non-governmental organisations (Cohen, 2011) and the Seals Convention was then adopted by the (at that time) 12 Consultative Parties in 1972. The Convention on the Conservation of Antarctic Marine Living Resources became the third framework adopted by the Antarctic Treaty Consultative Parties in 1982, which was encouraged and advised by the Scientific Committee on Antarctic Research that had outlined potential threats to Antarctic marine ecosystems as a result of increased interest in Antarctic fisheries resources (Cohen, 2011). In the same year the Commission for the Conservation of Antarctic Marine Living Resources (CCAMLR) was added to the Antarctic Treaty System. The 25 international members of the CCAMLR (as of April 2016) continuously update conservation measures that determine the use of marine living resources in the Antarctic. The focus on marine living resources in the ATS is said to have "led to growing involvement by both intergovernmental and nongovernmental organizations in providing expert analyses, participating as observers in ATS meetings and participating as members of national delegations to Antarctic Treaty meetings" (Cohen, 2011, p. 2).

The debates on marine living resources also encouraged discussions on the potential of mineral resources in Antarctica and at the eleventh ATCM in 1981 it was agreed to negotiate a legal instrument on their exploitation. The Convention on the Regulation of Antarctic Mineral Resource Activities (CRAMRA) was signed by the consultative parties in 1988 but never entered into force as it was blocked by "leading Antarctic states such as Australia and France" after its creation (Liggett, 2015, p. 63). Instead, at the fifteenth ATCM in 1989 France and Australia introduced joint proposals for the establishment of a protection regime in Antarctica and in 1991, the latest ATS Protocol on Environmental Protection (the so-called Madrid Protocol) was signed, which, however, "contained many of the environmental impact assessment tools elaborated in CRAMRA" (ibid.). With the purpose of providing protection for the Antarctic environment, the Madrid Protocol prohibits any mineral resource extraction for commercial purposes and provides regulations on environmental impact assessments, marine pollution, fauna and flora, waste disposal, protected areas and liability (cf. Secretariat of the Antarctic Treaty, 2013). Nevertheless, scholars, journalists and politicians often refer to the "uncertain future of the environmental protection framework offered by the Madrid Protocol" (Degeorges and Ali, 2015, p. 157) – an understanding that also empowers the "polar orientalist" (cf. Dodds and Nuttall, 2016) representation of geopolitical interests often ascribed to China (cf. The Guardian, 2013; Einhorn, 2014; Tadjdeh, 2016), South Korea, Pakistan and India, which are seen as "just exploiting the poles for national gain" (Bennett, 2014). Particularly in view of the unlikelihood of the Madrid Protocol's not being renewed after 2041, at this point in time these seem to be unfounded fears. Although, hypothetically speaking, mining could be allowed after 2048, this is very unlikely as the ATS is a consensus-based system, thus all signatory states would need to vote against the continuing application of the Madrid Protocol (see also Interview, January 2015). On the occasion of the 25th anniversary of the signing of the Madrid Protocol, the Consultative Parties to the Antarctic Treaty further reaffirmed their commitment to continue banning mining in the Antarctic (Santiago Declaration, 2016).

Following a provision of the Madrid Protocol, the Committee for Environmental Protection (CEP) was established in 1998. It functions as the ATS's primary advisory body on environmental questions. The work of the Committee, like that of ATCMs, is further facilitated by work conducted in intersessional contact groups. After a long-standing disagreement on its location between the Consultative Parties Argentina and the United Kingdom, the Antarctic Treaty Secretariat was created in Buenos Aires in 2004 (43 years after the ratification of the Antarctic Treaty). Under direction of the ATCM, the secretariat has administrative responsibilities and provides access to scientific findings and transparency in the ATS. In formal terms, the Antarctic Treaty Secretariat is an Argentine organisation with international status.

The debate on mining in particular exposed the ATS to greater public scrutiny in the 1980s (Dodds, 2010) and interest in Antarctica from outside the ATS has continuously increased. Several non-member states were of the view that mineral resources in Antarctica should be treated as a common heritage of mankind, just as the United Nations Convention on the Law of the Sea (adopted in 1982) had provided for minerals located off the seabed beyond national jurisdiction. As a consequence, in 1983 the question of Antarctica was discussed in the United Nations General Assembly and became a regular agenda item until 2005. In this regard also, the aforementioned criticism of the closed nature of the ATS put the system under pressure and "served as a catalyst for opening the Antarctic Treaty meetings to non-Consultative Parties and to international organizations, including nongovernmental" (Cohen, 2011, p. 3). These debates further resulted in the amendment of the Rules of Procedure at the fourteenth ATCM in 1987, which provided for participation by representatives of the Scientific Committee on Antarctic Research (SCAR) and the CCAMLR Secretariat as observers, and by representatives of several international organisations as experts. An interviewee stated in this regard that at present

> the system is a lot more open than it was. I think when ASOC started the meetings we had all closed doors. [. . .] So just the fact that there are multiple non-state organizations that are allowed to come into the meetings I think is an acknowledgement that outside groups have something to contribute.
>
> (Interview, April 2014)

Aside from the discussions on the role and geopolitical interests of the different actors participating in the ATS, climate change in particular can be regarded as an important and challenging external factor in the contemporary governance of Antarctica. While, for instance, decision-making by consensus and a growing number of parties underpin the stability of the ATS (as also outlined in the case of the AC), both impede its ability to reach agreements on measures to react to the pace of environmental changes in the Antarctic region in a timely fashion. Instead, in the ATS "decision making can be glacially slow" (Liggett, 2015, p. 68). Likewise, ratification processes take longer as more countries need to take the agreement through national legislation (Interview, January 2015). Despite past institutional changes, the inclusion of more actors and greater transparency (although still very much limited to participating actors, see also Cohen, 2011), which are all widely perceived as successes (Hemmings, 2009; Liggett, 2015), the ATS is still seen as needing to adapt its structure, in light of environmental concerns to address newly emerging and ever more complex issues of concern in a more expedient manner (as is the AC). Further, as Dodds (2010, p. 115) emphasises, the "managerial challenges pertaining to the Antarctic can be summarized as being largely a function of the region's diminishing isolation in a variety of political, scientific, commercial, cultural and environmental contexts". In this regard again, the questions "Who decides about the setting of priorities in the politics of the Polar Regions?" and "How are these priorities backed up?" come into play. As with the observer debate in the AC, it is still often feared that membership of the ATS is primarily motivated by the aim of securing future claims in Antarctica and not encouraged by a common interest in protecting the Antarctic environment. Paradoxically, particularly the submissions made to the Commission on the Limits of the Continental Shelf by the polar-rim states indicate that the former aim is a driver of their positioning as well.

Overall, the AC and the ATS are considered to be the main regional polities for policy making in the Arctic and in the Antarctic. Formed at different times and in distinct global settings (the AC in 1996 after the end of the Cold War ended; the ATS after the signing of the Antarctic Treaty in 1959 during the Cold War), they were established to meet different aims: The AC was formed as a non-binding soft-law forum to enhance international cooperation with regard to sustainable development and environmental protection. The ATS, on the other hand, is based on legally binding agreements such as the Antarctic Treaty, which was adopted to avoid an escalation of conflicts over territory in Antarctica. The institutional structures of the AC and ATS have diversified since their establishment to meet the complexity ascribed particularly to the changing Polar Regions. Both have also expanded the number of actors involved in policy making at times when global interests in the regions grew and when the legitimacy of both polities was increasingly challenged. In both the AC and the ATS, however, all policy making has remained under the guidance of states (the member states of the Arctic Council and the Antarctic Treaty Consultative Parties).

3.1.3.3 Actors participating in the Arctic Council[6]

The debate on a "Globalized Arctic" (cf. Heininen, 2013) is also linked to the growing interest shown by various actors (also non-Arctic and non-state actors) in participating in the AC and to the renewed discussions on the inclusion of more non-Arctic (predominantly Asian) states (cf. Claes and Moe, 2014; Koivurova, 2009; Shadian, 2010). However, all actors participating in AC subsidiary bodies are appointed by the Arctic 8 and PPs. Members of working groups and task forces are usually representatives from sectoral ministries, government agencies and researchers who are either affiliated to the Arctic 8, the PPs or to non-Arctic states and international organisations with observer status at the AC. Experts with a corporate background can also contribute to the negotiations if they are invited. All the following actors are entitled, AC Ministers, SAOs and delegations of Arctic states, representatives of PPs, Observers or Invited Experts, and they all have different rights and duties when participating in the Council.

Observer status can be given to non-Arctic countries, intergovernmental organisations, inter-parliamentary organisations and to non-governmental organisations that the Council "determines can contribute to its work" (Arctic Counil, 2013b, p. 5). While the means to contribute to policy making in the AC differ between states and non-state actors, the diverse characteristics of non-state actors with observer status in the Arctic Council make it necessary to further differentiate among this very heterogeneous group that comprises transnational advocacy networks, international non-governmental organisations, science networks, inter-parliamentary and intergovernmental organisations. The – to date – 13 intergovernmental and inter-parliamentary organisations as well as the 13 non-governmental organisations and the 13 states with observer status "have different internal structures, differ in access to knowledge and material and therefore also vary regarding their own possibilities of influencing policy-formulation processes at the Arctic Council" (Wehrmann, 2017, p. 191). Despite this heterogeneity, it is notable, however, that observer states and non-state observers have the same formal status at the Council (cf. Duyck, 2015).

Thus, all observers have similar rights and duties and can be invited to all AC meetings. According to the Rules for Observers, their primary role is to "observe the work of the Arctic Council" and "to make relevant contributions [. . .] primarily at the level of Working Groups" (Arctic Council, 2013b, p. 5).[7] Once invited to participate, observers are allowed to "provide views on the issues under discussion" and may submit documents (Arctic Council, 2013b, p. 8); they can propose projects through an Arctic state or a permanent participant. At ministerial meetings, observers are able to submit written statements. Observer status once granted, all actors retain it unless and until they are suspended. Up to now, however, no observer has been suspended from the work of the Arctic Council, which can only be done if the ministers no longer agree on the inclusion of the specific actor or if the observer "engages in activities which are at odds with the Ottawa Declaration or with the Rules of Procedure" (Arctic Council, 2013b, p. 5). Observers are required to report their activities and contributions pertinent to the work of the AC and to the Arctic region 120 days before a ministerial meeting (cf. Annex 2, Rules of Procedure, Arctic Council, 2013a)[8] and are "responsible for all costs associated with their attendance at a meeting of a subsidiary body" (ibid., p. 9). In sum, observers are mostly perceived as actors that "facilitate positive outcomes, acting as bridge builders between various national positions or by researching and proposing political options" (Duyck, 2012, p. 101) but they are also seen as "the least influential players within the Arctic Council" (Nord, 2016, p. 40).

Since the formation of the AC, no corporate actor has received observer status. However, representatives of the International Oil and Gas Producers Association and the International Association of Drilling Contractors, for example, have been able to participate on an ad-hoc basis as invited experts. According to the Rules of Procedure (Arctic Council, 2013a, p. 9), as with all other decisions, only Arctic states participating in the AC or in any of its subsidiary bodies can decide whether or not an individual or organisation is invited to participate as an invited expert. Any such invitation needs to be renewed for each new meeting. Particularly given their resources, which can exceed those of states, large corporate actors "are often seen as players in their own right whose activities should be monitored and regulated in different ways" (Koch, 2011, p. 202). Overall, however, industry representatives are regarded as having "little impact on the work of the Arctic Council" (Duyck, 2015, p. 31). Individuals are further able to collaborate in delegations from Arctic states. Although members of delegations usually represent government agencies, they are sometimes also affiliated with corporate actors or with environmental non-government organisations (ENGOs). In the past, for instance, the U.S. occasionally included representatives of ENGOs in their delegations. In some cases this was done to circumvent the official exclusion of ENGO observers from specific task force meetings as articulated by task force chairs and to deal with domestic pressures from environmentalists demanding a greater say in AC policy making (Interview, August 2015; see also Wehrmann, 2017). The same applies for permanent participants who are able to designate representatives and delegations to all AC meetings: The co-chair of the ICC himself, for instance, used to manage the Arctic Slope Regional Corporation.

Although these explanations show that how much say different actors have depends very much on their classification as a member state, permanent participant, observer or invited expert, the influence exercised by actors also differs among these various actor groups. When one looks, for instance, at their attendance (and thus, their possible participation) in Arctic Council meetings, these differences can be easily identified. Although attendance at SAO and ministerial meetings is obviously not necessarily accompanied by the introduction of official statements by the participants, the participation lists[9] reveal that the Arctic Council member states and permanent participants always formed the largest groups at both SAO and ministerial meetings. Thus, besides their particular voting and consultation rights, these two actor groups have the greatest presence in both AC organs. Of all the member states and permanent participants, the U.S. and the ICC were notable for being represented most at these meetings.

It is surprising that, particularly in the early years of the AC, the U.S. sent the largest delegations. Scholars often point out that the U.S. had a minor interest in the Council during and after its establishment (cf. i.e. Nord, 2016). Canada, the country that initiated the Council's formation sent the second largest delegations to all SAO and Ministerial Meetings, while the lowest number of delegates represented Iceland, Russia and Sweden. The difference in the numbers of delegates from these member states underpins an explanation introduced before and given by an interviewee (August 2015), who highlighted the fact that it is not solely the interest of actors but, in particular, their different financial resources and capacities to send experts to numerous meetings (which often take place in remote Arctic locations) that have an impact on the number of delegates they send.

This explanation also applies to other actor groups. Among the group of non-governmental organisations, the Association of World Reindeer Herders sent most representatives to ministerial and SAO meetings. The ENGOs the Circumpolar Conservation Union (CCU) and World Wildlife Fund (WWF) sent the second largest number of representatives (together with the University of the Arctic) to ministerial meetings and the third-largest number to SAO meetings. From the other group with observer status, the intergovernmental and inter-parliamentary organisations, most representatives were sent to both AC organs by the Nordic Council of Ministers and by the Standing Committee of the Parliamentarians of the Arctic Region.

Among the group of invited experts and ad-hoc observers – the category of ad-hoc observers no longer exists in the institutional structure of the AC – the Association of Oil and Gas Producers (OGP) was the only industry actor that participated in SAO and ministerial meetings. In total, however, OGP sent only one representative to the second SAO meeting in 2012 and three representatives to the ministerial meetings in 2009 and 2013. Interestingly, representatives of actors that also have a say in Antarctic politics (the British Antarctic Survey, the International Polar Year and also the International Maritime Organization) only attended SAO meetings, not ministerial meetings. Argentina, on the other hand, sent two governmental representatives as invited guests to the ministerial meeting in 2009. This indicates that representatives preferred to attend SAO meetings, presumably because they believed they either had a greater say in SAO meetings[10] or obtained better insights into the work of the AC, as operational matters in particular are more often discussed in SAO meetings in which reports from all the AC's subsidiary bodies are introduced as well.

Representatives of non-state actors have sometimes also formed part of national delegations. Apart from to which delegation the representatives belong, however, their affiliation to organisations is often not made explicit in the list of participants. From those cases where it is, it is apparent that, for instance, scientists and representatives of subnational governments participated in the U.S. and in the Canadian delegations to ministerial meetings. On the other hand, representatives from WWF-US and Greenpeace-USA were included as members of U.S. delegations to SAO meetings. The much larger number of representatives of non-state actors who formed part of U.S. and Canadian delegations in SAO meetings again underpins the impression that, in that setting, the member states seem to have been less protective and to have offered a greater say to non-state actors than in ministerial meetings.

An explication of these formal settings should not neglect the fact that observers as well as invited experts and other state and non-state actors also maintain informal relations with the AC and within the subsidiary bodies, which also differ in their intensity and impact and are also determined by the settings in which meetings of the different subsidiary bodies take place. Task force meetings, for instance, are perceived as being more formal than working group meetings, but the same participants are involved in them over the years, besides maintaining their relationships outside the AC and often discussing matters informally during coffee breaks as well (Interview, August 2015). Furthermore, actors who are not invited to participate in AC meetings can, of course, visit these gatherings and try to make their voices heard through demonstrations, as, for instance, Greenpeace and other environmental groups and representatives of Indigenous peoples have regularly done at past ministerial meetings (see also Dodds, 2015; Nord, 2016).

The new Arctic forums that have been recently founded also constitute additional means to influence AC policy making (indirectly). Forums such as the Arctic Economic Council (AEC), the Arctic Coast Guard Forum (ACGF) and the Arctic Offshore Regulators Forum (AORF) are not directly related to the Arctic Council and its projects and programmes, but are associated with them. Moreover, they are increasingly seeking to influence developments in the Arctic region and add to the overall "Arctic complexity" (Stokke, 2015, p. 333). The three mentioned serve as forums for selected actor groups (particularly for state representatives and corporate actors). The AORF, for instance, was intended as a technical forum for regulatory agencies from Arctic Council member states. It aimed to solve two fundamental problems related to work in AC subsidiary bodies, namely the lack of expertise sometimes experienced in negotiations and the tight time schedule hindering in-depth discussions (Interview, August 2015). The "perhaps idealistic idea behind the AORF", as an interviewee explained,

> has been that if regulators, who have actual expertise, experience and responsibility for carrying out policies and making decisions, had their own forum and got together more frequently, they would develop a mission to protect the Arctic and to share information confidentially with the possible effect that they would strengthen their hand against industry.
>
> (Interview, August 2015, cf. Wehrmann, 2017, p. 199f.)

Although the AORF operates outside the Arctic Council structure and was not formed as a related subsidiary body, the results of its deliberations are also introduced and considered in AC working groups and task forces, in which members of the AORF participate, too. Various interviewees from observer organisations regard this as problematic, as so far AORF deliberations have been conducted behind "closed doors". They demand more transparency through the publication of AORF meeting agendas and reports at least.

Based on their exclusive voting rights, their power to set the agenda, their general oversight and the size of the delegations they send to ministerial and SAO meetings, member states (and, to a lesser degree, permanent participants) remain the main steering-actors in the AC. Although, generally speaking, observers have more means to participate in the AC than invited experts, in concrete terms the say of the different actors classified as such varies on the basis of their attendance rates and also with regard to the different types of meetings. Observers and invited experts have been much more interested in participating in SAO meetings than in ministerial ones. In those cases in which non-state actors formed part of U.S. and Canadian delegations, scientists and representatives of subnational governments were included in delegations to both ministerial and SAO meetings, while ENGOs only participated in the latter.

The Council, however, strives for a broad acceptance of its institutions and adopted agreements (Heininen, 2004). In the past decade, it has, accordingly, diversified its structure and allowed various new bodies (states and non-state actors) with interests in the Arctic region to participate (Duyck, 2012). Both the structural changes and the inclusion of new actors are intended to facilitate the co-production of knowledge (Kankaanpää and Smieszek, 2014) and account for a shift from the former traditional "one-way" flow of information from member states and PPs to other actors affected by the changing Arctic. Although overall the "knowledge production and information sharing [. . .] have further strengthened the foundations for regional stability" (Graczyk and Koivurova, 2015, p. 321), this shift has not altered the Council's hierarchical governance structure, as Arctic states continue to have the final say in negotiations. Instead, the growing formalisation of institutional practices and increasing number of regulations adopted to improve the Council's efficiency and effectiveness are regarded as counteracting this shift,[11] since they enforce top-down steering such as the empowerment of the SAOs to monitor and to review all AC activities. This growing political oversight has recently caused concern that the subsidiary bodies "may lose more and more of their independence and autonomy and become a playing field for politicised contentions" (Knecht, 2016a, p. 4). Overall though, and arguably also as a consequence of these transformations, in the past 20 years, the AC has clearly increased its relevance in Arctic affairs and beyond (Kankaanpää and Young, 2012). The inclusion of new actors, however, has also caused much debate on whether or not new actors counteract the aim of the Arctic states (and permanent participants) to maintain exclusive control of the AC and the region (cf. Lackenbauer and Manicom, 2015; Graczyk and Koivurova, 2014; Humrich, 2017).

Although any further inclusion of new actors in the Arctic Council is constantly under discussion (as is the intensifying internationalisation of the Arctic region in general, cf. Heininen and Southcott, 2010; Humrich, 2017; Koivurova and VanderZwaag, 2007; Young, 2010), interviewees stressed that the very complex topics of concern require the inclusion of many different actors to arrive at an adequate outcome, thereby confirming the observation that the transformation of the AC is motivated by the need to include more expertise, to take advantage of a "cross-pollination of ideas" (Charron, 2014) and to acquire additional material resources. The structural adaptations of the AC and its shift towards a co-production of knowledge can thus be identified as strategies to manage the complex processes under way in a changing Arctic (cf. Wehrmann, 2017) and link them with a much discussed phenomenon inherent in global governance. In cases where a problem cannot be managed by a single state or several states alone, it has to be addressed at a level beyond the state that comprises a variety of state and non-state actors (Koch, 2011). As Lackenbauer and Manicom (2015, p. 529) illustrate, however, this development can also be framed quite positively: "As the Arctic changes, its challenges will be met with technologies, ideas and insights from the entire world. For this, the Arctic countries should be grateful".

3.1.3.4 Actors participating in the Antarctic Treaty System

While the Arctic Council emerged at a time in which "the international community progressively acknowledged the role of nonstate actors in environmental governance" (Duyck, 2015, p. 14), the Antarctic Treaty was established at the onset of decolonisation and in the first phase of the Cold War, an era predominantly shaped by the fear of open conflict between states (see also Dodds and Nuttall, 2016). This state-centred formation phase and the legal framework the ATS is grounded on, to which only states can become parties, explain why, until the 1980s, non-state actors played a very subordinate role in policy-making processes in regional Antarctic governance. The criticism of closed ATCMs articulated at the UN, on the other hand, arguably contributed to the opening of meetings and the inclusion of new actors (cf. Herr, 1997). The ATS, however, still functions as a tiered operational system, in which the Consultative Parties have the exclusive right to vote (see also Zebich-Knos, 2015).

In ATCMs, the representatives of the 24 non-Consultative (state) Parties (as of March 2018) and of three observers – the Commission for the Conservation of Antarctic Marine Living Resources, the Scientific Committee on Antarctic Research and the Council of Managers of National Antarctic Programs – contribute to the negotiation processes through participating in ATCMs, formal committee and working group meetings and are allowed to submit information documents to the Secretariat for distribution prior to ATCMs. The observers further present reports during ATCMs (ATCM Rules of Procedure, 2015, Art. 31–35). At the end of ATCMs, the meeting decides on the international organisations (also NGOs) that are to be invited to send experts to the forthcoming ATCM, although they are not necessarily invited to attend the whole meeting and "indeed until the late 1990s experts were asked to leave the meeting during discussion of at least one agenda item" (Cohen, 2011, p. 3). Experts are further allowed to submit documents to the Secretariat to inform other meeting participants of their position prior to the ATCM in question (ATCM Rules of Procedure, 2015, Art. 39–45).

All parties to the Madrid Protocol are entitled to send delegations to the annual Committee for Environmental Protection (CEP) meetings. Observer status is open to all contracting parties of the Antarctic Treaty that are not parties to the Madrid Protocol and to the heads of the three ATCM observers. After approval by the ATCM also "other relevant scientific, environmental and technical organisations which can contribute to the work of the Committee" are allowed to participate in CEP meetings (CEP Revised Rules of Procedure, 2011, Rule 4). Observers are encouraged to submit papers (Rule 12) and to participate in discussions, but have no decision-making rights. On an ad-hoc basis, experts are also invited to participate in CEP meetings. The CEP can also establish subsidiary bodies, when approved at ATCMs, such as informal open-ended contact groups (Rules 9–10).

It is noteworthy that, up to the present day, only observers with a particular focus on Antarctic science have been included in the ATS. The non-governmental organisation the Scientific Committee on Antarctic Research (SCAR) is a committee of the International Council for Science (ICSU) that coordinates scientific research carried out by different institutions and countries in Antarctica. CCAMLR, as an intergovernmental organisation, and COMNAP, as an international association of government employees, are made up of the contracting parties' scientific representatives to the CAMLR and the Antarctic Treaty. Among the group of experts, on the other hand, are three non-governmental organisations, seven intergovernmental organisations and several that represent environmental and corporate interests independently of the United Nations. Like observers invited to participate in the Arctic Council, the capacities of actors classified as experts able to contribute to ATS policy making differ. Overall, these observer and expert organisations are, however, regarded as providing

> important information and advice without which the Antarctic Treaty Parties could not effectively or efficiently manage Antarctica. In other words, if these organisations did not provide certain necessary information, the Parties would have to develop that information themselves. I refer in particular to the scientific advice that is received, for example from SCAR, ASOC, IHO, IOC, IUCN, SCAR, UNEP and WMO.
>
> (Cohen, 2011, p. 3)

Among the group of observers, the SCAR, founded in 1957 and representing government and private sector scientists (see also Zebich-Knos, 2015), was the only non-state actor to receive recognition and a monitoring role in the ATS even before the opening of the System in the 1980s. The 1972 *Convention for the Conservation of Antarctic Seals* already lays down that all contracting parties should report annually on their activities to SCAR. Based on SCAR reports on permissible catch limits (provided also in collaboration with the UN Food and Agriculture Organization) contracting parties are further requested to "take appropriate measures to prevent its nationals and vessels under its flag from kill-ing or capturing seals of that species" (ibid.). The 1982 CAMLR Convention, however, did not continue "this approach of delegating tasks to a specific nongov-ernmental scientific organization" (Duyck, 2015, p. 20).

In the late 1970s the ENGO umbrella organisation, the Antarctic and Southern Ocean Coalition (ASOC, created in 1978), demanded greater participation of civil society in the governance of Antarctica and advocated stronger environmental protec-tion in the Antarctic. ASOC pushed these two emerging issues of concern in the ATS and, although it did not receive invited expert status in CCAMLR meetings before 1987 and in ATCMs until 1991, representatives were occasionally included in national delegations from supportive governments even before then (see also Duyck, 2011). To date, ASOC and the International Union for Conservation of Nature and Natural Resources (IUCN) are the only ENGOs that contribute to the ATS officially. Applica-tions by the ASOC member organisation Greenpeace, for instance, were rejected.

The International Association of Antarctic Tour Operators (IAATO), created in 1991 to advocate the interests of the tourism industry at a time when the ATCPs started to discuss possible regulation of Antarctic tourism, was the first corporate actor to receive invited expert status to ATCMs in 1992. Prior to the seventeenth ATCM, at which tourism regulations were supposed to be discussed, the IAATO adopted self-regulating guidelines to limit the environmental impacts of tour-ism on the Antarctic, which resulted in the abandonment of further regulating mechanisms on the part of the ATS and showed that IAATO had a special say in this regard. This implementation of industry standards has been constantly under discussion (see Liggett, 2015), but was also regarded as the most effective means of control because of the "complicated nature of jurisdiction over tourism operators – most notably the issue of flags of convenience" (Duyck, 2015, p. 23). More recently, this self-regulatory approach has, however, been challenged by the adoption of regulations through the International Maritime Organization (IMO):

> In 2010, the Marine Environment Protection Committee amended Annex I of the International Convention for the Prevention of Pollution from Ships (MARPOL) in order to bar the use and carriage of heavy-grade oil below 60 degrees latitude south from August 1, 2011. Implementation of this rule has led to the first drastic reduction in pollution through regulation affecting tourism in the region. In addition, the IMO is currently developing its Interna-tional Code of Safety for Ships Operating in Polar Waters (Polar Code). [. . .] The entry into force of the code will further impact Antarctic tourist operators, limiting the capacity of the industry to self-regulate in order to avoid manda-tory standards. In this context, the ability of IAATO to avoid international regulation of Antarctic tourism activities seems to fade as the stakes are raised by the growing scale of the activity.
>
> (Duyck, 2015, p. 23)

Nevertheless, the IAATO is still considered a very influential non-state actor as the Polar Code is not specifically aimed at managing tourism but intended to prevent ship-related environmental disasters, which leaves the regulation of tourism in the Antarctic in the hands of the industry and "[u]nlike the domestic arena, there are not many business sector NGOs competing over tourism-related issues in the Antarctic, which gives IAATO an advantage when offering input into the policy-making and regulatory process" (Zebich-Knos, 2015, pp. 98 and 107). With the adoption of the CAMLR convention that also regulates fisheries in the Southern Ocean, the interests of other private entities came to the fore, which had been "absent from the governance of the Antarctic" before the 1980s (Duyck, 2015, p. 22). In 1987, for instance, the Coalition of Legal Toothfish Operators (COLTO) was founded. It advocated particularly against illegal unregulated and unreported fishing and obtained observer status to CCAMLR (to which it also contributed financially, Duyck, 2015) in 2003.

While overall, and as with the AC, policy-making power in the ATS lies in the hands of states (particularly of the Consultative Parties to the Antarctic Treaty), "NGOs like IAATO certainly play a valuable role as disseminators of information and translators of policy and subsequent regulations to their members" (Zebich-Knos, 2015, p. 101). They also provide information to a wider and more general public (Cohen, 2011). The means by which non-state actors with expert status (ASOC, the IAATO and COLTO) can influence policy making in the ATS are similar to those provided to non-state actors with observer status in the AC: they range from attending ATS meetings, monitoring official decisions, submitting of papers on issues of interest to meetings with government officials in order to represent their perspectives and influence national policies domestically. As mentioned before, national delegations to the ATS, like those to the AC, are sometimes composed of non-state actor representatives (Cohen, 2011) and non-state actors can, informally, submit position papers to national delegations with similar interests, hoping that the latter will draw on their arguments and positions in meetings that they are not allowed to attend themselves (see also Hemmings, 2015).

Although, generally speaking, the different classification of actor groups indicates their particular say in the ATS, their actual influence in policy making differs among the heterogeneous actor groups and also from meeting to meeting.[12] The participation lists reveal that among the group of consultative parties (and also overall) the U.S. sent the largest and Chile the third-largest delegations to past ATCMs. Thus, these two American polar-rim states have been the actors with the greatest presence at these meetings. Canada acceded to the Antarctic Treaty in 1988, attended an ATCM for the first time in 1989, and became the non-Consultative Party represented by the fourth-largest number of delegates sent to ATCMs during the period of investigation. Among the group of observers, SCAR has been the only subsidiary organisation that has contributed as an observer in all past ATCMs. ASOC, however, has sent almost twice as many delegates to ATCMs as SCAR since 1990 (in that year ASOC was invited to participate as an expert in ATCMs for the first time). The IAATO, meanwhile, is the only industry actor that has participated as an expert in ATCMs. Besides Belarus and Malaysia (both became non-Consultative Parties later), the Arctic Council is the only organisation that has sent delegates as invited guests to ATCMs. The IPCC was invited to contribute to past ATCMs (e.g. in 1989 and 1991) but did not attend the meetings. As in the AC, non-state actor representatives have sometimes taken part as members of national ATCM delegations. Representatives from the ENGOs Greenpeace (in 1992) and ASOC (in 1997 and 2002), for instance, were members of the U.S. delegation, and an IAATO representative also participated in the U.S. delegation in 2002. At special ATCMs (SATCMs) the CEP observer organisations (such as ASOC, IAATO, IUCN) that usually have the status of experts at ATCMs are also accredited as observers. During the period under investigation, two SATCMs took place: In 1990, the SATCM focused on the negotiation of the Environmental Protocol while the 2000 SATCM was shaped predominantly by discussions on the establishment of the Antarctic Treaty Secretariat, which had been identified as a necessary step (a "pressing need", ATCM, 1992) after the implementation of the CEP.

In total, the number of delegates sent by Consultative Parties far exceeds (by more than ten times) the number of other delegates that have participated in past ATCMs and the ratio of state to non-state actor delegates is similar. Although ASOC and IAATO sent more delegates to ATCMs in the 2000s and 2010s than before, in view of the overall participation numbers (and as with AC meetings examined before), state representatives have formed the clear majority of delegates in ATCMs. The first meeting in which non-state actors participated was the 1987 ATCM to which the IUCN, WMO (World Meteorological Organisation) and SCAR were invited by the host country Brazil. At that meeting too, the Rules of Procedure were adopted that regulated the attendance of experts for the first time and in accordance with which non-state representatives were only allowed to participate for the discussion of specific agenda items. As was pointed out in the first section of this chapter, it was a common practice until the late 1990s for experts to be asked to leave the meeting. Against this backdrop, although the representation of non-state actors' interests seems to have been rather weak in view of the smallness of the change in participation numbers during the period of investigation, the general expansion of parties participating in ATCMs and the inclusion of non-state actors already provide evidence for a changing understanding and acknowledgement of contributions made by non-state actors to the politics of Antarctica.

In view of the intensity of the issue-related and institutional challenges that the ATS has to deal with, Hemmings concluded in 2009 that "Antarctica appears to be significantly more unstable than at any previous time since the adoption of the Antarctic Treaty" (p. 56). To meet these challenges, scholars have recommended a "greater internationalization of Antarctica and the adoption of a common heritage of mankind regime" (Liggett, 2015, p. 70) based on agreed principles to protect the environment. In this regard they have argued for

> [r]ecalling the key role of science in Antarctica and with respect of Antarctic governance [. . .] an increase in funding for science, including for basic research, would support continued public participation in the diffusion of

knowledge about Antarctica and its role in global physical processes, including biological, geochemical and environmental.

(Cohen, 2011, p. 7)

These recommendations are particularly underpinned by increasing knowledge of the global impacts of the changing environment in Antarctica (and in the Arctic), which encourages an understanding of Antarctic governance as being a global concern.

To sum up, the ATS and the AC were formed to address distinct purposes at times when global settings differed significantly, which also resulted in distinct institutional structures: a binding treaty system in which states (and scientific organisations) have been regarded as primary actors, on the one hand, and a non-binding intergovernmental forum primarily composed of "Arctic states" (and permanent participants), on the other. Despite the numerous differences, many of the formal settings for negotiations and participation provided by both the AC and the ATS have evolved in a similar manner although at different times. In both cases, their diversification and greater inclusion of non-state actors were triggered by growing complexities related to environmental changes in the Polar Regions on the one hand and the growing significance of the Polar Regions in global and future terms being translated into criticism regarding the legitimacy of both polities on the other. Over the past decades, consequently, the interests of (similar) state actors in the Arctic and Antarctic regions who demand to be included in policy making although they are still most often regarded as "outside actors" have multiplied (Bratspies, 2015, p. 176).

Against this backdrop, cooperation between state and non-state actors in regional polar governance has also been growing. This collaboration, however, has not followed a linear progress but was initiated at different points in time. In the ATS, scientific organisations were already playing an important role in the 1970s, ENGOs contributed increasingly in the 1980s and industry representatives from the IAATO and COLTO have participated since the 1990s. In Arctic regional governance, on the other hand, Indigenous peoples' organisations received a greater say in the early 1990s, and other non-state actors have been increasingly included since the beginning of the 2000s. Few industry representatives, however, have participated in Arctic Council ministerial or SAO meetings.

In both regional governance settings, the number of state actors and non-state actors involved in policy-making processes is said to have grown "to a larger extent than is commonly the case at the global level" (Duyck, 2015, p. 31). As Duyck points out, in view of global environmental governance this development is, however, not regarded as being inspired by any institutional structures unique to the AC and the ATS but as following general trends. On the one hand, the inclusion of more actors has been motivated by the aim of achieving a broad acceptance of policies adopted by the AC and ATS and underpinning their overall legitimisation as in other international settings; see, for instance, the 1992 UN Conference on Environment and Development whose final declaration stated that "environmental issues are best handled with participation of all concerned citizens, at the relevant level" (Art. 10). On the other hand, the knowledge, resources and expertise provided by (non-state) observers and experts are mostly regarded as contributing significantly to the work of the ATS in particular, but also that of the AC. Their general say is, however, much more limited in AC policy making, although in quantitative terms the number of observers and invited experts is larger than in the ATS. In both the ATS and the AC, voting rights have remained in the hands of state parties.

The institutional contexts in which state and non-state actors interact differ in the AC and ATS and also in the respective subsidiary bodies. The increase in the number of their actors and their institutional diversification, however, is seen as not only contributing to but also challenging their effectiveness and efficiency. Against this backdrop, new Arctic forums are being set up, though these are often perceived as problematic "closed clubs" by those who are not included. The revised rules of procedure adopted to strengthen the effectiveness of the AC, on the other hand, are seen as enforcing top-down steering and weakening the say of non-state actors. As an alternative way of including more views without diminishing the effectiveness and efficiency of both bodies, for instance, the elimination of the consensus system is increasingly (and controversially) being discussed as well as their transformation into international organisations. Having said this, however, both the ATS and the AC need to be perceived as non-rigid entities that have transformed themselves over the years to respond to numerous similar institutional challenges – particularly in terms of effectiveness and efficiency and their overall legitimisation.

3.2 Geopolitical reasoning in the regional politics of the Polar Regions

In trying to better understand why certain patterns of interpretation have become dominant in the politics of the Polar Regions one may ask oneself to what extent the views introduced in the aforementioned regional settings by state and non-state actors from the American polar-rim states correspond and to what extent they differ. Before comparing the reasoning introduced at regional level, this section

1 explores the significance ascribed to the changing Polar Regions in the geopolitical reasoning introduced at different times by the different actors under analysis, and
2 evaluates the degree to which different views are considered in the regional politics of the Polar Regions.

The previous section found that the say of actors in regional policy formulation processes has differed not solely among the different actor groups (classified as member states/parties, permanent participants, observers, invited experts and guests), but also within these groups. The attendance lists have further shown that the possibilities for introducing differing views have also varied from meeting to meeting and within the different subsidiary bodies. Interviewees (February and

August, 2015) confirmed that, in this regard, much depends on the overall constellations of meetings, the agenda items under discussion and the time available to negotiate. In order to explain why certain patterns of interpretation have become dominant in the regional politics of the Polar Regions, it is thus necessary to consider the contexts within which state and non-state actors from the American polar-rim states operate when representing a specific view in the AC and ATS. Shedding light on the overall contexts such as the type/year of meeting, and the agenda item under which the respective geopolitical reasoning was introduced thus allows one to examine whether or not policy making in both the AC and ATS is based on a similar reasoning, to compare how views (and the imaginaries that strengthen these views) have changed, and whether this change is likely to have been inspired by positions introduced by non-state actors. While this chapter focuses particularly on corresponding and conflicting views represented by state and non-state actors in the AC, the ATS and, in the next section, the American polar-rim states, Chapters 4 and 5 consider the compatibility of these views with those predominantly promoted by newspapers and non-governmental theorists in formal and popular geopolitics.

Multilateral negotiations are, without a doubt, shaped by complexity, as many actors address numerous issues, and have (strategic) preferences that also relate to different policy levels. Against this backdrop, the actors participating in the AC and ATS and their disparate views need to be considered as exemplary. In accordance with the theoretical approach to this study, it is important to keep in mind that neither the ATS nor the AC operates "in a vacuum" (cf. e.g. Smieszek and Kankaanpää, 2015; Berger, 2015). Any assessment of the general leverage of participants needs to include discourses conducted at other (e.g. national, international) levels and to consider the significance of events that are also likely to shape participants' positions and views.

3.2.1 Discursive entanglements? Topics addressed in the Arctic Council and in the Antarctic Treaty System

During the period of investigation, topics prominently placed on the agendas of AC and ATS subsidiary bodies ranged from rather broad issues such as "international cooperation" to very specific topics of concern such as "waste management" and the "Arctic Council structure"/"the future of the CEP". "International cooperation", for example, is a topic that has been placed continuously on the agendas of all four subsidiary bodies as has "environmental concerns". Most of the specific issue-related topics prominently addressed in the AC and in the ATS, however, differ, underlining the disparateness of the processes and management tasks related to the changing Arctic and Antarctic (cf. Boxes 7.2–7.5, Annex).

Despite the different topics prioritised in the AC and ATS, in both settings discussions on the changing environment are entangled with the majority of specific issue-related topics (e.g. "protected areas" and the "conservation of Antarctic flora and fauna" in the ATS and "energy" in the AC). Moreover, seemingly "distinct concerns" often reappear in different contexts in the other regional setting: in AC subsidiary bodies, for instance, "human development" and "human conditions" are recurring topics placed prominently on agendas for ministerial and SAO meetings (2007–2013 in SAO meetings; 2002–2013 in ministerial meetings). Human needs are also continuously discussed in the ATS, however, in a different context – almost exclusively in relation to scientific operations conducted in Antarctica (e.g. under the ATCM agenda item "education and training"). Human impacts on the Arctic and Antarctic environment, on the other hand, are prominently addressed in both bodies, particularly with regard to pollution caused by shipping and tourism.

As exemplified by the different viewpoints linked to "classical" and Critical Geopolitics, opinions may differ on whether these similarities result from discursive entanglements shaped by the actors engaged in policy making or from the "nature" or physical environment of the Polar Regions and the challenges that arise from it. From the point of view of "classical Geopolitics" it is no surprise that in the unique conditions of the Arctic and Antarctic environmental concerns in particular are dominant issues addressed in regional policy making. Critical Geopolitics, however, helps to explain why the topics addressed differ in this regard. In the AC, for example, climate change was discussed as a trigger of both environmental and institutional changes, as the involvement of new actors was generally argued to be a necessary step towards including more expertise and resources to better understand the changing Arctic. In ATS negotiations, the changing environment and related processes in both Polar Regions, on the other hand, constituted a separate agenda item: "Relevant developments in the Arctic and in the Antarctic" were continuously discussed in ATCMs during 1995 and 2006, which can be explained by the historical significance of the International Polar Year (IPY) for the ATS and the relevance also ascribed to the 2007–2008 IPY (by contrast, the IPY, preparations for it and its "legacy", never acquired a prominent place on the AC agenda).

The next section builds on the viewpoint promoted by Critical Geopolitics and considers the different say of state and non-state actors in the AC and ATS to explore the transnational (and transregional) discursive entanglements in more detail. My purpose is to explain how specifically distinct geopolitical views and the meaning ascribed to recurring topics and concepts changed and thus contributed to the prioritisation of different issue concerns in the politics of the Polar Regions.

3.2.2 Geopolitical reasoning introduced in the AC and ATS

While state representatives have dominated policy making in the AC and ATS, topics related to "environmental concerns" in the Arctic and Antarctic have dominated the agendas of their subsidiary bodies investigated in this book. As "environmental concerns" and "international cooperation" are topics that have been prominently and continuously addressed in those subsidiary bodies, focusing on the geopolitical views of the state and non-state actors from the American polar-rim states makes it possible

1   to identify conflicting and corresponding views and the geopolitical imaginaries that are used to empower these views,

2　to classify whether these views and imaginaries have shifted and changed over time and among the different actor groups in and across both Polar Regions.

> The fact that Saami and other Arctic indigenous peoples' areas continue to be exploited without their consent implies that the colonisation of Sápmi and other indigenous peoples' territories is *not a practice of the past but very much a continuing process of today.*
>
> (Javo, on behalf of the Saami Council, on the occasion of
the Eighth Ministerial Meeting of
the Arctic Council, 2013, emphasis added)

> I call upon all those responsible for our planet to protect and preserve nature created by God. This responsibility has been broadly demonstrated for three decades within the Antarctic System. We are lawyers and politicians, scientists and explorers, but, *above all, we are men responsible for the birth of a new southern world, five centuries after Columbus.* On the eve of the third millennium, *this is the last continent at man's disposal in an increasingly populated, eroded and polluted planet.* I repeat we Chileans cannot remain indifferent to the importance of a clean Antarctic. Not only do *we live in America, bordering on the Antarctic, but we also permanently inhabit the southern continent itself.*
>
> (Vargas Carreño on behalf of the Chilean delegation
on the occasion of the Tenth Special Antarctic Treaty
Consultative Meeting, 1990, emphasis added)

As these two statements from the years 2013 and 1990 exemplify, it is not only the environment of the Arctic and Antarctic that is considered as impacting the rest of the world in numerous ways. The management of both regions is closely bound up with geopolitical understandings that are shaped by special relationships with geographical spaces, by emerging responsibilities (in view of the future of the planet and of the growing populations inhabiting the planet), and by experiences (of domination). These factors have, however, been prioritised differently throughout the period of investigation and in the reasoning introduced by the different actor groups engaged in policy making at the AC and ATS.

3.2.2.1　*Representations of geography and the environment in the Arctic Council*

In the AC, the special geographical relationships that people have with the Arctic differ particularly between member states, permanent participants and observers who have continuously regarded the Arctic either (1) as a national territory due to annexation, (2) as a homeland because of their cultural entanglements with the Arctic environment and territory or (3) as a common heritage because of its global significance.

AC member states like the U.S., for instance, declare that they identify as "Arctic nations" because of their acquisition of territory located in the Arctic (see, for instance, Kerry, 2013). Permanent participants, on the other hand, representing peoples that have lived in the Arctic for several thousand years, often stress their understanding of "the Arctic, our joint homeland" (Javo on behalf of the Saami Council, 2015) in order to argue that "[t]he Arctic [. . .] remains a territory of untouched nature and the living home of indigenous peoples" (Sulyandziga on behalf of RAIPON, 2013). Permanent participants often emphasise their perception of the Arctic in a more personalised way as well when demanding that it should be protected. For them, it is "not just an important international theatre of events; this is my home" (Gundersen on behalf of the Aleut International Association, 2011). This understanding of an Arctic home/homeland (also emphasised in the Arctic Human Development Report, 2004) is further portrayed as being inherently interrelated with the identity of Indigenous peoples, as permanent participants often emphasise that their "home" is also associated with their way of life: "The Aleut people are supremely adaptable; they have survived invasion, disease, forced relocation, war, internment, and nuclear testing on their land, all *while living in a tremendously challenging physical environment*" (see Gundersen, 2011, emphasis added). Or argue that "cultural heritage, sacred sites and cultural landscapes" (Olli on behalf of the Saami Council, 2013) are located in areas that the "indigenous peoples of the Arctic have managed and been an integral part [. . .] for thousands of years" (Nuorgam on behalf of the Saami Council, 2002).

As these examples show, the imaginary of an "Arctic homeland/home" is often used to emphasise the significance of the Arctic for Indigenous peoples and as a means to defend the use of this geographical space against the interests of others. It is also an imaginary that can be regarded as mirroring a nationalism of those who feel oppressed, who use this imaginary to relate to past struggles over territory and to emphasise their attempts to become empowered and thus taken into account by others. The meaning ascribed to the homeland imaginary differs, however, within the groups of Indigenous peoples (and also with respect to the "planetary home" imaginary that is introduced in regard to the Antarctic).

The differing views of the Arctic as a "homeland", "national territory" or "common heritage" also contribute to controversial understandings regarding the question of "who has a legitimate say". While AC member states emphasise their authority (and also refer to their special responsibility for "their nations"), permanent participants demand an exceptional say equal to those of states. Although the member states implemented the unique permanent participant status in the AC to acknowledge the special heritage of Indigenous peoples in the Arctic, environmental changes and the growing global interest increasingly conflict with this understanding of the Arctic being a unique place for Indigenous peoples as the theme of the U.S. 2015 chairmanship "One Arctic" most notably exemplifies: The theme was presented as being based on the understanding of a "shared Arctic not just for the nations that touch it, but for the way that what happens here, [. . .] *touches every single person around the world and our way of life*" (Kerry, 2013, emphasis added). The theme also emphasised shared responsibilities, since "the consequences of *our nations'* decisions don't stop at the 66th parallel" (ibid., emphasis added).

The reasoning promoted by the U.S. Secretary of State thus stresses the global significance of the Arctic region. In contrast, when the Inuit Circumpolar Conference introduced the theme "One Arctic – one future" in 1992 at their Sixth General Assembly (20 years before it was reproduced by the U.S.), they referred to "global significance" in a different manner (as Kerry also notes), arguing that "the entire world shares a responsibility to protect, to respect, to nurture, and to promote the region" (ibid.) for the sake of the people who live in the Arctic and are particularly affected by the changes taking place there. While Kerry attempted to motivate "every single person around the world" to reflect on the consequences of their daily actions for the Arctic in times of climate change, representatives of Indigenous peoples still often apply the former reasoning when demanding an exclusive say: "From research to implementation, they [Indigenous peoples] should be the ones to make all decisions" (Sulyandziga on behalf of RAIPON, 2013; see also Olli on behalf of the Saami Council, 2013).

This example illustrates how different actors use similar reasoning for different purposes that (in the case of the permanent participants) has not changed much over time but has been introduced in negotiations of different concerns. Kerry was speaking about environmental changes in the Arctic in relation to climate change, which he depicts as a "threat" and as "one of the most obvious shared challenges on the face of the planet today" (Kerry, 2013). The ICC also referred to environmental changes in 1992. Sulyandziga, the representative of RAIPON, on the other hand, stressed the say of Indigenous peoples particularly in view of the regulation of economic activity in the Arctic ("the planet's resource base of the 21st century", ibid.). Further, while Kerry ascribes a specific significance to the Arctic because of climate change impacts, more recently permanent participants have most often perceived not climate change but economic interests as the main driver for the growing global interest in the region, as the following exemplary statements introduced by the Saami Council representatives at different Ministerial meetings show:

The Arctic, our homeland, has again become an attractive region for people from outside. [. . .] Oil and gas activities particularly are major drivers of social, economic and environmental changes in the Arctic.
(Kobelev on behalf of the Saami Council, 2006)

First, increased access to non-renewable resources in our homelands has created a "race to the Arctic" and a change in land-use.
(Åhren on behalf of the Saami Council, 2009)

Although other non-state actors, such as the inter-parliamentary organisation the Standing Committee of Conference of Parliamentarians of the Arctic Region (SCPAR) also relate the growing global spotlight on the Arctic to interest in energy resources, when stating that "many countries [have] a strong and increasing interest in the Arctic region. This interest is due, not least, to the expected substantial quantity of energy resources and other natural resources in the Arctic", they also acknowledge that "climate change in the region and its projected wider impact has also contributed to this focus" (Solberg on behalf of SCPAR, 2006). In contrast to the view often promoted by permanent participants and like the U.S. Secretary of State, environmental organisations such as the WWF regard climate change as the prime driver of the growing global significance ascribed to the Arctic (WWF, 2011). Quite obviously, for those without territorial rights in the Arctic, this understanding of climate change and the significance ascribed to the Arctic are the most powerful arguments for demanding a say in regional policy making, while the permanent participants need to emphasise their territorial rights in the face of growing economic activities so as not to weaken their special say at the AC. This is not to disregard the fact that the permanent participants also perceived climate change as a driver of change when they stated, for instance: "We face many challenges related to the climate change" (Åhren, 2009); "We who live in the Arctic are impacted in our daily life by the effects of climate change" (Lynge, 2013). And back in 2002, permanent participants emphasised the global significance of climate change in the Arctic and the related understanding of the Arctic as a barometer: "The Arctic is maybe the most important indicator of the global environmental status. The developments we observe here are in fact signals of what to expect globally in a couple of decades" (Nuorgam on behalf of the Saami Council, 2002). In ministerial meetings, however, all permanent participant representatives demanded a particular say in the regulation of oil and gas development, something that has not been urged so far with regard to climate change.

Generally, permanent participants do not argue against the development of oil and gas resources (as most environmental organisations obviously do). Like AC member states, they often regard growing industries as beneficial to the North (see, for instance, Olli on behalf of the Saami Council, 2013). But in consideration of their colonial past, they air their fears of experiencing domination (cf. e.g. Nuorgam on behalf of the Saami Council) and exploitation of their resources:

> While the wealth of these activities flows south, we are left with the environmental, social, and cultural impacts of the exploration, construction and shipping activities. [. . .] We will not accept outsiders coming to the Arctic, taking out the resources and degrading our land, without even having the common courtesy to ask.
>
> (Kobelev on behalf of the Saami Council, 2006)

> The fact that Saami and other Arctic indigenous peoples' areas continue to be exploited without their consent implies that the colonisation of Sápmi [Saami territory/DW] and other indigenous peoples' territories is not a practice of the past but very much a continuing process of today.
>
> (Javo on behalf of the Saami Council, 2013)

> Industry, Arctic states and others cannot simply sail into the Arctic and take what they want for their benefit.
>
> (Lynge on behalf of the Inuit Circumpolar Council, 2011)

As these examples show, perspectives differ particularly with regard to the question of what issues are to be regarded as global and as local concerns. With regard to resource development and climate change these perspectives are further intertwined and conflicting as climate change is commonly understood as facilitating resource development in the Arctic and resource development is perceived as enforcing the impacts of climate change. In other words, they merge under the so-called "Arctic Paradox" (see e.g. Palosaari, 2016): Both determine each other.

Also the "victimization" of permanent participants has been viewed differently by Arctic Council member states as most notably the statement by the Canadian Minister of Foreign Affairs during the 2002 Inari Meeting exemplifies, who relates environmental problems to the management of resources and land-use in the North and thereby included Indigenous peoples:

> The Arctic Council's earlier work on contaminants depicted the North as a "victim" of problems originating in the South. However, increasing resource development and industrial activity in the North itself is also a source of the problems we are facing, and this increases the pressure on the Arctic environment.
>
> (Graham, 2002)

Particularly with regard to climate change, however, representatives of permanent participants depict themselves mostly as victims (see e.g. Eegeesiak on behalf of the Inuit Circumpolar Council, 2015 or Cochran on behalf of the Inuit Circumpolar Council, 2009) and use this victimisation as well as the territorial rights granted them to demand more political and decision-making rights:

> Today, when it comes to combating it [climate change], we are still often left out. While we are now named as stakeholders, I would remind those here that we are more than that. We are rights holders, Mr. Chair. We are land owners, resource owners; we have settlement and treaty rights and *it is our right to be at the table on all matters related to the Arctic.*
>
> (Cochran on behalf of the Inuit Circumpolar Council, 2009, emphasis added)

The ownership of resources is thus also perceived as a factor providing a particular say in Arctic Council policy making, as a factor that gives permanent participants an equal standing with Arctic Council member states. With regard to the management of resources, this understanding includes the right to determine "if and how the minerals, fish, marine mammals, tourism, and other things are developed in the Arctic" (Cochran, 2009). It is also noteworthy that the ICC representative invokes settlement and treaty rights, and stresses that they are "land owners" and "resource owners" in order to claim the right to be included in policy making. This reasoning is introduced to emphasise the differences between permanent participants and other non-state actors and as a reason for the demand for a greater legitimacy in policy making. In a similar vein (and in line with the Arctic home/homeland imaginary), the Saami representative also emphasises the Saami's "inherent right to our territories", protesting "for large parts of the Saami area, our land and governance rights are still not respected" (Kobelev, 2006).

As the impacts of change are manifold in the Arctic, the statements introduced at ministerial meetings often make it clear that impacts and consequences of the changing Arctic are perceived differently. Some permanent participant representatives, for instance, expressed a greater fear of losing territorial rights than of suffering from climate change ("We believe that the Saami people can adapt to climate change. But there is nothing we can do if you take our land from us", Åhren on behalf of the Saami Council, 2009). In contrast to this perspective of the Saami representative and underpinning again the differences regarding the uses of territory and cultural distinctions among Indigenous peoples in the Arctic, others, such as the representative of the Inuit Circumpolar Council relate to the "Arctic Paradox" when stressing to perceive the melting sea ice as a great challenge:

> sea ice has helped sustain Inuit for thousands of years. And now it is thinning and melting. We need the ice to access our resources and to sustain us. Ironically, and perhaps tragically, others need the ice to melt so that they can access easier travel routes and resources found deep beneath our world, the Inuit homelands.
>
> (Cochran, 2009)

Both statements relate to the "elephant in the room": discussion of the impacts of the changing Arctic is almost always coupled with discussion of rights and sovereignty not solely with regard to territory but also with regard to the power over interpretation that unite in the question of "Who has the right to decide about actions with implications for the future in (and beyond) the Arctic?" In this discussion, permanent participants such as the ICC make their different views explicit and hint at the problem of obtaining a legitimate say in Arctic politics when stating that while "the Arctic has increasingly become the focus of states, industry, academics, and others, we have had to address for ourselves the questions of 'who owns the Arctic?' (Cochran on behalf of the Inuit Circumpolar Council, 2009). The Saami representative, on the other hand, does not highlight the need to cooperate with others in order to deal with impacts of climate change but stresses the demand to respect their territorial rights.

Another example of clashing perspectives is the significance ascribed to "environmental health". While member states such as the U.S. and Canada have often emphasised the global significance of environmental health in the Arctic (e.g. the protection of its biodiversity), the ICC demanded that the well-being of the people living in the Arctic should be prioritised:

> Some say the best indicator of a healthy Arctic can be found by measuring the effects of climate change. Others say the best indicator can be found by measuring the health of Arctic biodiversity. Still others say the best indicator is related to resource and economic development. Inuit believe, however, the

key indicator that encompasses these and other measurements of a healthy Arctic environment is the state of well-being of Inuit and other Arctic peoples.

(Lynge, 2011)

In this regard, permanent participants also stated that they experience how this view (the prioritisation of human beings living in the Arctic and their needs over environmental issues) conflicts with the positions introduced by ENGOs at the Arctic Council and have often demanded respect for their way of life (see also Cochran, 2009):

> We experience that environmental and animal rights organizations have a tendency, when they speak about the Arctic, to forget about the human beings living here. It is a daily struggle for the Indigenous Peoples' Organizations present, to explain our ethical and moral values to those organizations.
>
> (Lynge, 2011)

Overall, the special geographical relationships said to link them to the Arctic differ particularly between member states, permanent participants and observers but also among the different actor groups. In relation to the question of "who has a legitimate say" in AC policy making, the representations of geography and the environment are in all cases instruments to increase influence. The related geopolitical views, as summarised next (see Figure 3.3), were introduced in different contexts (most notably in view of the emerging possibility of developing hydrocarbons in the Arctic), in which most often "local" and "global" perspectives conflict. In the examples mentioned earlier, however, member states mostly share similar views. Yet, these often conflict with positions expressed by permanent participants (e.g. regarding their authority in Arctic politics, collaboration with businesses, the causes and solutions of climate change). Particularly with regard to the impacts ascribed to climate change and mitigation actions such as

Self-Understanding	Understanding of the Arctic	Understanding of Actions Needed
Arctic Nation (Arctic identity)	As a place of global interests (accessibility to resources)	Everywhere (shared responsibility)
Rightsholder (Arctic homeland)	As a place of global significance (climate change)	By the primary polluters (outside the Arctic)
Observer (common heritage)		

Figure 3.3 Summary – conflictive entanglements among representation of the geography and the environment in the Arctic Council

investments in renewable energy, the perspectives expressed by observers, on the other hand, often correspond to those of member states: Concerning the perceived impacts and consequences of the changing Arctic, for example, member states as well as observer states often refer to climate change as being the greatest challenge/threat. With respect to the perception of the Arctic as a common heritage (an understanding expressed by numerous observers), however, the perspective of observers differs significantly from the views expressed by permanent participants and member states.

3.2.2.2   *Representations of international cooperation and environmental changes in the Arctic Council*

In spite of the different views expressed at ministerial meetings and of the particular say demanded by permanent participants, all statements considered emphasise cooperation as the best strategy for dealing with the changing Arctic. As the representatives of the Aleut International Association and the Saami Council and the U.S. Secretary of State highlighted, this cooperation is also regarded as a result of the changes taking place:

> I want to see my home prosperous, protected, and filled with laughter of healthy children. How can we achieve this? How can we identify threats and take timely and adequate measures to assure that my children and your children will inherit a sustainable Arctic? How can we be better prepared for new opportunities that are knocking at our doors? *It is only our collective efforts and good will that could pave the way for finding true solutions.*
>
> (Gundersen, 2011, emphasis added)

> Our pledge to you all is that we need to safeguard the unique work of the Arctic Council. We need to continue to cooperate as *one Arctic family learning from each other and respecting each other.* That is our responsibility and is important for sustainable development and well-being of all.
>
> (Javo, 2015, emphasis added)

> But there is nothing that should *unite us* quite like *our concern for both the promise and the challenges of the northern-most reaches of the earth.*
>
> (Kerry, 2013, emphasis added)

It is noteworthy that, particularly in recent years, this cooperation has also related to the inclusion of actors from outside the Arctic. With regard to the debate on expanding membership in the Arctic Council, for instance, representatives of the AIA have stressed that "change has often been a part of the Arctic Council, and that change has nearly always resulted in a stronger Council, better able to fulfil its mandate" (Gundersen on behalf of the AIA, 2013). It is the original mandate to facilitate circumpolar cooperation (Art. 1a) that Gundersen is referring to and assessing positively and as necessary to strengthen the Council. This is particularly remarkable because what she is saying goes against the perception that permanent participants believed the inclusion of more actors would weaken their position (cf. e.g. Lackenbauer and Manicom, 2015). She further explains this need to collaborate with the new challenges related to the changing Arctic, arguing that the expansion of the AC would contribute to "helping it be better able to respond to an more rapidly changing Arctic and an increasing focus on the Arctic globally" (Gundersen, 2013). The AAC representative, Michael Stickman, likewise emphasises that environmental changes are caused by "countries outside our region. So, we need to talk to these countries" and in particular asks "Can we take some of the politics out of this, and talk about how best to engage non-Arctic states with Arctic interests? We should use the Arctic Council to engage non-Arctic states" (Stickman on behalf of the Arctic Athabaskan Council, 2013). In a similar vein, other permanent participants also expressed concern "that geopolitical issues not related to the Arctic might threaten the discourse on Arctic issues" (Aleut International Association, 2015), and argued during the Ukraine crisis:

> International co-operation in the Arctic is important to those of us who live in the North. Decision-makers in Washington DC, Ottawa and Moscow, and our Asian and European observers, should understand this. We are not naïve, but this Council and its individual members should shield our cooperation from broader political and geopolitical rivalries.
>
> (Stickman on behalf of the Arctic Athabaskan Council, 2015)

In concrete terms, in the Arctic Council circumpolar cooperation is accordingly promoted:

1   to support and strengthen the work of the Arctic Council in view of environmental changes, and
2   to strengthen the voice of the permanent participants.

By applying the former reasoning, the WWF too demanded the inclusion of knowledge provided by environmental organisations such as itself. Having reviewed AC project reports in the past, it offered its expertise to refine the reports or share knowledge generated in own assessments (see WWF, 2011). Permanent participants have also argued that, in order to strengthen the AC, the sharing of knowledge "can help build a good platform for decision makers and those who live in the Arctic" (Olli on behalf of the Saami Council, 2013). While SAOs (representatives of member states) had demanded back in 2007 that observers should, in general, be restricted, permanent participants often argued in favour of including non-Arctic actors in the AC. Similar to the "moral pledge" often referred to in the Antarctic Treaty System, a representative of the AAC, for example, stressed the principle of circumpolar cooperation as having been promoted in the founding phase of the Arctic Council and recommended the inclusion of non-Arctic actors when outlining the need to increase "the role and participation of observer states in activities of the council, and opening a dialogue with China, Japan, India and

Brazil about their potential involvement" (AAC, 2007, p. 10). This recommendation obviously conflicted with those who believed that expanding the number of participants in the AC would drown out their voices (a view often ascribed to permanent participants, cf. Steinberg et al., 2015) or would compromise "the efficiency of the Arctic Council process" (SAO Meeting, 2007/2). The reasoning provided by the AAC is also noteworthy as it underpins how much the challenges ascribed to climate change encourage cooperation in the Arctic:

> Decisions made in non-Arctic states and by global institutions have a growing influence on the well-being of Athabaskans who continue to adjust to a rapidly changing world. Adapting to the impacts and effects of climate change is becoming a central task. In this rapidly changing world our objective is for the Arctic Council to effectively address the international dimension of economic, environmental, social, cultural and other issues of concern to Athabaskans, and to other Indigenous peoples in the circumpolar world.
>
> (AAC, 2007, pp. 3–4)

Again, entangled issues of concern at the international level, climate change and the need for the AC to improve its effectivity are stressed and at the same time this statement illustrates how much the discussion on dealing with climate change and the opening of the Arctic Council to "outside actors" are interrelated. This nexus has also been made explicit in conferences such as the one organised by the Arctic Council Observer Nordic Council of Ministers (NCM) on which the NCM reported in the second SAO meeting in 2008. The conference was entitled "Common Concern for the Arctic", and its purpose was to inform non-Arctic stakeholders about a variety of current Arctic issues while simultaneously recognising that "the transformation of the Arctic is and will be an event of global significance" and thus also acknowledging the interests of stakeholders who were not from the Arctic region: As the representative of the NCM declared:

> Mr Chairman, it can be argued that *the whole world is a stakeholder in our Polar Regions*. Therefore, we too believe that it is important to take a holistic view of the Arctic region. [. . .] The participation of non-Arctic states in Arctic science is not the only aspect through which they qualify as responsible stakeholders. There are many examples that demonstrate that specific and unusual 'polar capabilities' can be found in non-polar nations. State Observers to the Arctic Council do provide added value.
>
> (The representative of the NCM at the SAO
> Meeting, 2008/2, emphasis added)

Overall, international cooperation in the Arctic Council is mostly described positively, as a solution that allows the sharing of resources and knowledge and encourages a better understanding of changes in the Arctic. In that sense, cooperation is not perceived as a reaction that primarily aims to weaken possible international conflicts over the Arctic but as a necessary means to deal with climate change.

3.2.2.3 Representations of geography and the environment in the Antarctic Treaty System

As the introductory quote already hinted, given the changes under way in the Antarctic, the perception of emerging responsibilities is reflected in the geopolitical reasoning introduced in regional policy making in the ATS too. The consultative parties often emphasise "emerging responsibilities" to argue in favour of multilateral cooperation while at the same time strengthening their territorial claims. The latter is very explicitly exemplified by the tensions and conflicting geopolitical understandings among different (state) actors in regard to the implementation of the Antarctic Treaty Secretariat. Most often, however, exemplification is more indirect and implicit, occurring when the consultative parties reproduce and strengthen the representation of a consistent and stable ATS in their statements, aiming to protect the unique Antarctic environment and thereby steal the thunder of critics of the legitimacy and functioning of the ATS. In regard to the latter, for example, even before the Cold War ended, Roberto Daverede (on behalf of the Argentine delegation, ATCM, 1989) referred to the "ties of harmony and co-operation which have been a feature of relations between the Contracting Parties" and the unity expressed in "the same ideas and goals dedicated to the Antarctic Treaty". The chairman of the Chilean delegation, similarly, addressed all the delegates as "dear friends" (Vargas Carreño, SATCM, 1990) and spoke of "a new fraternal spirit". This family analogy was also used by the representative of the Chilean delegation in relation to shared responsibilities and when arguing that

> in the face of an increasingly complex Antarctic world, we must be understanding and flexible in order to coordinate our presence on the vast Polar Continent and the neighbouring archipelagos so that science and technology, operations and logistics, do not remain in ignorance of one another but enjoy a close, brotherly understanding with the Consultative Meetings, the fundamental structure of the whole Antarctic system.
>
> (ATCM, 1995, Pinochet de la Barra on behalf of the Chilean delegation)

The representative of the Argentine delegation likewise spoke of its "sister Republic of Chile" (ATCM, 1991, Fleming on behalf of the Argentine delegation). Both consultative parties thus qualified cooperation in the Antarctic as being based on very intense, quasi-familial relationships, grounded on the principles of "solidarity" and "peace". Andrew Fleming (on behalf of the Argentine delegation, ATCM, 1991) further related the "spirit of cooperation" among Antarctic Treaty parties to the "lofty moral pledge to ensure peace". This moral pledge in particular was described by the Chilean representative as the reason why the implementation of CRAMRA failed and the negotiation of the Environmental Protocol succeeded, reminding all parties of the "very best Antarctic tradition":

> At the end of the first thirty years, we were sorely tempted by the mineral exploitation issue, yet we found a way to emerge victorious and united from

the trial. The temptation was difficult to withstand, since reconciling science and business, peace and economic performance in the remote world of the Antarctic seemed so attractive a proposition. When, after quietly considering the proposals in a spirit of self-criticism, we came to realise how dangerous they were for the environment and for peace, we reacted by taking appropriate steps at Viña del Mar and Madrid. Thus it was that the Protocol on Environmental Protection came into being in the very best Antarctic tradition of understanding and co-operation.

(ATCM, 1991, Pinochet de la Barra, on behalf of the Chilean delegation)

Individual geopolitical and economic interests, as Óscar Pinochet de la Barra outlines, were thus set aside in this particular case and instead the "moral pledge to ensure peace" united all parties, resulting in an intensification of cooperation and encouraging the understanding of a shared responsibility. In retrospect, this "moral pledge" has lasted as the Argentine representative outlined 18 years later:

I remain convinced that the best way to protect the Antarctic, its pristine environment and the entire planet is through solidarity, peace, teamwork and cooperation. In this light, my country reaffirms its commitment to furthering scientific knowledge, protecting the environment and promoting close cooperation with other countries in order to achieve this goal.

(ATCM, 2009, Mansi on behalf of the Argentine delegation)

Here, Ariel Mansi even transfers the "moral pledge" inherent in Antarctic cooperation to the global level, which, on the one hand, can be understood as an acknowledgement of the global significance ascribed to the Antarctic but, on the other, implies a worldview according to which the "entire planet" can be protected "through solidarity, peace, teamwork and cooperation". Substantiating the latter, disputes over the status of the Islas Malvinas/Falkland Islands did not hinder cooperation among Argentina and the United Kingdom which tabled a number of joint proposals to ATCM meetings.

The negotiations on the establishment of the Antarctic Treaty Secretariat, however, provided an important exception and revealed how – despite the continuous cooperation in view of other issue areas in the ATS – geopolitical reasoning conflicted between Argentina and the United Kingdom and was closely entangled with their shared past. Further, the statements introduced in this regard illustrate how the "moral pledge" was used as an instrument to strengthen geopolitical views.

During the negotiations for the Environmental Protocol, Argentina proposed Buenos Aires as the future location for the secretariat that all consultative parties saw as a necessary organ to manage the work of the numerous ATS-subsidiary bodies and of the Committee for Environmental Protection (at that time agreed on but still to be formed). The United Kingdom opposed the location of the secretariat in Buenos Aires, arguing that it had not offered to host the secretariat itself due to its overlapping (but "frozen") territorial claims in Antarctica. Instead, the United Kingdom stated that it would accept any location other than Buenos Aires. Later statements provided evidence that the United Kingdom was particularly doubtful of Argentina's geopolitical interests in Antarctica, as most of Argentina's activities in Antarctica were conducted by military forces and it was further argued that "a country known for aggressive military records is not suitable for hosting an agreement dedicated to secure peace" (Interview, January 2015). In 1994, the U.S., the other candidate country to be the location of the Antarctic Treaty Secretariat and depositary of the Antarctic Treaty withdrew and appointed Buenos Aires as an appropriate candidate to achieve "a weighted geographic balance as regards the headquarters of the various components of the Antarctic Treaty System" (see ATCM, 1994, Scully on behalf of the U.S. delegation). In view of almost unanimous support, Argentina argued officially it was not "willing to analyse any other alternative regarding the geographic location of the Secretariat" (ATCM, 1997, Solari on behalf of the Argentine delegation). In 1998, however, Australia put forward Hobart as an alternative proposal, a move that was highly criticised by Argentina, which perceived this application as counteracting the interest of several countries in achieving a prompt solution and unjustified in view of the almost unanimous support that Argentina's proposal had received for the preceding five years. To strengthen its position, the representative of the Argentine delegation referred to the "moral pledge"[13] that arguably forms the base of Antarctic cooperation and classified the opposition of the United Kingdom as undermining the consensus principle by stating:

> It is therefore in no way beneficial for the consolidation of the spirit of harmony and co-operation which has always prevailed in the Consultative Meetings that, in this case, a single reservation, which throughout the past five years has not obtained any support, be enabled to thwart the will of the vast majority of the Consultative Parties. The acceptance of this position certainly indicates an unusual use of the consensus mechanism which could imply potentially hazardous consequences for the correct operation of the Antarctic Treaty System in the future. Of the implied motives for the formulation of a reservation against the Argentine proposal, we could infer a discrimination between States Party to the Antarctic Treaty which goes clearly against both its text and its spirit, and affects one of the Treaty's fundamental pillars such as Article IV. In effect the word and the spirit of the Antarctic Treaty have allowed for its effective operation despite global, regional or bilateral conflicts which have occurred outside its field of application.
>
> (ATCM, 1998, Solari on behalf of the Argentine delegation)

As a reaction to the Hobart proposal, the U.S. delegate renewed the "continued support [of the U.S.] for the establishment in Buenos Aires of a modest, cost-effective Secretariat to assist the work of the Antarctic Treaty System" (ATCM, 1998, Scully on behalf of the U.S. delegation). In 2001, the United Kingdom agreed to the secretariat's being established in Buenos Aires. This decision came after Argentina had reorganised the structure of its Dirección Nacional del Antárctico (DNA) with a civilian rather than military leadership, strengthened its science division and enacted the Argentine Antarctic policy in Argentine legislation that included the implementation of the Environmental Protocol and its Annexes (ATCM, 2001, Juanarena on behalf of the Argentine delegation). According to an interviewee, this change of authorities was a crucial step that also ensured that in Argentina the military has only had an advisory role in terms of Antarctic logistics up to the present (Interview, January 2015).

3.2.2.4 Representations of international cooperation and environmental changes in the Antarctic Treaty System

In the ATS international cooperation is thus considered to be a long-lasting "Antarctic tradition" shaped by shared responsibilities. In the wake of more rapid environmental changes international cooperation is further accompanied by a feeling of shared overburdening, as numerous representatives of consultative parties such as Secretary of State Hillary Clinton (on behalf of the U.S. delegation) pointed out 18 years after the signing of the Environmental Protocol:

> Exploring our planet, protecting its future, is *too large a task for any one country to undertake*. And of course, no country owns the market on good ideas. Breakthroughs can and should come from anywhere and everywhere, especially when genuine collaboration and teamwork are involved.
>
> (ATCM, 2009, emphasis added)

In contrast to "the Antarctic tradition", which builds on the feeling of a shared responsibility and was mostly related to *state actors* in the ATS, the framing of a "shared overburdening" specifically included "representatives from governments, academia, science, and the private sector" with whom, cooperation is more often envisioned "to meet the challenges of this time" (ibid.). Earlier, in the 1990s, representatives of the Chilean and U.S. delegations, amongst others, supported the inclusion of non-state actor representatives in the ATS and in the intersessional work. In contrast to Clinton's vision of their inclusion in joint activities, however, in the 1990s monitoring functions and expert knowledge were what was ascribed to non-state actors and perceived as beneficial for the ATS (cf. ATCM, 1995, 1996).

As in the statements introduced in the Arctic Council, therefore, environmental changes were used in the ATS as a stepping stone to argue in favour of more collaboration with non-state actors. In addition to this more inclusive and global understanding, which was also promoted by others, Clinton also referred to another dimension that has become recognised as significant in the present politics of the Polar Regions and that was already a strong argument during the negotiations for the Madrid Protocol: time. When speaking – like others before her – of "succeeding generations" Clinton not only extended the principle of a shared responsibility to other countries (in accordance with the aim of securing peace), she also considered the consequences that are likely to materialise in the future. This dimension of time, however, includes the challenge of uncertainty.

In particular, this uncertainty links back to the initial and most prominent characteristic ascribed to Antarctic governance right after peaceful cooperation (Art. 1) in the Antarctic Treaty, that is, scientific cooperation (Art. 2). Fifty years after the signing of the Antarctic Treaty, the representative of the Argentine delegation (Taiana, ATCM, 2009) built on the challenges addressed by Hillary Clinton and (again) put particular emphasis on international and "polar scientific cooperation" as a means of providing more reliable scientific results, pooling ideas and efforts and meeting complex, risky and costly challenges, arguing that the evolution of the Antarctic climate over the next 100 years represented a major scientific challenge. Joint scientific collaborations had often before been perceived (by representatives of the American polar-rim states among others) as illustrating the fundamental need for and success of international cooperation in the Antarctic:

> Antarctica has long been described as a continent for science and more recently as a *land of science*. [. . .] Antarctic science also has a critical role extending far beyond the Antarctic. It has been long known that the Antarctic offers unique opportunities for research in a variety of disciplines which contribute to understanding problems outside the Antarctic. In recent years it has also come to be accepted that research in the Antarctic, including the Southern Ocean and the Sub-Antarctic islands, is crucial in its contributions to understanding global change, development which affect all human beings. [. . .] The subject has a relevance far beyond the confines of the Antarctic. It includes global warming and the thinning of the ozone layer.
>
> (SATCM, 1990, Laws on behalf of SCAR, emphasis added)

> In a world of complex interdependence and scarce resources, the preservation of the Antarctic as a natural reserve, devoted to peace and science, with values and interests shared by all Parties, must strengthen the commitment of each Party to the System, and favor the common interest over the interests of each individual State. Chile considers that the Antarctic Treaty and its System must be broadened and reinforced, as appears in our recently approved Antarctic Strategic Plan 2011–2014. In the face of global pollution and climate change, the Antarctic must be safeguarded, and preserved as our gift to future generations.
>
> (ATCM, 2011, Moreno Charme on behalf of the Chilean delegation)

In the wake of environmental changes, the significance ascribed to science, moreover, connects collaborative efforts in the Antarctic with the work conducted under the auspices of the Arctic Council, which is often summarised as "polar scientific cooperation": In 2009, for example, with the conclusion of the predominantly science-based International Polar Year and on the occasion of the 50th anniversary of the Antarctic Treaty, the first ever joint session of the Antarctic Treaty Consultative Meeting and the Arctic Council ministerial meeting was held. But the Canadian delegation had revealed these science-based interlinkages 10 years earlier when arguing that "both the principle of arriving at major decisions in the polar regions by international consensus, and an increasing number of scientific and environmental monitoring activities undertaken through the Council, have relevance to Antarctic decision-making and science" (ATCM, 1999, opening statement by the Canadian delegation).

Beside the fact that this interpolar entanglement is based on shared environmental and scientific challenges, the institutionalised means of international cooperation embodied in the Antarctic Treaty System were considered during the formation of the Arctic Council. Mary Simon, Canada's first Ambassador for Circumpolar Affairs and a lead negotiator during the formation of the AC, for example, refers to the "systematic exchange between the intergovernmental organizations dealing with the Arctic and the Antarctic" that had been "welcomed" by the ministers "of the eight Arctic countries" who "agreed to reciprocate" (ATCM, 1996). She thus clarifies that actors with a say in both Arctic and Antarctic politics exchanged their knowledge. Moreover, she reveals that the representatives of circumpolar countries also discussed "the experience of the Antarctic Treaty" during negotiations leading to the creation of the Arctic Council, particularly when dealing with the question of how to address "both national and international issues in polar regions" (ATCM, 1996), and thus considered the governance of Antarctica as a frame of reference. After its formation, the Canadian delegation further introduced the AC as "the appropriate body to provide liaison with the Antarctic Treaty on matters of bi-polar significance" (ATCM, 1997, opening address by the Canadian delegation) and, specifically, the implementation of the consensus principle in AC policy making is depicted as being inspired by the ATS:

> The Treaty System is also a successful example, although not without difficulties, of the development of consensus-based international policy to carry out a shared responsibility and foster a sense of caring for all parts of our planetary home.
>
> (ATCM, 1997, opening address by the Canadian delegation)

As this statement further reveals, the Canadian delegation relates the consensus principle to the understanding of "a shared responsibility" for "our planetary home", thus providing evidence that identical norms are shared in Arctic and Antarctic politics, and at the same time making use of a rhetoric usually ascribed to common heritage discourses. As in the discourses conducted in the Arctic Council a decade later, both the Argentine and the U.S. representatives substantiated this particular understanding by outlining the significance of the Polar Regions in the context of climate change:

> The Arctic and Antarctic play a primordial role in many issues that are of crucial importance to humanity, such as global warming, climate change and rising sea levels. It is at the poles, more than anywhere else in the world, that we can best observe the huge environmental impact that climate change is having.
>
> (ATCM, 2009, Taiana on behalf of the Argentine delegation)

We need to increase our attention not only to the Antarctic but to the Arctic as well. As a senator, I travelled to the Arctic region, both in Norway and Alaska. I saw for myself the challenging issues that the region is facing today, especially those caused by climate change. This too provides an opportunity for nations to come together in the 21st century, as we did 50 years ago in the 20th century. We should be looking to strengthen peace and security, and support sustainable economic development, and protect the environment.

(ATCM, 2009, Hillary Clinton on behalf of the U.S. delegation)

In general, the recognition of interrelationship between the AC and the ATS is particularly reflected in the regular ATCM agenda item "relevance of developments in the Arctic and the Antarctic" that is not limited to the scientific dimension. The related reports show that interpolar entanglements have not only existed through the participation of individuals, organisations and states in both regional governance settings but also in the "increasing number of the issues and topics important to international governance" that are of "bi-polar and global relevance". As was specified, for instance, in the opening address by the Canadian delegation to the ATCM of 1997 "[e]nvironmental protection and sustainable management of resources are priority interests in both areas". Not only Canada but also Chile (as another American polar-rim state), moreover, emphasised "the need for a bipolar perspective" back in the 1990s.

Overall, the views presented previously reflect the fact that the actors under analysis consider international cooperation in the AC and ATS as being particularly entangled with environmental concerns and increasingly with climate change during the period under investigation. More recently, representatives from the American polar-rim states have increasingly emphasised the need for cooperation with non-state and nonpolar actors too, who are perceived as providing significant knowledge that is needed to deal with environmental challenges in both regional settings. By the same token, in the AC and particularly in the ATS, scientific initiatives are highlighted as means to protect the "planetary home". This particular imaginary can be regarded as a variation of the understanding of the "Arctic home" but, unlike the statements introduced in the Arctic Council, this imaginary is used to argue in favour of inclusivity by being linked to the notions of "shared responsibilities" and "shared overburdening" in dealing with climate change. Moreover, in both the AC and the ATS the "moral pledge" of the latter as well as the "principle of circumpolar cooperation" in the founding phase of the former were emphasised by the actors under analysis when demanding an intensification of cooperation.

3.3 Geopolitical reasoning in the domestic politics of the Polar Regions

By following the understanding promoted by scholars of Critical Geopolitics that "power [is exercised] through the ability to impose order and meaning upon space" (Ó Tuathail, 1996, p. 1), this book has already outlined inter- and transregional entanglements that have contributed to the dominant interpretations of the changing Polar Regions that have influenced the governance of the Arctic and the Antarctic and its implications beyond both regions. But to what extent do the geopolitical views introduced by the American polar-rim states in the AC and ATS build on or conflict with their positions in domestic politics?[14]

3.3.1 Classification and contextualisation of the polar policies of Canada, the U.S., Chile and Argentina

3.3.3.1 Canada

In general, Canada's Arctic policy (like the mapping of its continental shelf) is coordinated by the Department of Foreign Affairs and International Trade (DFAIT, Bergh, 2012; Conley et al., 2013). DFAIT is headed by Canada's Senior Arctic Official who took over "much [of] the workload of the Circumpolar Ambassador, a position that was abolished by the [former] government in 2006 in an effort to cut costs" (Bergh, 2012, p. 9) and who meets with SAOs from other Arctic-rim states in the AC on a regular basis. Related to the new joint policy *Shared Arctic Leadership Model* (2016) announced by U.S. President Barack Obama and the Canadian Prime Minister Justin Trudeau to build a sustainable northern economy and reduce climate change emissions taking traditional knowledge into consideration (Indigenous and Northern Affairs Canada, 2016), the position of Canadian Minister's Special Representative (MSR) was given to Mary Simon in March 2016. Simon is responsible "for leading an engagement and providing advice on the development of a new shared Arctic leadership model" (Indigenous and Northern Affairs Canada, 2016). At the national level, policy advice is provided by numerous other state agencies, for instance, Transport Canada and Environment Canada. Canada also used to have a Canadian Polar Commission and the Canadian High Arctic Research Station initiative of Aboriginal Affairs and Northern Development Canada. Both were combined and replaced in 2015 by the federal research organisation Polar Knowledge Canada (POLAR) that focuses particularly on advancing knowledge of the Arctic. POLAR moreover sends representatives to meetings conducted under the auspices of the Antarctic Treaty System (e.g. to SCAR and to COMNAP).

Particularly in the past decade under the former government of Prime Minister Stephen Harper who placed "the Arctic at the top four of our foreign policy agenda" (Wright, 2013, p. 104), various policies that relate to the Arctic were released in Canada, most notably *Canada's Northern Strategy – Our North, Our Heritage, Our Future* from 2009 and the 2010 *Statement on Canada's Arctic Foreign Policy – Exercising Sovereignty and Promoting Canada's Northern Strategy Abroad*. In 2009 the Canadian Standing Senate Committee on Fisheries and Oceans published the *Rising to the Arctic Challenge: Report on the Canadian Coast Guard*. Back in 2007, on the subnational level, the Northwest Territories, Yukon and Nunavut had introduced the first joint policy document, *A Northern Vision: A Stronger North and a Better Canada* (followed by four other jointly adopted documents in 2011, 2014 and in 2016), on which the 2011 *Pan-Territorial*

Adaptation Strategy: Moving Forward on Climate Change Adaptation in Canada's North builds.

Numerous studies have already found that Canada's interest in its Arctic territories has greatly depended on the government in power (e.g. McRae, 2007) even though the country's relationship with the territories located in its North is based on a long historical development (Bartsch, 2015). Despite their sparse population, these Canadian territories have been regarded as of particular significance throughout history and particularly since European colonisation because of the desirable goods and resources developed north of the Arctic Circle (such as furs, coal and gas, ibid.). More recently, the significance of the people who lived in Canada's North even before European colonisation, who "have occupied Canada's Arctic lands and waterways for millennia" as Canada's Minister of Indian Affairs and Northern Development put it (Government of Canada, 2009, p. 3), has been emphasised in Canada's Arctic policies. But before as well, in the early days of the AC, for instance – the Canadian government was a driving force in the Council's establishment and in building up its international influence – Canada's Arctic interests were much shaped by domestic processes. At that time, for example, "indigenous peoples had been successful in claiming and asserting their right to participation in the governance of their Arctic homeland" (Scrivener, 1996 in Humrich, 2016). As a result, Canada encouraged the inclusion of Indigenous peoples in the AC, and Canadian representatives from the Inuit Circumpolar Council and the Gwich'in Council International have been able to participate as permanent participants in the Arctic Council ever since.

Also in regard to its second chairmanship of the Arctic Council (2013–2015), Canada's agenda was grown from domestic needs (English, 2013) and reflected the significance ascribed to Indigenous peoples in Canada's North: Canadian officials, for example, travelled "to the north of Canada to listen to the Northerners' own view of what was important to focus on during the chairmanship period" (Berger, 2015, p. 65). The main priorities outlined in Canada's chairmanship program *Development for the People of the North*, were moreover represented by the Minister of Health and Social Services Leona Aglukkaq who became the head of the Canadian delegation (the Minister for the Arctic Council), having been born in Inuvik (Northwest Territories) and raised in Nunavut herself.

As far as the ATS is concerned, Canada is a non-Consultative Party and has not released an Antarctic policy. As outlined previously, however, Canada contributed actively to the Antarctic Treaty Consultative Meetings from 1989 and its interests were represented by the fourth-largest number of delegates sent to ATCMs during the period of investigation. Canada further participates in the Committee for Environmental Protection and has adopted CCAMLR and the various Climate Change Agreements.

3.3.3.2 The U.S.

In the U.S., the number of federal agencies with responsibilities in the U.S. Arctic has grown far beyond 20,[15] which marks a sharp contrast to 1971 when the first U.S. Arctic policy was released and only seven U.S. state agencies had coordinating tasks (Conley et al., 2013). Until very recently, the number of state agencies involved in policy formulation processes concerning the Arctic did not mirror the general interest in the region shown in the U.S. In 2010, for instance, the U.S. was still described as being rather "slow to implement its articulated interests and define a comprehensive and assertive strategy in the region" (Conley and Kraut, 2010, p. 26) even though the *NSPD-66/HSPD-25 Arctic Region Policy* (2009, National Security Presidential Directive and Homeland Security Presidential Directive) had already been released at that time. This changed with the introduction of numerous policies in the 2010s, the preparation of the second U.S. Arctic Council chairmanship and the new position of the State Department's Special Representative for the Arctic given to Admiral Robert J. Papp, Jr. in 2014.

The *NSPD-66/HSPD-25 Arctic Region Policy* introduced by President George W. Bush was the first official "update" of the 1994 United States Policy on the Arctic and Antarctic Regions developed under the Democratic government of President Bill Clinton. As in Canada, however, in recent years under President Barack Obama numerous other policies were adopted, most notably the federal *National Strategy for the Arctic Region* (The White House, 2013b), the *Implementation Plan for The National Strategy for the Arctic Region* (2014) and the *Executive Order – Enhancing Coordination of National Efforts in the Arctic* (2015). In 2015, also on the subnational level, the Alaska Arctic Policy Commission published its final report. Other agencies, such as the U.S. Coast Guard, the Department of Defense and the National Ocean Council also introduced the *United States Coast Guard Arctic Strategy* (2013), the *Arctic Strategy* (2013), and the *Strategic Action Plan – Changing Conditions in the Arctic* (2011).

In contrast to Canada, the U.S.'s increased interest in the Arctic is fairly new, and particularly "the decreasing ice cover and increasing global attention given to the region" are often said to have encouraged "the Arctic [. . .] becoming more firmly established on the US Government's agenda" (Bergh, 2012, p. 10). Before, regardless of the political party in power, U.S. Arctic policy "remained constant" (Arnaudo, 2013, p. 85) – in the sense that the Arctic was described as having had a "salient position" in the U.S. foreign policy (Westermeyer and Shusterich, 1984).

With increased international and multilateral cooperation in the Arctic, particularly under the auspices of the AC, the U.S.'s priority interests in the Arctic have continued to evolve around national security interests, scientific research, preservation of the principle of freedom of the sea and the environment. During the formation process of the Arctic Council, the U.S. was viewed as a "reluctant collaborator" (Huebert, 2009a), and in more general terms, it is often said that after the Cold War the Arctic region was "overlooked in U.S. national politics" (Bergh, 2012, p. 10). With respect to the Arctic Council, Nord (2016, p. 18) explained this reluctance thus:

American objections to the Arctic Council took a variety of forms. Most prominent of those voices during the period of the George H.W. Bush Administration was a concern over how the operation of a multifaceted Arctic Council might interfere with Washington's established orientation toward the region. The Arctic policy of the United States had long given a priority to military and defense issues in this area. Even with the apparent end of the Cold War, most Washington policymakers were reluctant to replace their strategic vision of the region with one that gave equal priority to environmental, economic and social concerns. Equally significant was the unwillingness of American policy analysts to abandon their decidedly hegemonic vision of Arctic decision making.

As far as Arctic research was concerned, however, the U.S. had already taken the lead in the Arctic Council in the late 1990s and initiated the Arctic Climate Impact Assessment (ACIA). Nevertheless, the ACIA initiative is said to have "correspond[ed] well with the climate policies of the Clinton administration" (Smieszek and Kankaanpää, 2015, p. 10). Similarly, in "One Arctic", the imaginary and theme of the second U.S. Arctic Council chairmanship programme highlighting the exceptionality of the region and the collaborative actions desired, climate change was highlighted as a phenomenon to be dealt with internationally by all Arctic-rim states – who were all described as being affected by the challenges related to it. English (2013, p. 295) explains the contrast between this focus promoted by the U.S. and that of Canada's prior chairmanship by pointing to the policy tradition of both countries, arguing that "Canada's Arctic policy grew from domestic needs, whereas American policy was shaped by global geopolitical concerns". Similar geopolitical concerns are also said to have turned the U.S. into "an Arctic nation" (U.S. Arctic Council chairmanship program, 2001) in the first place. With respect to the purchase of Alaska in 1867, however, it was also Russia that – "as a byproduct of European balance-of-power politics" – preferred to sell Alaska to the U.S. instead of to its principal rival Great Britain (cf. Zellen, 2009a, p. 74).

In the Antarctic, on the other hand, the U.S. had already introduced the aforementioned policy from 1994, though it "[did] not yet make a distinction between the Arctic and the Antarctic as spheres of interest to U.S. Foreign policy" (Pedersen, 2012, p. 152). This changed in 1996, with the *Presidential Decision Directive/National Security Council-26 U.S. Antarctica Policy* (U.S. Department of Defense, 1996), which focused particularly on budgetary decisions regarding the U.S. Antarctic program and on the renovation of the South Pole Station. Nevertheless, the U.S. has had a significant say in the Antarctic region since the 1950s. The Antarctic Treaty was signed following an invitation by the U.S. to the Soviet Union and all other countries claiming territory in Antarctica to participate in The Washington Conference on Antarctica in 1959. The U.S. is also the depository of the Antarctic Treaty and thus perceived as its "guardian" (Interview, January 2015). As an original signatory to the Treaty, the U.S. also adopted all other instruments under the ATS and sent the largest delegations to the ATCMs in the period under investigation. The U.S. also supported the adoption of CRAMRA but – in retrospect – acknowledged the opposition provided by representatives from Australia and France at these times as a "perhaps heroic" intervention that "led to something better" – the Environmental Protocol (Bloom, 2016).

The U.S. position in the ATS, like its position regarding the Arctic, has been particularly driven by scientific interests that are rooted in the Antarctic Treaty and increasingly relate to climate change. In contrast to the numerous different state agencies involved in Arctic science, however, all Antarctic research activities and programs are managed by the National Science Foundation (The National Science Foundation, 2011, p. 1f.).

3.3.3.3 Chile

In Chile the Dirección de Antártica [Directorate of Antarctica] and the Consejo de Política Antártica [Antarctic Policy Council] are the core actors responsible for Chilean Antarctic policy and also for Chile's cooperation in the ATS. Since 1978 and in accordance with *Decreto N° 161*, the Council, in particular, has had the function of determining national actions in the Chilean Antarctic Territory. The Directorate operates under the Department of Foreign Affairs (Ministerio de Relaciones Exteriores) and cooperates with numerous other state agencies, particularly the Chilean Antarctic Institute (Instituto Antártico Chileno, INACH) and the Direction of Borders and Limits of the State (Dirección de Fronteras y Límites del Estado, DIFROL), as well as with various ministries (the Ministry of Defense, the Ministry for the Environment, the Subsecretariat for Tourism, and the subnational government of the XII Región de Magallanes y Antártica Chilena, cf. Ministerio de Relaciones Exteriores de Chile, 2016b). In 2000, the Dirección de Antártica released the *Chilean Política Antártica Nacional* (Decreto N° 429, 2000) on behalf of the Ministry of Foreign Affairs and under President Ricardo Lagos. This National Antarctic Policy has been updated since via strategy plans (Planes Estratégico Antártico). The newest strategy plan is the first long-term vision paper (it applies until 2035) and the most detailed paper so far (it runs to 63 pages). In accordance with the National Antarctic Policy all strategy plans highlight the history and geography of Chile in regard to Antarctica, its understanding as a "país-puente a la Antártica" [bridge country to the Antarctic] and refer to the guiding principles of the Antarctic Treaty, for instance, when outlining international scientific cooperation as a main priority (cf. e.g. Ministerio de Relaciones Exteriores de Chile, 2013).

In the late 1970s, Chile and Argentina intended to substantiate their territorial claims in Antarctica, but during the period of investigation Chile's interests in Antarctica revolved mainly around environmental concerns. Chile was, for instance, a leading party during the negotiation of the Madrid Protocol and hosted the related Special Antarctic Treaty Consultative Meetings in 1990 in Viña del Mar. The National Antarctic Policy, however, also highlights the role of Punta Arenas as an "Antarctic city" – a logistic centre for national and international activities in Antarctica. To enforce this understanding, the Chilean Antarctic Institute (INACH) became the first federal agency whose headquarters were located outside the capital Santiago when it was transferred to Punta Arenas in 2003 (also as a way of establishing closer relations with the distant Magallanes region; Interview, February 2015). The INACH is responsible for "planning, co-ordinating, directing and controlling officially authorized scientific and technological activities of the

Chilean government and private organizations in Antarctica" (COMNAP, 2016). Besides this international orientation directed towards the support of activities in the Antarctic and its endeavours during the negotiation of the Madrid Protocol, as a claimant state and original signatory of the Antarctic Treaty, Chile has also been very active in the ATS and sent the third-largest delegation to past ATCMs in the period under investigation.

Chile does not have an Arctic policy and did not participate in the AC during the period of investigation.

3.3.3.4 Argentina

In addition to the Ministry of Foreign Affairs and Worship, the National Antarctic Commission (La Comisión Nacional del Antártico) and the Argentine Antarctic Institute (Instituto Antártico Argentino – IAA), which falls under the National Directorate for the Antarctic (Dirección Nacional del Antártico, DNA), shape Argentina's Antarctic policies. Members of the DNA also represent Argentina internationally at COMNAP and the Reunión de Administradores de Programas Antárticos Latinoamericanos (RAPAL) meetings. In April 1940, the National Antarctic Commission was founded with the mission to consider Argentine interests in Antarctica (Genest, 2004). The IAA, established in 1951 under *Decree Nº 7338*, which focuses on academic research carried out in and relating to Antarctica, also participated in the National Commission of the International Geophysical Year, and is known as the first organisation in the world to be exclusively devoted to Antarctic research. In 1969 the DNA was created, its mission being to regulate and control Argentina's activities in Antarctica in accordance with the nation's objectives, policies and strategies. Further, the DNA is responsible for releasing the *Plan Anual Antártico* [Annual Plan for Antarctica] and coordinates the participation of private entities in Antarctic activities. According to *Decreto N° 1041/95* adopted by the Ministry of Defense in 1995 the DNAntártico has the primary responsibility for realising

> programación, planeamiento, coordinación, dirección, control y difusión de la actividad antártica argentina, a fin de lograr el cumplimiento del objetivo, políticas y prioridades de la Política Nacional Antártica, contribuyendo a la permanente actualización de la misma en concordancia con la dinámica del quehacer en la región, para afianzar la eficacia del accionar argentino en la material. [the programming, planning, coordination, direction, control and diffusion of Argentina's activities in the Antarctic, in order to accomplish the objective, policies and priorities of the National Antarctic Policy, contributing to the permanent actualisation of the same in accordance with the regime of work in the region, to strengthen the effectiveness of Argentina's activities there].
>
> (Ministerio de Defensa, 1995)

Formerly subordinate to the Ministry of Defense, the DNA was transferred to the Ministry of Foreign Affairs in 2004. As mentioned before, this transfer of responsibilities was a prerequisite for the location of the Antarctic Treaty headquarters in Buenos Aires and was also (more indirectly) outlined as such in the *Decisión Administrativa 509/2004* (Dirección Nacional del Antártico, 2004).

Argentina's National Antarctic Policy (Política Nacional Antártica) was released in 1990 under the Peronist government of President Carlos Menem and – like Chile's – has been updated since via national action plans developed by the Argentine Antarctic Institute under the auspices of the National Directorate for the Antarctic. Building on this policy also, Law N° 307 was adopted, which relates to Argentina's sub-Antarctic province of Tierra del Fuego, Antarctica and the South Atlantic Islands.

During the Argentine dictatorship, activities promoted under Argentina's Antarctic policy are described as having had "una connotación predominatamente 'territorialista'" [a predominantly "territorial" connotation] (Colacrai, 2012, p. 272). Despite its continuing claim to the Islas Malvinas/Falkland Islands referred to before (and supported by Chile), this connotation has diminished. Since the end of the dictatorship, Argentina's position towards Antarctica has been particularly characterised by a desire to strengthen the ATS. Like Chile, Argentina has focused mainly on scientific activities, collaboration with other members of the ATS and using its geographic proximity for the international management of Antarctica. Although, in comparison to other original signatories of the Antarctic Treaty, Argentina has been represented by smaller delegations at ATCMs, during the period of investigation its delegation contributed actively to negotiations in these meetings and also supported the negotiation of the Madrid Protocol.

After both countries had returned to democracy, Argentina collaborated particularly with Chile in Antarctica. Interviewees said that this cooperation was also seen as building on the "historias antárticas [. . .] muy similares en ambos países" [very similar Antarctic histories in both countries] (Interview, February 2015). This collaboration was formalised with the *Declaración Conjunta sobre la Antártida* [Joint Declaration on the Antarctic] (1990) signed by Chilean President Aylwin and Argentine President Menem, in which they declared their intention to cooperate in activities in Antarctica. In 2009, this collaboration was expanded by the *Tratado de Maipú* [Treaty of Maipú] signed by Presidents Bachelet of Chile and Kirchner of Argentina in which they formalised the collaboration between two particular scientific bases in Antarctica. In 2012, moreover, President Kirchner and Chile's President Piñera signed an agreement to create an ad-hoc committee aimed at promoting joint positions within the Antarctic Treaty System.

Argentina does not have an Arctic policy and has not been regularly active in the Arctic Council in the period of investigation, although two governmental representatives attended the Ministerial Meeting in 2009 as invited guests.

3.3.2 Entangled representations of environmental changes and international cooperation

So how do the polar policies of the American polar-rim states relate to and represent the topics (environmental changes and international cooperation) most prominently addressed in the AC and ATS? And do these representations correspond or conflict with the dominant understandings and positions regarding these topics at regional level?

Despite the differences between the Arctic and Antarctic and as with policy making at the regional level, the national Arctic and Antarctic policies formulated by the four American polar-rim states show very similar concerns as their priorities. Although, irrespective of their different dates, all national policies highlight the exercise and defence of their sovereignty and national security as their main objective, they also prioritise international cooperation and the protection of the Arctic and Antarctic environment. Nonetheless, the significance they ascribe to the AC, the ATS and to partnerships with neighbouring states differs. Likewise, the changes taking place in the Arctic and Antarctic and the protection of the environment are addressed differently in policies of the different states and reflect their particular interests.

3.3.2.1 Entanglements and disentanglements in the policies put forward by the American Arctic-rim states

As with the positions introduced in the AC outlined before, it has been the representation in the American Arctic-rim states of climate change as a phenomenon caused "inside" and/or "outside" the Arctic that has been used to empower a "local" or "global" perspective in the geopolitical reasoning by actors with different political interests. Canada's *Northern Strategy* (Government of Canada, 2009, p. 24) for instance states:

> Facing the challenges and seizing the opportunities that we face often require finding ways to work with others: through bilateral relations with our neighbours in the Arctic, through regional mechanisms like the Arctic Council, and through other multilateral institutions. The United States is our premier partner in the Arctic and our goal is a more strategic engagement on Arctic issues.

Similarly, the U.S. *National Strategy for the Arctic Region* (The White House, 2013b, p. 8f.) declares:

> What happens in one part of the Arctic region can have significant implications for the interests of other Arctic states and the international community as a whole. The remote and complex operating conditions in the Arctic environment make the region well-suited for collaborative efforts by nations seeking to explore emerging opportunities while emphasizing ecological awareness and preservation. We will seek to strengthen partnerships through existing multilateral fora and legal frameworks dedicated to common Arctic issues. We will also pursue new arrangements for cooperating on issues of mutual interest or concern and addressing unique and unprecedented challenges, as appropriate.

In view of changing conditions in the Arctic and related challenges, therefore, both policies represent international cooperation in multilateral institutions as being of crucial significance and the Canadian *Northern Strategy* ascribes a particular priority in this regard to the Arctic Council and to its partnership with the U.S. (Government of Canada, 2010). Whereas the U.S. *National Strategy*, however, considers international cooperation as a means of raising ecological awareness beyond the Arctic region, Canada's *Northern Strategy* promotes an understanding of strategic cooperation to deal with "Arctic issues" in the region. Moreover, while the U.S. *National Strategy* represents changes and activities in the Arctic as having an impact beyond the region, Canada's *Northern Strategy* promotes the opposite perspective when referring to impacts on the Arctic caused by the phenomenon of climate change originating "outside the Arctic":

> The Arctic environment is being affected by events taking place far outside the region. Perhaps the most well-known example is climate change, a phenomenon which originates outside the Arctic but is having a significant impact on the region's unique and fragile environment.
>
> (Government of Canada, 2010, p. 16)

The perspective promoted by the Canadian government is also taken in Alaska's *Arctic Policy* (Government of Alaska, 2015a), which uses this understanding of climate change as a phenomenon caused by others as a stepping stone to justify economic development in the Arctic:

> Climate change is a global challenge and Alaska's citizens and its economy should not bear the consequences of mitigation. Economic development provides funding for needed infrastructure that will empower Alaskans to adapt, respond and plan *for changes that may result from sources beyond its jurisdiction.*
>
> (Government of Alaska, 2015a, p. 9, emphasis added)

Unlike the U.S. federal government, which places its approach to the Arctic in a national and a global context – when stating, for instance, "The U.S. approach to the Arctic region must reflect our values as a nation and as a member of the global community" (The White House, 2013b, p. 10) – the members of Alaska's Arctic Policy Commission do not consider it to be their responsibility to adjust economic activities in Alaska to mitigate climate change in the Arctic. Instead, the policy reproduces the understanding of citizens in Alaska of their being victims of climate change (as also emphasised by Indigenous peoples' representatives on the Arctic Council), and argues that economic development is needed as a means of empowerment in view of climate change impacts in the Arctic.

In a similar vein, Alaska's policy – although entitled *Arctic Policy and Climate Change* – focuses particularly on economic questions. Prior to recognising climate change, it states, for instance, "[m]any of us live in the Arctic, and all of us depend on its resources" (Government of Alaska, 2015a, p. 1). In following this reasoning through and accusing "others" of domination in this regard, Alaska's policy statement demands that the state take a leadership position in national and regional politics of the Arctic, declaring:

> Alaska is America's Arctic, and the Arctic is a dynamic region that is changing rapidly. We cannot let the perceptions of others – who might not understand its value or its people – determine Alaska's future. Alaska's future in the Arctic demands leadership by Alaskans.
>
> (Government of Alaska, 2015b, p. 2)

It is notable that, although Alaska's *Arctic policy* is not a foreign policy but primarily addresses domestic audiences, it urges "[i]ncreasing involvement [by Alaskans] and prominence in all areas of Arctic governance, with particular focus on the Arctic Council" (Government of Alaska, 2015a, p. 2). This special emphasis on the AC relates specifically to the U.S. chairmanship (2015–2017), which is seen as "creat[ing] opportunities for the State [of Alaska/DW] to demonstrate leadership and bring expertise to decisions that affect our residents". It is made explicit that the State of Alaska aims not only at shaping domestic politics but also at influencing international cooperation and the politics of the Arctic on the regional level, as policy making on both levels is understood as "determin[ing] Alaska's future". In contrast, in the subnational policy developed by the Canadian Northwest Territories, Yukon and Nunavut in 2007 the international dimension is only slightly touched upon and it is almost exclusively the domestic context that is addressed. In view of the changing Arctic, the international dimension is only considered when the policy emphasises the need to "[s]peak [. . .] with a Northern voice" and, corresponding to Alaska's Arctic Policy insofar as a particular say is ascribed to the people inhabiting Canada's Arctic territories: "Northern issues must be addressed by Northern voices" (Northwest Territories, Yukon, Nunavut, 2007). Similarly, Canada's *Northern Strategy* emphasises Canada's special "commitment to protect our North" when stating "[c]ooperation, diplomacy and respect for international law have always been Canada's preferred approach in the Arctic", but continues "[a]t the same time, we will never waver in our commitment to protect our North" (Government of Canada, 2010, p. 27, emphasis added). In line with the Alaskan subnational policy, therefore, the understanding of territory in Alaska and the Canadian North as being "owned" by Canadian and Alaskan citizens and needing to be defended against any possible domination by others is reinforced. Unlike in U.S. federal policies, it is not the significance of the Arctic for the planet that is stressed.

The demand for leadership in Alaska's *Arctic policy*, on the other hand, has historical roots. As explained before, during the formation of the Arctic Council in the 1990s, the U.S. showed little interest in the establishment of the intergovernmental forum. This lack of interest in the Council and in the Arctic in general has been explained by the geographical distance that separates the "lower 48" from Alaska, the only state through which the U.S. can be identified as an "Arctic nation" (Government of Alaska, 2015a, p. 1). In the 1990s, this lack of interest gave Alaska a special say in Arctic regional politics; it was authorised by the federal government to develop the U.S. Arctic Council chairmanship program, thus to decide the priorities of the Arctic policy that the U.S. presented during its first chairmanship in the years 1998–2000 (Interview, August 2015).

During national chairmanships, AC member states are said to have significant agenda-setting power (Berger, 2015) in this regional intergovernmental forum. The special say ascribed to Alaska actually provided the state with a leadership role in international Arctic affairs. This leadership role was, however, taken back by the federal government itself as the preparations for the second chairmanship, which launched in 2015, were conducted in Washington D.C. Thereby the federal government not only took away Alaska's former international leadership role in Arctic affairs, but its chairmanship program also marked a significant shift away from the Arctic issues prioritised before and also conflicted with primary interests traditionally promoted by the State of Alaska. The chairmanship theme developed under the federal government (*One Arctic: Shared Opportunities, Challenges, and Responsibilities*) highlighted global climate change as a particular challenge and emphasised the Arctic's role in global ocean and climate systems which the program outlined as the reason why "the Arctic Council seeks to educate and inform the public worldwide that the Arctic should matter to everyone" (U.S. Department of State, 2015). The objective of raising global ecological awareness was thus operative at the national and regional levels.

By contrast, in the State of Alaska, the development of oil resources had consistently been prioritised by all politicians in previous decades. Even at the time of the Exxon Valdez oil spill, which coincided with negotiations on the opening of the Arctic National Wildlife Refuge (ANWR) to oil development, Alaska's Governor Hickel continued to promote drilling in Alaska and accused the U.S. government in Washington (that opposed the opening) of turning Alaska into "a state on the brink of colony status once again – because we have let our rights be taken away by the Federal Government" (Egan, 1991). Building on the perception that their interests were subordinated to federal priorities, Alaska's *Arctic Policy* objectified climate change and emphasised the prioritisation of the human dimension in the politics of the Arctic:

> [W]e value our federally-protected wilderness and marine areas, but Alaskans should decide for ourselves whether we want any more; and we are concerned with climate change and want to partner with the federal government to adapt, *rather than endure any federal attempts to solve world climate change on the backs of Alaskans.* [. . .] We are concerned that Alaskans will not be able to develop our economy in a way that will allow us to respond to, and prosper in the face of, change. All levels of government can work together to empower Alaskans to adapt and promote resilient communities. *We believe that people should come first.*
>
> (Government of Alaska, 2015b, p. 3, emphasis added)

This emphasis is remarkable as it underlines the perception of being dominated by federal policies even though the U.S. *Arctic Policy Implementation Plan* of 2014 also drew particular attention to "Alaskan residents [who] are experiencing the impacts in the Arctic" (The White House, 2014, p. 4) and highlighted that "[o]ur highest priority is to protect the American people, our sovereign territory and rights, and the natural resources and other interests of the United States" (The White House, 2014, p. 5). The *Executive Order* of 2015, moreover, demanded the establishment of an Arctic Executive Steering Committee to facilitate the interagency coordination also "where applicable, with State, local, and Alaska Native tribal governments and similar Alaska Native organizations, academic and research institutions as well as the private and nonprofit sectors" (The White House, 2015).

Similar to Alaska, a state that often feels "cut off from the lower 48" (Interview, January 2014), the Canadian North is located remote from most Canadians. In contrast to the global perspective and responsibility emphasised in federal U.S. Arctic policies that conflict with Alaska's interests, Canada's Northern Strategy predominantly highlights the significance of the North for Canada and promotes an understanding of needs and responsibilities that incorporates the subnational interests formulated there. Correspondingly, instead of highlighting climate change as a driving force, the 2007 policy formulated by the governments of the Northwest Territories, Yukon and Nunavut takes the experience of "massive social, political, environmental and economic change" (Northwest Territories, Yukon, Nunavut, 2007) as a starting point. So, when speaking about the "[d]ramatic transformation [that] is taking place in the North" (Northwest Territories, Yukon, Nunavut, 2007), references are made to settlement agreements, to the transfer of province-like powers and particularly to the "booming" of "renewable and non-renewable resource industries". The transformation is depicted as being of relevance not only at the subnational level but also to Canada at large, and it is clear that this transformation is thus particularly related to the economic gains made in the Arctic/Northern territories and not to environmental changes.

This understanding is also picked up in the national 2009 *Northern Strategy*, which states "Canada's future is intimately tied to the future of the North" (Government of Canada, 2009, p. 39) and refers to the "unlocked" economic potential in the North that is perceived as providing "unprecedented opportunities" (Government of Canada, 2009, p. 5). Although Canada's 2009 and 2010 policies highlight "rapid changes" as the backdrop to the formulation of these policies and, as opposed to the 2007 subnational policy, link these changes to the impacts of climate change, as in the 2007 subnational policy, neither climate change nor environmental changes are highlighted as areas of concern, instead the spotlight falls on "exercising sovereignty, promoting economic and social development, protecting our environmental heritage, and improving and devolving Northern governance" (Government of Canada, 2010, p. 3). Likewise, when addressing the international dimension, climate change is understood as a driver but not the predominant driver of changes in the Arctic, and what is emphasised is not the Canadian leadership's responsibility to protect the Arctic environment but the importance of Canada's exerting "effective leadership both at home and abroad in order to promote a prosperous and stable region responsive to Canadian interests and values" (Government of Canada, 2009, p. 5).

It is notable that, unlike in the U.S., the priorities outlined in Canada's national and subnational Arctic policies correspond to each other and indicate that they are based on a shared perspective of the North. Moreover, the geopolitical reasoning implicit in Canadian policies is also entangled with the understanding promoted in Alaska's *Arctic Policy*, as it is particularly the well-being and empowerment of people in the North and improving and developing northern governance that are outlined as significant drivers to meet the objective of a strong North. Unlike Alaska's policy, however, according to which empowerment is necessary to deal with the impacts of climate change, the Canadian national policies understand a strong North primarily as a means of defending sovereignty and thus stress again that the environmental impacts caused by climate change are not perceived as a primary concern.

More recently, however, the significance ascribed to climate change and to sovereignty on the subnational level has changed, which is likely to have affected the former shared prioritisation of objectives between the subnational and national level in Canada. In contrast to the 2007 policy formulated by the three subnational governments in Canada's North, the 2011 strategy focuses predominantly on dealing with climate change. Its executive summary starts right away with "Climate change impacts vary widely in nature and magnitude across Canada's North" and similarly, the first sentence of the introduction to the policy is "Climate change is affecting Canada's North" (Northwest Territories, Yukon, Nunavut, 2011, p. 9). In contrast to the Canadian policies discussed before, moreover, the strategy understands the transformations taking place in the North as driven by climate change and promotes a more global perspective when concluding:

> Climate change is a reality in our world. We need to look ahead, to be thoughtful, comprehensive and action-oriented when addressing the issues and making plans to adapt to a changing environment. We believe that future success in this area depends on collaboration, support and integration of our efforts with respect to the urgent concerns and interests of Northerners. We have committed to working closely with our local, territorial, national, Aboriginal *and international partners* to share climate change adaptation knowledge and practices in order *to develop collaborative activities*.
>
> (Northwest Territories, Yukon, Nunavut, 2011, p. 29, emphasis added)

Through highlighting climate change and no longer economic development as the main driver of the changing Arctic, the strategy introduced by the Canadian subnational governments thus reveals a growing entanglement with the perspective promoted by the U.S. government discussed earlier. This shift is also illustrated as the strategy no longer solely speaks about the Canadian North but about "Northern" in general and no longer exclusively addresses the Canadian government but speaks of "governments" in the plural that are said to need "to manage risks and ensure that Northern infrastructure, ecosystems and cultures are resilient to future changes" (ibid.). Accordingly, and as a means to that end, it is no longer empowering the "Northern voice" that was exclusively emphasised before but collective actions in the areas of knowledge sharing, the protection of people and property in the North, the strengthening of Northern economies and the maintenance of food security. In a similar vein, the meaning ascribed to adaptation (a concept that is – unlike in U.S. policies – explicitly defined in both documents) has also changed: The 2007 policy states that adaptation "means taking actions that can prevent or reduce the negative impacts of climate change, and build on the positive impacts" (Northwest Territories, Yukon, Nunavut, 2007). The 2011 definition of adaptation centres less on possible economic benefits arising from climate change but on the risks related thereto:

> Adaptation means accepting that climate change is happening and taking action to prevent or reduce potential harm while building on new opportunities. To be effective, adaptation requires governments to assess vulnerabilities and manage risks to ensure that Northern infrastructure, economies, ecosystems and cultures are resilient to future changes.
>
> (Northwest Territories, Yukon, Nunavut, 2011)

The 2011 subnational strategy, however, is the only policy introduced in Canada representing this shift in providing a greater weight to climate change.

Similarly, it is also observable in the U.S. that the understanding of climate change represented in Arctic policies has changed. In 2009, for instance, *NSPD-66/HSPD-25* referred to sustainable development of resources in the Arctic, which is considered as important "in meeting global energy demand" even though it recognises that "[i]ncreased human activity is expected to bring additional stressors to the Arctic environment" (The White House, 2009). Accordingly, it is stated that the U.S. will "seek to balance" both (ibid.). In the 2013 National Strategy for the Arctic Region, on the other hand, it is explicitly the changing Arctic that is identified as the reason why "[w]e must proceed, cognizant of what we must do now, and consistent with our principles and goals for the future" (The White House, 2013b, see also National Ocean Council, 2011). In this regard and more explicitly, the 2015 Executive Order states

> [a]s a global leader, the United States has the responsibility to strengthen international cooperation to mitigate the greenhouse gas emissions driving climate change, understand more fully and manage more effectively the adverse effects of climate change, protect life and property, develop and manage resources responsibly, enhance the quality of life of Arctic inhabitants, and serve as stewards for valuable and vulnerable ecosystems.
>
> (The White House, 2015)

As these examples show, the representation of climate and environmental change in the Arctic has also shifted over time in U.S. Arctic policies and, as the previous quote illustrates, the new significance ascribed to climate change resulted in an understanding that recognises a greater responsibility. Another implication arising from this new understanding of climate change is revealed in the acknowledgement of failure in dealing with climate change, as both are used as a stepping stone to emphasise the need of "collaborative efforts by nations seeking to explore emerging opportunities while emphasizing ecological awareness and preservation" (The White House, 2013b, p. 8f.). Building on this understanding of a shared challenge, the 2013 strategy speaks of a "spirit of trust, cooperation and collaboration" that "[t]he United States and its Arctic allies and partners seek to sustain" (The White House, 2013a, p. 2). Although depicted as such, it becomes clear that this spirit is seen as a necessary condition for dealing with environmental and climate change when it is compared to the "common spirit and shared vision of peaceful partnership [that] led to the development of an international space station" (The White House, 2013b, p. 4). The 2014 *Implementation Plan* reproduces this understanding as well when it refers to "works with international partners to pursue global objectives of addressing climate change" (The White House, 2014, p. 1).

Despite this new significance ascribed to climate change, however, the understanding of international cooperation as being particularly significant in view of environmental concerns is nothing new. In earlier polar policies introduced by the U.S., international cooperation is already presented as a prioritised means to deal with environmental matters, which were continuously understood as being a "shared concern" with other state and non-state actors: Although in 1994, for instance, the United States Policy on the Arctic and Antarctic Regions (a document that dedicated four pages to the Arctic and one to Antarctica) referred particularly to "post-Cold War national security and defense needs" (The White House, 1994, p. 2) as a central area of concern, the policy presented international cooperation and environmental protection as being entangled objectives. Also, domestically, the 1994 policy promotes a greater inclusion of actors "where appropriate on U.S. delegations to relevant international meetings and in domestic decisions" and encourages "federal agencies to explicitly involve the State of Alaska and Alaskan indigenous peoples [. . .] in policy making regarding this region" (The White House, 1994, p. 4). Thus, as in Arctic policy making on the regional level, particularly Indigenous peoples organisations were considered legitimate non-state actors who could be included in the policy negotiations at that time, while others – ENGOs and industrial actors – are not explicitly outlined as non-state actors contributing to the production of knowledge in this matter. Sixteen years later, however, the Strategic Action Plan released by the National Ocean Council in 2011 illustrates that inclusion of a broader range of actors is perceived as being of benefit when encouraging "work [. . .] *with all stakeholders*, including the State of Alaska, and Alaska Native communities, [. . .] to sustainably manage and encourage use of the Arctic while also continuing to protect it" (National Ocean Council, 2011, emphasis added). Also in 2009 *NSPD-66/HSPD-25* addressed "positive results" that the collaboration with non-state actors and particularly with the Arctic Council "has produced [. . .] for the United States by working within its limited mandate of environmental protection and sustainable development" and describes the forum as a "beneficial venue for interaction with indigenous groups" (The White House, 2009). As in policy making in the Arctic Council, the inclusion of actors is, moreover, represented as the preferred means for dealing with the changing Arctic as well as "updating the structure of the Council" particularly with respect to its subsidiary bodies (ibid.). More fundamental changes, on the other hand, such as the regulation of the Arctic under an "Arctic Treaty" or the transformation of the Arctic Council "into a formal international organization" (The White House, 2009) are opposed in the U.S. policy as well as in the 2009 Canadian *Northern Strategy*.

With respect, more specifically, to international cooperation as a means to react to the changing Arctic, it is notable that, as in regional policy making, all Arctic policies and strategies introduced in Canada and the U.S. prioritise "science-based decision making" (see e.g. the White House, 2015), the need to improve scientific knowledge of environmental and climate change in the Arctic and ascribe "scientific leadership" to the countries in which such improvements are made. Due to the growing significance ascribed to environmental and climate change effects in the Arctic, this scientific leadership implies a desire to gain and maintain an international reputation on the one hand, while, on the other, it is also entangled with the power of ordering and interpretation not only of scientific information but also in regard to international cooperation. In dealing with environmental changes in the Arctic, it is the sharing of information and cooperative research that all policies particularly emphasise, when, for instance, promoting better coordination of (collaborative) research among "all Arctic nations" and "sharing access to research platforms" (The White House, 2009). More specifically, it is stated that research will focus on providing "[a]ccurate prediction of future environmental and climate change on a regional basis, and the delivery of near real-time information to end-users" (ibid.). Scientific leadership thus also enables Canada and the U.S. to make important research decisions with political, economic and social implications that, to a certain extent, can be at least regional in scope. Moreover, through deciding on funding and the design of research projects "scientific leadership" offers additional room for manoeuvre, for instance, regarding the inclusion or exclusion of state and non-state actors in and beyond the Arctic-rim states. In this regard, when Alaska's *Arctic Policy* recognises the significance of science for the future of the Arctic, it also outlines science as an instrument to increase its say in the politics of the Arctic. Moreover, by placing science on the same plane as "community input", the State of Alaska underpins its demands for empowerment and inclusion in national and regional policy making (cf. Government of Alaska, 2015a, p. 2). However, although scientific findings contribute to the meaning and significance ascribed to priorities in political processes and to their ordering, in the case of climate change mitigation these conflict with interests represented by the State of Alaska and contribute to the perception of its being dominated.

Underlining the significance of scientific research also in regard to the representation of the Arctic as a region of potential future conflict, the Canadian Coast Guard report and the U.S. Coast Guard Arctic Strategy quote and refer to the four-page *U.S. Geological Survey* published in 2008 to emphasise the "vast and largely untapped natural resources" (Canadian Standing Senate Committee on Fisheries and Oceans, 2009, pp. vii and 11) and the general resource potentials in the Arctic Ocean (United States Coast Guard, 2013, p. 17f.). In contrast, the 2010 *Statement on Canada's Arctic Foreign Policy* also refers to the perception of conflicting geopolitical interests. It speaks directly against the "widespread perception that the region could become a source of conflict", which is often said to be associated with "[t]he increasing accessibility of the Arctic" (Government of Canada, 2010, p. 25). With regard to territorial disputes, the 2009 Northern Strategy depicted Canada's sovereignty in the Arctic as being "undisputed" – the only exception being Hans Island, a dispute, however, described as being "on a diplomatic track" and as only being "about the island, not about the waters, seabed, or the control of navigation" (Government of Canada, 2009, p. 13). All other disagreements, such as the maritime boundary in the Beaufort Sea, are said to be "well-managed and pose no sovereignty or defence challenges for Canada" and are described as not affecting collaboration with "Arctic neighbours on issues of real significance and importance" (ibid.). In its 2013-*Arctic Strategy*, the U.S. Department of Defense represented the potential for future conflicts in the Arctic as being overestimated, and classified "[p]olitical rhetoric and press reporting about boundary disputes and competition for resources" as a challenge that "may inflame regional tensions" (U.S. Department of Defense, 2013, p. 13). More specifically, it is said that "[e]fforts to manage disagreements diplomatically may be hindered if the public narrative becomes one of rivalry and conflict" (ibid.), and it is against this backdrop that the Department of Defense outlines one of its objective to be the "mitigat[ion of] this risk by ensuring its plans, actions, and words are coordinated, and when appropriate, by engaging the press to counter unhelpful narratives with facts" (U.S. Department of Defense, 2013, p. 13). To counteract the idea that military activities in the Arctic "may lead to an 'arms race'" (ibid.) the Department recommends building "trust through transparency" and, more explicitly, enabling the sharing of information concerning the "intent of our military activities" (U.S. Department of Defense, 2013, p. 13).

3.3.2.2 Entanglements and disentanglements in policies introduced by the American Antarctic-rim states

Returning now to the American *Antarctic*-rim states, Chile's Antarctic Policy (*Decreto Supremo N° 429*) from the year 2000 emphasises a special responsibility ascribed to environmental concerns in the Antarctic when it states that the Madrid Protocol marked the beginning of a "nueva etapa" [new stage] in the Antarctic Treaty System in the first place because it caused the formulation of a new Antarctic policy in which national priorities regarding Antarctica were redefined "bajo esta nueva perspectiva" [under this new perspective] (Ministerio de Relaciones Exteriores de Chile, 2016a, p. 4). It is particularly notable that Chile's Antarctic Policy described this redefinition of national priorities as being a consequence of a change in the regional governance under the ATS, as it illustrates the scope of the binding legislation enacted under the regional framework. Like the policies of all American polar-rim countries, Chile's prioritises Chilean sovereignty interests. Unlike the national and subnational Arctic policies introduced by Canada and the U.S., which outlined those countries' interests and objectives concerning the Arctic region, Chile's Antarctic Policy predominantly represented sovereignty as being entangled with its participation in the ATS. Consequently, the Antarctic policy states that Chile envisions strengthening its influence in the ATS (priority 2), participating effectively in the System (priority 3), preserving peace, promoting scientific activities and protecting Antarctica as a natural reserve (priority 5, Ministerio de Relaciones Exteriores de Chile, 2016a). Similar to Canada's representation of international cooperation in the AC, the prioritisation of these "regional" objectives is related to the understanding that collaboration in the ATS provides Chile with a significant opportunity to promote its views and interests internationally (see also Ministerio de Relaciones Exteriores de Chile, 2015). The policy, moreover, describes this format for international collaboration as being particularly significant in view of "grand emerging tendencies" and refers in this context to the globalisation of science, the protection of the environment and the economic and technological management of Antarctica (Ministerio de Relaciones Exteriores de Chile, 2016a). Like the *Política Antártica Nacional* [National Antarctic Policy] of 2000, the *Plan Estratégico Antártica* [Strategic Plan for Antarctica] 2015–2019 published by the Consejo de Política Antártica in 2014, which builds on the policy from the year 2000, prioritises this regional perspective, when outlining the aim of obtaining a relevant position for Chile in the ATS and when relating Chile's national interest to "las nuevas tendencias globales" [the new global trends] (ibid.). The strategic plan, however, does not specify what is exactly meant by new global trends.

   Although climate change is not specifically mentioned among these emergencies, Chile's *Antarctic Policy* addresses the need to improve research on climate change (that is related to "el estudio de problemas globales de interés para toda la humanidad" [the study of global problems of interest to humanity at large] (Ministerio de Relaciones Exteriores de Chile, 2016a) when specifying its objectives relating to environmental concerns (among them better control of tourism and intensified conservation of marine living resources are also listed). The policy further summarises measures demanded that are "national" in scope, such as support for national Antarctic institutions that coordinate Chile's activities in Antarctica. In putting particular emphasis on Chile's continental link as "país-puente" [bridge country] to Antarctica, the policy moreover states the objective of improving the infrastructure of the Región de Magallanes y de la Antártica Chilena particularly in view of increased tourism and scientific activities in Antarctica (Ministerio de Relaciones Exteriores de Chile, 2016a, priorities 7 and 11). Like the territories in the North American Arctic, the latter region is difficult to access (one has to drive through Argentina to reach Punta Arenas). The geographic proximity of Punta Arenas and the Región de Magallanes to the Antarctic peninsula is moreover understood as an important national and regional connection to Antarctica that serves the needs of international Antarctic programs and the tourism industry, since there are many hotels in Punta Arenas, the flight connection to the Chile's capital Santiago is reckoned to be good and, compared to other "gateways to Antarctica", operating costs in Chile and Punta Arenas are low (Ministerio de Relaciones Exteriores de Chile, 2015, p. 31). In view of the risk to losing its competitiveness with respect to other emerging Antarctic cities, the long-term strategic plan recommends constructing an international Antarctic centre in Punta Arenas (focusing on science and tourism), promoting sustainable tourism activities given the growing numbers of tourists, improving the port infrastructure and digital connectivity and finding a solution to the high living costs and absence of goods and services in Punta Arenas (ibid., p. 32f.).

The latter objective is also reproduced in the Chilean *Plan Estratégico Antártica 2015–2019* released 14 years later that, however, speaks of the objective of empowering the "Región de Magallanes y Antártica Chilena como puerta de entrada a la Antártica" [the region of the Magallanes and Chilean Antarctica as a gateway to the Antarctic] (Consejo de Política Antártica, 2014). The strategic plan more explicitly emphasises the aim of empowering Chile's southernmost region (e.g. it demands the establishment of "Centro Antártico Internacional en Punta Arenas" [an International Antarctic Centre in Punta Arenas]) while the *Antarctic Policy* of the year 2000 outlined this objective as being inspired by the need to meet the challenges and opportunities arising from its location and obligations as "país-puente" (e.g. search and rescue activities). Corresponding to the *Antarctic Policy*, however, the strategy plan also demands the improvement of the operational and logistic capacities of the state of Chile, and the strengthening of national Antarctic operators for cooperative polar scientific endeavours (Consejo de Política Antártica, 2014).

Particularly in view of the latter, Chile's *Antarctic Policy* encourages cooperation with nations that also share permanent interests in Antarctica for historical and geographical reasons (Ministerio de Relaciones Exteriores de Chile, 2016a, priority 5). In this regard, and also as with the Canadian and U.S. Arctic policies, the Chilean *Antarctic Policy* and the succeeding strategy plans emphasise cooperation with neighbouring states ("en particular con países de la region" [particularly with countries in the region], Consejo de Política Antártica, 2014) and with states that are regarded as sharing a similar (and more legitimate) relationship with Antarctica. An interviewee highlighted, for instance, that due to the geographic proximity and its (frozen) territorial claim, Chile perceives itself as a more legitimate actor involved in the politics of the Antarctic ("de alguna manera nosotros reconocemos mucho más legitimidad a parte obviamente de la reivindicación chilena a Argentina que a otros actores, no?" [in some ways, also because of the Chilean claim obviously, we recognise Argentina as having much more legitimacy than other actors, don't we?]) and this understanding is also projected onto other countries from Latin America ("no cierto del America de Sur en el continente Antártico"). Although the interviewee also speaks about a "cierta rivalidad" [certain rivalry] between Chile and Argentina with respect to Antarctica, it is this shared understanding as more legitimate actors and the geographical position that "explica digamos que en términos de cooperación practica la cooperación mayor que nosotros tenemos es con Argentina" [explains why in practical terms the country we mainly cooperate with is Argentina] (Interview, February 2015).

In Chile's *Antarctic Policy* reference is also made to political cooperation among the "países-puente" (Argentina, Australia, Chile, New Zealand and South

Africa) that are known as the Group of countries from the Southern Hemisphere/ Group of Valdivia, formed in 1995 in the Chilean city Valdivia, that focus on environmental questions relating to climate change, the depletion of the ozone layer, desertification and the conservation of biodiversity. Logistic cooperation, on the other hand, is outlined as being provided particularly among countries from Latin America who are members of RAPAL. In addition to scientific cooperation under the auspices of the Scientific Committee on Antarctic Research, it is notable that all these forms of cooperation are represented as supporting political cohesion among its members without adversely affecting Chile's territorial rights in Antarctica (Ministerio de Relaciones Exteriores de Chile, 2016a). This emphasis illustrates that, despite the general commitment to international cooperation, cooperation is particularly motivated by the aim of maintaining Chile's say in Antarctica. Accordingly, it is also the potential conflict "entre los páises antárticos y otros Estados Miembros de la comunidad internacional" [between Antarctic countries and other states that are members of the international community] that is identified as a "mayor riesgo" [major risk] affecting international cooperation in the ATS caused by overlapping territorial claims and by countries that do not recognise "soberanías" [sovereignties] in Antarctica (Ministerio de Relaciones Exteriores de Chile, 2016a, p. 4). Accordingly, Chile's long-term strategic plan (*Plan Estratégico Antártico – Visión al 2035*, Ministerio de Relaciones Exteriores de Chile, 2015) considers the growing number of consultative parties to constitute a risk. Thus, like the position often ascribed to Canada in the Arctic Council, the increase in the number of parties involved is understood as diminishing Chile's influence in the ATS as it is said to be more difficult to find the necessary consensus (Ministerio de Relaciones Exteriores de Chile, 2015, p. 37).

On the other hand, however, the ATS is regarded as enforcing bilateral and multilateral initiatives with non-state actors advocating the protection of the environment that are recognised with it. The ATS is accordingly perceived as providing an important opportunity for public–private cooperation in shared areas of interest (ibid.). The strategic plan further implies that this kind of opportunity is missing at the national level and also draws attention to other problems to be dealt with at the national level that relate to Chile's engagement in the Antarctic such as: the small number of professionals who are able to respond to the agendas of ATCM, CEP and CRAMRA meetings, the absence of Chilean regulation addressing environmental concerns in Antarctica, other national institutions' lack of knowledge about the norms promoted in the ATS, the principles represented by the Chilean delegation to the ATS and the functioning of the ATS in general (Ministerio de Relaciones Exteriores de Chile, 2015, p. 37).

As a means of maintaining its special say, the long-term Antarctic strategy plan recommends strengthening the legitimacy of the ATS and, for instance, providing better accessibility to scientific data (also to counteract the tendency to commercialise scientific information) and conducting investigations taking account of environmental imperatives (Ministerio de Relaciones Exteriores de Chile, 2015, p. 58). It is notable that, unlike the Arctic policies examined previously, the strategic plan explicitly emphasises the need to preserve "la libertad y la accesibilidad de la investigación científica [que] resulta fundamental para un país con derechos soberanos e intereses permanents en el Continente Antártico" [the liberty and accessibility of scientific investigation that is of fundamental significance for a country with sovereignty rights and permanent interests in Antarctica] (ibid.). As with the Arctic policies introduced in Canada and in the U.S., scientific cooperation is thus seen not only as a means to gain more knowledge of environmental concerns but is also explicitly represented as providing the opportunity to gain and maintain political influence. Accordingly, the long-term strategic plan further highlights the need to maintain a national Antarctic scientific program, which is said to be of particular relevance in domestic and international terms and – as an interviewee outlined – also presented by the INACH as "major forma de 'hacer soberanía' en la Antártica" [main means to 'maintain sovereignty' in Antarctica] (Interview, February 2015). In this regard, the plan refers to the highly qualified "recursos humanos" [human resources] who are capable of investigating and operating in the extreme Antarctic environment as a strength (ibid.). Similarly, the network of Chilean Antarctic bases in regions of high scientific interest to others besides Chile and the interest of "nuevos países sudamericanos, asiáticos y europeos por desarrollar investigación científica en la Península Antártica y zonas marítimas aledañas" [new countries from South America, Asia and Europe that aim to conduct scientific investigation in Antarctica and nearby maritime zones] are emphasised as another strength that supports Chile's relationships with these countries, allows the exchange of scientific knowledge and is described as having resulted in greater international visibility (Ministerio de Relaciones Exteriores de Chile, 2015, p. 25). Again, international cooperation is described as being beneficial to Chile's national, regional and international position.

Argentina's position with respect to Antarctica is – like the Canadian one – often described as being based on geopolitical and "nation-building" interests and its involvement in the Antarctic Treaty System is particularly inspired by its determination to defend its sovereignty over the Islas Malvinas. However, Argentina's four-page national Antarctic Policy (*Decreto 2316/1990*) reveals numerous similarities to the Chilean policy adopted 10 years later: it not only highlights the expansion of Argentina's sovereignty rights in Antarctica as a main objective (Art. 1.I), but also aims to do so via the ATS. Similarly, strengthening the ATS is set down as the first priority, followed by an interest in increasing Argentina's influence in it. Argentina, like Chile, also refers to other Latin American countries and claims to envision intensified cooperation with respect to Antarctica, primarily in logistical terms, in order to enhance the effectiveness of the Argentine presence in Antarctica. In accordance with this priority, Argentina also initiated the establishment of the Union of Latin American Antarctic Tour Operators, through which this kind of cooperation has been managed since.

In accordance with Argentina's *National Antarctic Policy* 1990, the National Directorate for the Antarctic provides information on the various scientific programs conducted at Argentina's Antarctic bases on an annual basis (detailing their maintenance, specific aims and planned activities). Until 2014, these annual reports always stated the general objective of Argentina's activities in Antarctica to be:

> To strengthen Argentina's sovereignty rights in Antarctica, to deepen the scientific and technical activities needed to achieve a complete knowledge of the natural world of the Antarctic especially in areas related to the country's priorities, to support the conservation and preservation of fish and mineral resources, the protection of the environment, the integration of Latin America in activities in Antarctica and the supply of services.
>
> (cf. e.g. Dirección Nacional del Antártico, 2006)

This statement of objectives thus explicitly refers to the aim of reinforcing cooperation with other countries from Latin America in Antarctica and supporting a stronger regional entanglement by integrating Latin American countries in general in activities there.

Most often, however, the annual plans refer to international cooperation in the shape of collaboration on scientific programmes at bases located in Antarctica, for instance, on climate change with Australia, Germany and the U.S. (Dirección Nacional del Antártico, 2010). The annual plans always refer to the Madrid Protocol at the outset to emphasise Argentina's obligation – as a party to the Protocol – to protect the Antarctic environment and its dependent ecosystems. Like Chile's policy statements, they also highlight an understanding of Antarctica as an area dedicated to peace and science, which harks back to the Antarctic Treaty and is described "as the reason why certain activities are prohibited and why others are dependent on prior permission" (Dirección Nacional del Antártico, 2006).

More specifically, climate change and its effects are to be researched and monitored at the Marambio base amongst others (Dirección Nacional del Antártico, 2006). Investigations into climate change are also conducted under the auspices of the IAA at the Larsen Shelf, and the annual reports recognise climate change as the reason for the collapse of sections of the Larsen (Dirección Nacional del Antártico, 2010). The reports also acknowledge climate change to be a global phenomenon that is particularly notable in the Polar Regions without, however, explicitly referring to the Antarctic as a climate change barometer (ibid.). The glaciers located in West Antarctica are assessed as "responden en forma directa al cambio climático de las últimas décadas" [responding directly to climate change over the last few decades] (ibid.) and also the "región occidental de la península Antártica, ha sido la mas afectada por el calentamiento climático, registrándose un incremento de 2.5 oc desde la década de 1950" [western region of the Antarctic peninsula has been most affected by climatic warming, registering an increase of 2.5°C since the 1950s] (Dirección Nacional del Antártico, 2010, p. 82; see also Dirección Nacional del Antártico, 2014, p. 47).

Although not developed by the subnational government but by the Argentine government itself, *Ley N° 307* on the "Antártida Argentina: Política Provincial" [Argentinian Antarctic: provincial policy], issued in 1996, applies particularly to the province of Tierra del Fuego, Antarctica and the South Atlantic Islands.

This law repeated the position outlined in Argentina's *National Antarctic Policy* 13 years earlier and expanded its interests to cover Ushuaia. The law's overriding object was to initiate actions to consolidate Ushuaia "como la puerta de acceso a la Antártida" [as the gateway to Antarctica] (Art. 2). As Chile did with Punta Arenas, Argentina aimed to improve Ushuaia's infrastructure with respect to communication, tourism, scientific activities and logistics. In 2014, a law was proposed in the Chamber of Deputies (the lower house of the Argentine National Congress) to – also similar to Chile – move the national institutions most involved in the national politics of Antarctica (the DNA and IAA) to Ushuaia. The proposal, however, did not receive the necessary support from those entitled to vote.

Overall, the positions expressed in the Argentine and Chilean national Antarctic policies, strategy and annual plans correspond in the sense that they all prioritise the need to defend their sovereignty in Antarctica via the ATS, which they envision being strengthened, and that this understanding has not changed during the period of investigation. In both countries, the principles of the Antarctic Treaty are continuously described as guiding national activities in Antarctica and are thus accepted as characterising the supranational framework. It is also highlighted that both countries are original signatories to the Antarctic Treaty and that both shaped all the other legal instruments subsumed under the ATS whose content is thus not questioned but seen as legitimate. Further, international cooperation in Antarctica is represented as the most important means to protect international peace and the environment there while, at the same time, Chilean and Argentine collaboration in the ATS is regarded as the only means to defend sovereignty claims in Antarctica. Moreover, all policies and strategic and annual plans considered emphasise international cooperation as necessary to conduct research in Antarctica. Both Chile and Argentina, however, aim to strengthen inter-American cooperation in Antarctica as well and to include more countries from Latin America in the ATS (even though Chile has expressed a fear that a growing number of consultative parties to the ATS would limit the Chilean say due to the consensus principle). While Chile has particularly focused on international cooperation among the "países-puente" [bridge countries] with regard to environmental protection, Argentina established RAPAL to encourage logistical cooperation among countries from Latin America in Antarctica. In both countries, these alliances with neighbouring states are envisioned and described as being based on shared continental and historical entanglements. Cooperation with other allies is further outlined as a means to strengthen Chile's and Argentina's say in and beyond Antarctica. As another means to this end, both countries put particular emphasis in their policies on the development of infrastructure and institution-building in their "gateway cities to Antarctica" Ushuaia and Punta Arenas (while Chile moved its Antarctic institutions to Punta Arenas, in Argentina the corresponding proposal for Ushuaia failed). Chile's long-term strategic plan further demands better education and more information on its role in Antarctica to encourage a stronger national identification with Antarctica that would also be conducive to legitimising the necessary funding for Antarctic research to remain competitive. So, as in the Canadian and Alaskan perspectives, the potentials ascribed to increasing human activities in Antarctica are seen as beneficial for Chile and Argentina and as encouraging region-building in their southernmost areas.

It is notable that in the Antarctic policies introduced by Argentina and Chile it is not the changing environment that is depicted as encouraging international and inter-American cooperation, but an understanding of Antarctica as a place for scientific cooperation (in accordance with the Antarctic Treaty), which also allows both countries to maintain a say there. Unsurprisingly, in both the Chilean strategy and the Argentine annual plans, the harsh environment and high operational costs are also mentioned as encouraging this kind of cooperation. The Chilean long-term strategic plan further lays down that new relationships with other countries will be built on the basis of scientific information about Antarctica that will not only allow an exchange of scientific knowledge but also contribute to a greater acknowledgement of Chile internationally. In this regard it is noteworthy that Chile's long-term strategy also involves non-state actors, collaboration with whom is said to be encouraged under the auspices of the ATS. In both countries, scientific cooperation is, however, not only perceived as a means to gain more knowledge about environmental concerns but also as a way to increase their international influence.

3.3.3 Conflicting and corresponding positions in national and regional discourses

In both regional and national settings, all American polar-rim states emphasise that the protection of the pristine polar environment is necessary, as it is perceived as being endangered by activities possibly causing pollution and by climate change impacts. My analysis of national polar policies and strategies has, however, illustrated, that – not solely among states but also between the federal and subnational levels – perspectives differ particularly concerning the concrete measures to be taken and with respect to those "legitimised" to decide on them (as illustrated most notably by the comparison of Alaska's *Arctic Policy* with the policies introduced by the U.S.). Like permanent participants in the AC, the State of Alaska and the subnational governments representing the Canadian North emphasise that their voices need to be considered in policy making on the Arctic. They oppose the dominance they perceive to be exercised by the federal government (in the U.S.) and by non-Arctic actors (in Canada), insofar as the latter decide on measures to be taken to protect the Arctic (for instance, in the AC). Their opposition is based on the reasoning that pollution and climate change particularly affect the territories they inhabit as well as having implications for their way of living. Consequently, as at the regional level, to counteract this perceived domination they continue to demand greater inclusion in policy making.

Particularly with respect to the Arctic and also corresponding to shifts at the regional level, understanding of the measures needed and their prioritisation have changed over time. This is most notable in the Canadian policies under analysis in which, for instance, the definition of adaptation has been renewed as a result of a growing acknowledgement (and experience) of climate change impacts in the Arctic. Against this backdrop and as with Alaska's *Arctic Policy*, the "development" and "empowerment" of the North are identified as measures needed in order to "adapt to" environmental changes in the Arctic. In the case of the U.S., on the other hand, the recognition of rapid environmental changes (and their economic implications) became a driver for federal engagement in the Arctic and the federal government's interest in policy making on "responsible measures" in the first place. Based on the "global" reasoning promoted by the U.S., measures negotiated in the Arctic are placed in global perspective, which is, however, also inspired by the self-understanding of the U.S. as a "global leader" with special responsibilities (particularly in raising ecological awareness). In this regard, it is notable that the U.S. federal government's interest in the Arctic was "absent" at the national and regional level before, indicating that environmental concerns as such are not the reason for its increased attention to the Arctic. In the 1990s, for instance, when environmental concerns relating to the pollution in the Arctic Ocean were addressed in the AC and also picked up in the U.S. *Arctic Policy*, the negotiation of international measures did not constitute a significant interest for the U.S. Instead, it depicted pollution in the Arctic Ocean as a "local" concern, and accordingly policy-making power at the national and regional levels was predominantly placed in the hands of the State of Alaska. This former lack of interest in environmental concerns helps to explain why the State of Alaska perceives itself as being dominated by the federal government at present.

All the policies and positions considered that were introduced by the American polar-rim states in national and regional governance settings identify cooperation and increasingly and particularly scientific cooperation as the best means to deal with climate and environmental changes that are widely perceived as challenges. At the national level in particular, however, all American polar-rim states regard scientific cooperation not only as being entangled with the aim of gaining knowledge of environmental and climate change but also as determining their general influence in the politics of the Polar Regions. Chile and Argentina even represent scientific cooperation as a means to build international relationships, strengthen the Antarctic Treaty System and thereby maintain a say in the System in the first place. Canada and the U.S., on the other hand, claim to have scientific leadership and at the same time emphasise that they conduct science-based decision making with respect to the changing Arctic which will also be informed by non-state actors, particularly Indigenous peoples from the North. Scientific cooperation is thus also understood as providing non-state actors with a say in polar policy making.

More generally, all American polar-rim states look forward to cooperation with neighbouring states with whom they either already share a long-lasting relationship on which future cooperation can be built or whom they regard as more legitimate actors in the governance of the Polar Regions because of their geographical proximity to the Arctic or Antarctic. This understanding operates in the AC, but it is a particularly remarkable feature of the ATS. Despite the juridical bindingness of the framework (which is, moreover, perceived in positive terms in national policy making as the ATS is viewed as providing opportunities to encourage international cooperation on issues above and beyond those that concern the

Antarctic, growing cooperation with non-state actors and a better awareness on environmental concerns) and despite the influence of regional negotiations on national policy making (as illustrated by the redefinition of national interests as a consequence of the negotiation of the Madrid Protocol in the case of Chile and the change of competencies ascribed to governmental agencies focusing on the Antarctic in the case of Argentina), Argentina and Chile consider their participation in the ATS as a means of maintaining sovereignty, which is why they further wish to strengthen the intergovernmental framework and, in particular, to keep the consensus principle. This understanding is closely entangled with the reasoning adopted in the Antarctic Treaty (Antarctica's being of benefit to all human beings; the "freezing" of all territorial claims), which, however, does not allow the exercise of sovereignty in Antarctica beyond the maintenance of scientific stations but does provide for inclusion in the ATS as a consultative party. Thus, unlike in the Arctic, the international bindingness of the treaty framework is not understood as a limiting factor but as actually providing both countries with a particular say (even though, as "países-puente" and "gateways" to Antarctica, both perceive themselves to have greater legitimacy than others, which is why they increasingly envision cooperation with other rim states also outside the framework, as the Group of Valdivia and RAPAL exemplify). This reasoning, moreover, constitutes a major disentanglement in the politics of the Polar Regions, as Canada and the U.S. also recognise the AC as the most appropriate governance setting for the management of the Arctic although policy making in the Council is mostly non-binding. Even though climate change and its impacts are perceived as "transregional" challenges and increasingly also as being related to a shared (global) responsibility, in their national policies Canada and the U.S. oppose the drawing up of an Arctic Treaty similar to the Antarctic Treaty or the restructuring of the AC into an international organisation, as such actions would be understood as limiting their sovereignty rights.

To sum up: In *regional and national* policy making, environmental concerns affecting the changing Polar Regions are increasingly being addressed and identified as enforcing international cooperation, which is perceived particularly as being necessary to provide better knowledge of the causes and impacts of these changes. In this regard, scientific cooperation with non-state actors, other international fora and non-polar-rim states is increasingly being promoted in regional and national policy making by the American polar-rim states. This reasoning illustrates a remarkable interpolar entanglement in the politics of the Polar Regions as it not only promotes transnational policy making but also provides a legitimate say to actors formerly excluded from it. As all American polar-rim states, however, identify themselves as more legitimate actors in polar policy making thanks particularly to their geographic proximity to the Poles, they prioritise the maintenance of their sovereignty rights as well as the objective of not losing influence in political processes against the backdrop of the changing Arctic and Antarctic. Very different strategies, however, are pursued with respect to the Arctic and the Antarctic that exemplify a significant disentanglement in the politics of the Polar Regions that is also rooted in their disparate regional governance settings and their formation at different times. While Argentina and Chile envision the strengthening of the ATS as a legally binding framework and thus promote supranational decision making, Canada and the U.S. oppose the transformation of the Arctic Council into a Treaty System and instead wish to steer the regional governance of the Arctic in accordance with the prioritisation of objectives developed on the national level and to transform the structure of the AC respectively. To maintain the legitimacy of both regional governance settings and to counteract the perception of possible future conflict over territory and resources taking place in the Arctic and the Antarctic, however, national policies in both regions demand more transparency concerning scientific findings and policy making. Against this backdrop, it is also the "spirit of trust, cooperation and collaboration" (outlined in regional and national policy making in both Polar Regions) that constitutes an interpolar entanglement that will likely guide the politics of the Polar Regions in the future.

Overall, the entanglements identified in this chapter provide evidence of transnational and interpolar dynamics that have influenced polar policy making. These dynamics particularly relate to the understanding of international cooperation and the inclusion of actors in polar policy making, the significance ascribed to environmental and climate change in the Arctic and Antarctic. They have, moreover, shown how the politics of the Polar Regions are increasingly structured by the idea of a "shared responsibility" in view of environmental changes. Although this idea supports the inclusion of more actors and perspectives in the politics of the Polar Regions, at the same time it challenges the interests particularly of those who fear being dominated by a homogenising global perspective. This trend towards a more "global reasoning" may thus encourage processes of political disentanglement in the politics of the Polar Regions, as the formulation of Alaska's Arctic Policy most notably exemplified.

In the subsequent chapters the findings of this chapter are compared to the representation of the changing Polar Regions in newspaper reporting (Chapter 4) and by non-governmental theorists and strategists (Chapter 5) to also evaluate (dis) entanglements between practical, formal and popular geopolitical discourses and between the regional and national levels that help to explain why similar patterns of interpretations have evolved in the politics of the Polar Regions.

Notes

1 For a detailed explanation of the five maritime zones (territorial sea, contiguous zone, exclusive economic zone, continental shelf, the high seas) and the application of UNCLOS in the Polar Regions, see Hinz (2011, p. 88); Rothwell (2015); United Nations (2013).

2 In the Barents Sea and Arctic Ocean, Norway and Russia developed a cod stock management regime, Canada and Denmark negotiated the "Agreement for Cooperation Relating to the Marine Environment", and Canada and the U.S. collaborated in developing a joint marine contingency plan in the Beaufort Sea (cf. Koivurova et al., 2009, p. 274).

3 In its most general sense, a rightsholder is any person or organisation that owns legal rights to something. As mentioned before with respect to free, prior and informed consent, UNDRIP gives Indigenous peoples these rights in a more explicit sense and

specifies that "Indigenous peoples have the right to participate in decision-making in matters which would affect their rights, through representatives chosen by themselves in accordance with their own procedures, as well as to maintain and develop their own indigenous decision-making institutions" (Article 18). As in this article, throughout the declaration Indigenous peoples are considered as collective human rightsholders, given that there are "certain individual human rights that can only be enjoyed 'in community with others', which means that for human rights purposes the group involved becomes a rights-holder in its own right" (Stavenhagen, 2011, p. 161).

4 The nearest landmasses to Antarctica belong to Chile, Argentina, South Africa, Australia and to New Zealand, which are also called Southern Ocean Rim States (SORS). Like the Antarctic Treaty Secretariat (Interview, January 2015), Dodds (1997) also considers India to be a SORS because it is one of the six major southern nations involved in the Antarctic region.

5 Until the 1980s it was largely "wealthy" and "western" countries that held consultative member status in Antarctic Treaty Consultative Meetings. Uninvolved nations, in particular "emerging" nations stressed their impression that an exclusive, western-inaugurated "Antarctic-Club" had been established with the intention of dividing up Antarctic resources among its members. It was, for instance, argued that "some parts of the world, such as Africa, Asia, and the Middle East, remain under-represented in the ATS, thereby giving substance to enduring criticisms of the regime's exclusive character" (Beck, 2003, p. 211).

6 The main observations and arguments included in this section are also introduced in Wehrmann (2017).

7 At the SAO meeting in October 2015, the addendum to the Arctic Council Observer Manual was adopted, which defines in more concrete terms possible "meaningful contributions" Observers are expected to make with regard to intersessional communication, meeting participation and project contributions (Arctic Council, 2015a, p. 11f.). For a more detailed analysis of the practical implications of this addendum, see Knecht (2016a).

8 This new rule was adopted in 2013 and Graczyk and Koivurova (2014, p. 234) argue that although the report of non-state actors' activities might serve the purpose of excluding any actor that engages in activities that subvert the principles that the members of the Arctic Council have agreed on, it also strengthens the assumption that "existing observers, as well as candidates to the status, are being evaluated on the basis that they are not seen as a challenge to Arctic states' and PPs' regional interests".

9 No attendance lists were accessible for the 1998, 2004 and 2006 ministerial meetings and for the 2003/2, 2004/2, 2005/1, 2005/2, 2006 and 2013/1 SAO meetings. See also the "Stakeholder Participation in Arctic Council Meetings (STAPAC) Dataset" (Knecht, 2016b) for an overview on the attendance and non-attendance of formally accredited stakeholders at all meetings conducted under the auspices of the Arctic Council.

10 In task forces "stakes tend to be higher as participants are working on a written document, an agreement, which is highly political in nature, and governments are more protective of their role" (Interview, August 2015, in Wehrmann, 2017, p. 20).

11 With the 2013 Observer Manual (Section 7.3), for example, it is no longer possible for observers to assign or designate others to represent them at Arctic Council meetings.

12 With the exception of the 1999 CEP meeting, no participation lists are included in the reports published by the Antarctic Treaty Secretariat during the period under investigation, which is why no data on changing attendance patterns can be provided with regard to CEP meetings. The Antarctic Treaty Secretariat, however, states that "CEP meetings are also attended by various experts and observers" (ATS, 2016). These experts either form part of national delegations or of observer parties (contracting parties to the Antarctic Treaty and "scientific, environmental and technical organisations which can contribute to the work of the Committee", Revised Rules of Procedure, Rule 4).

13 Although it had been the main hindrance, during negotiations on the establishment of the Antarctic Treaty Secretariat, the Argentine delegation continued to argue in favour of the consensus principle: "It is important to persist in the effort to achieve a consensus within such a framework, and to avoid conflictive situations which can lead to impasse difficult to overcome in the near future" (ATCM, 1994, Rebagliati on behalf of the Argentine delegation).

14 This section focuses exclusively on the Arctic and Antarctic policies introduced by the American polar-rim states under different federal and subnational governments. The views expressed by non-state actors in Canada, the U.S., Chile and Argentina cannot be considered due to the absence of comprehensive documentation on national policy formulation processes that would allow for the systematic assessment of which non-state actors contributed to the formulation of the policy and strategy reports under analysis and how they did so. In those reports, moreover, the contributors are most often not identified. Some do, however, state that "[s]takeholder input will inform key decisions" (e.g. NSPD-66/HSPD-25, The White House, 2009) and demand a "future collaboration" with non-state stakeholders (e.g. La Política Nacional del Año 2000, Ministerio de Relaciones Exteriores de Chile, 2015), thus indicating that the direct involvement of non-state actors in these processes might not have been a common practice before. An interviewee confirmed this impression when stating: "No hay muchas organizaciones ambientales en Chile y tampoco en Argentina que tienen influencia en la política nacional y subnacional" [There are not many environmental organisations in Chile or Argentina that influence national and subnational politics]; (Interview, January 2015). Sometimes participation lists offer a possibility of identifying attendees at hearings. These, however, do not mirror how these attendees contributed to the hearings and the positions they represented. Information provided by the non-state actors themselves (in interviews, annual reports and on websites), on the other hand, are difficult to verify and biased. An evaluation of the say of non-state actors in the domestic policy making processes would thus be based on much uncertainty and findings would be rather speculative. When focusing on newspaper reporting and on scientific publications in the following chapters, however, the domestic non-state actors referred to in these discourses are described and an assessment is made as to how their say and positions are perceived and represented.

15 For an overview of the complex institutional structure, see Arctic Science Portal (2016), which was launched in 2016 and results from the U.S. Arctic science initiative that was also implemented under the U.S. chairmanship in the Arctic Council. For an analysis of the role and significance ascribed to these different actors, see Conley et al. (2013).

4 Popular geopolitics

The changing Polar Regions in newspaper reporting

Although state and non-state actors from the American polar-rim states have related differently to the changing Polar Regions in foreign policy, the meanings they have ascribed to environmental concerns and international cooperation in the Arctic and Antarctic have often been alike and have provided evidence of transnational and interpolar dynamics. But how have practical geopolitical discourses been viewed and evaluated by newspapers published in those same states? Have similar patterns of interpretation in newspaper reports empowered or challenged their understanding of the changing Polar Regions? And to what extent has newspaper reporting in the American polar-rim states contributed to the formation of recurring interpretations, dominant views and positions that have ultimately influenced polar politics? This chapter shows that dominant patterns of interpretation and imaginaries have indeed been enforced in the press by:

1 representations reproduced at different times (e.g. of Canada as an international leader in the Arctic or Argentina's and Chile's special relationship with the Antarctic because of their geographical proximity to it),
2 the representation of concerns as becoming entangled and bound up with one another as they have evolved (e.g. resource development, the protection of the Arctic environment and territorial claims),
3 the repetition of positions over time and beyond national boundaries at the regional and interpolar levels (e.g. positions on resource development and the colonisation and internationalisation of the Polar Regions).

Similarly, the say that politicians and scientists have had in newspapers in the American polar-rim countries strengthens the argument put forward in this book, namely, that polar politics has been shaped by many regional, transnational and interpolar entanglements and interconnected (practical, popular and formal) geopolitical discourses, which explain why similar patterns of interpretation became dominant.

4.1 Newspaper reporting in Argentina, Chile, Canada and the U.S.

Print newspapers such as the *Toronto Star* and *Globe and Mail* in Canada, and *USA Today*, the *New York Times* and the *Washington Post* in the U.S. have been

among the national and international daily newspapers with the highest circulation rates in their respective countries. Although the *Washington Post* is usually
not ranked among the top five U.S. newspapers, it is the most-read newspaper in
Washington D.C., and the particular attention it receives from politicians and policy advisors underlines its significance. Environmental organisations, for instance,
often try to publish editorials in the *Washington Post*, knowing that their views
will be more likely to receive attention from politicians (cf. Interviews, March
and April 2014). The same applies to *Página 12* in Argentina, which has a high
average circulation rate but cannot compete with the country's two most widely
read newspapers, *Clarín* and *La Nación*. However, *Página 12* became known as
the strongest medium of opposition to the government under the presidency of
Carlos Menem (1989–1999) and gained a high reputation through the quality of
its leading writers, of whom Eduardo Galeano was formerly one.

The aforementioned newspapers are published by different publishers and take
various political lines. Analysing them thus allows the inclusion of a wide and
diverse reading community. The *Globe and Mail* is classified as liberal conservative
with a political affiliation to the Liberal Party (and, until 2003, to the Progressive
Conservative Party of Canada), while the *Toronto Star* is regarded as left-liberal.
The *New York Times* is considered to be a liberal paper with a political affiliation
to the Democratic Party. *USA Today* is a conservative tabloid and the *Washington
Post* is regarded as favouring conservative positions (Editor and Publisher, 2013).
La Tercera (conservative) is one of the 14 leading newspapers in South America
and shares the highest circulation rate in Chile with *El Mercurio* (which is also
right-wing). These two newspapers were the only dailies allowed to be published
during the dictatorship of General Augusto Pinochet (1973–1990) and supported
the military regime in their news coverage (Rinke, 2008, p. 159). The *Clarín*
forms part of the largest Argentinian media group (Grupo Clarín) and its political
stance is centre-right, while *La Nación* is known as a conservative newspaper: it
supported the military dictatorship in Argentina (1976–1983). *Página 12*, on the
other hand, is considered to be progressive (Rojas-Kienzle, 2013).

News reporting has also been influenced to different degrees by political interventions and media regulations in the four American polar-rim countries. Furthermore,
a growing concentration of media ownership has been noticeable. The rise of new
media channels in particular has led to increased competition in the news market,
causing financial difficulties for the traditional newspaper industry worldwide. The
financial dependence of newspapers on investment from private companies has
been growing ever since, which has also encouraged a concentration of ownership.
As an example of company investment, the *Washington Post* was run by the Washington Post Company until 2013 when it was sold to Jeff Bezos, founder and chief
executive of Amazon.com. To illustrate concentration of ownership, one can point
to Torstar's former 20% stake in CTVglobemedia (Torstar owns the *Toronto Star*,
and the *Globe and Mail* belonged at that time to CTVglobemedia). Although, at
present, these particular newspapers are owned by different companies, a concentration of newspaper ownership has been visible in all the four countries in question
(Brede and Schultze, 2008; Stüwe, 2008), most notably in Chile, where 99% of all

subscription-based daily newspapers belong either to the Consorcio Periodístico de Chile Sociedad Anónima (COPESA) or to the El Mercurio group. In Argentina, the Grupo Clarín is regarded as the main media conglomerate. It is, however, less the concentration of newspaper ownership than the strength of the executive branch under the political system in Argentina, which has led to the press's position in Argentina being very much determined by the personality and interests of the president of the day (Hänsch and Riekenberg, 2008, p. 79).

In addition to their varying degrees of dependency on foreign news agencies, the quality of news reporting in the selected newspapers differs not least because of the nature of the legislation affecting the media in each country. On the World Press Freedom Index published by Reporters without Borders, which has marked a "worldwide deterioration in freedom of information", Canada was ranked highest of the four polar-rim states (in 20th place), the U.S. was ranked 32nd, Chile was ranked 60th and Argentina 54th (Reporter without Borders, 2013). This index gives a guide to the freedom of the press in the countries under analysis, even though, like other indexes, it does not offer full transparency with regard to how the information was collected, evaluated and quantified.

Despite the limited autonomy of newspapers outlined above and their diminishing significance given the rise of social media, within the time frame applied in this book focusing on newspapers allows for a systematic analysis of how media discourses on the politics of the changing Polar Regions transcend the local, national and regional spheres and of the significance ascribed to media representations and the actors shaping them. Print media in general do not aim to represent the plurality of perspectives in an equivalent manner and instead (consciously) prioritise them differently. Consequently, analysis of this communication medium offers the opportunity to recognise specific interpretations of events as well as recurring patterns of interpretation in dominant discourses that ultimately have an impact on the politics of the Polar Regions (see also Husseini de Araújo, 2012, p. 143).

4.2 Themes mainly addressed in newspapers

The themes proportionally and predominantly addressed in the newspapers mentioned above differed significantly. Like the topics placed on the agendas of the regional governance settings (practical geopolitics), newspaper coverage in the four American polar-rim states also concentrated on different issues and mirrored the disparities in the governance of the Polar Regions. In the mid-1990s, newspaper articles from Canada and the U.S., for example, mostly dealt with resources and the environment when focusing on the Arctic. By comparison, "territorial disputes" only received minor coverage. Conversely, newspapers from Chile and Argentina particularly addressed "territorial disputes" in the Antarctic in the 1990s and began to refer more often to the "environment" only after 1996. When one considers a longer period of time (in the case of this book the years 1989–2014), however, it is noticeable that newspaper reports in the American polar-rim states highlighted and referred to similar themes when focusing on the Arctic and Antarctic, though at different points in time. The following overview shows that overall, in the 7,329 articles examined that related to the Arctic and/or the Antarctic (out of 29,399 articles initially considered), the environment constituted the main theme continuously addressed by all the newspapers under analysis. The following figure also provides a more detailed overview of the absolute number of articles in regard to the themes prominently addressed.

Figure 4.1 shows that many events that, prior to analysis, I would have expected to receive exceptional attention in newspaper articles did not, in fact, receive it at the times at which they occurred (for an overview of the "crucial events" selected beforehand, see Annex). Instead, some of these events, such as the collapses of the Larsen Shelves, were only noted in retrospect. This indicates that the significance ascribed to particular events depends in part on the overall priority ascribed to the themes to which they refer. In the case of the Larsen Shelves, for example, climate change and global warming were topics that were increasingly discussed after 2005 when the theme of the "environment" was attracting more attention. The Larsen A and B Shelves, however, had already collapsed in 1995 and 2002 respectively.

Continuity of interest is also evident in newspaper reporting on Canada's Arctic identity and U.S. interests in energy resources. "Canada's North" was already receiving a great deal of attention at the end of the 1980s and in the early 1990s, not only in the context of the Nunavut land claims but also in connection with Canada's sovereignty and its military defence programme. Given the amount of attention that Canada's North was already receiving at that time, it is not surprising that more than a decade later Canada's Arctic strategy was still being much discussed. Similarly, U.S. and Canadian newspapers had made U.S. energy policy a topic of concern and focused on resources in the Arctic National Wildlife Refuge (ANWR) and Prudhoe Bay before.

Chile and Argentina's national policies towards the Antarctic, on the other hand, did not receive much attention from their own press and were also ignored by the North American newspapers under analysis. Likewise, international policy making in the Arctic Council and the Antarctic Treaty System was not picked up on to any great extent by the press in the American polar-rim states, the only exceptions being the negotiation and signing of the Madrid Protocol and the establishment of the Antarctic Treaty Secretariat.

In contrast to policy setting in the Polar Regions, international policy making on climate change has been of considerable interest to the newspapers under analysis, as the comparatively high number of articles relating to the Kyoto Protocol and the IPCC reports illustrates. Moreover, some events of particular significance for the Arctic and/or Antarctic received prominent attention in newspapers published in the countries of the opposite polar-rim, most notably the Russian flag planting at the North Pole, the extent of Arctic summer sea ice and the Larsen A and B collapses (in retrospect) as well as the Islas Malvinas/Falkland Islands conflict.

Numerous topics relating to the Polar Regions attracted attention in the press of all four countries at different points in time, for example the Nunavut land claim, the ANWR, the Islas Malvinas/Falkland Islands, the South Patagonian Ice Field, endangered species, search-and-rescue (SAR), the ozone layer, climate change

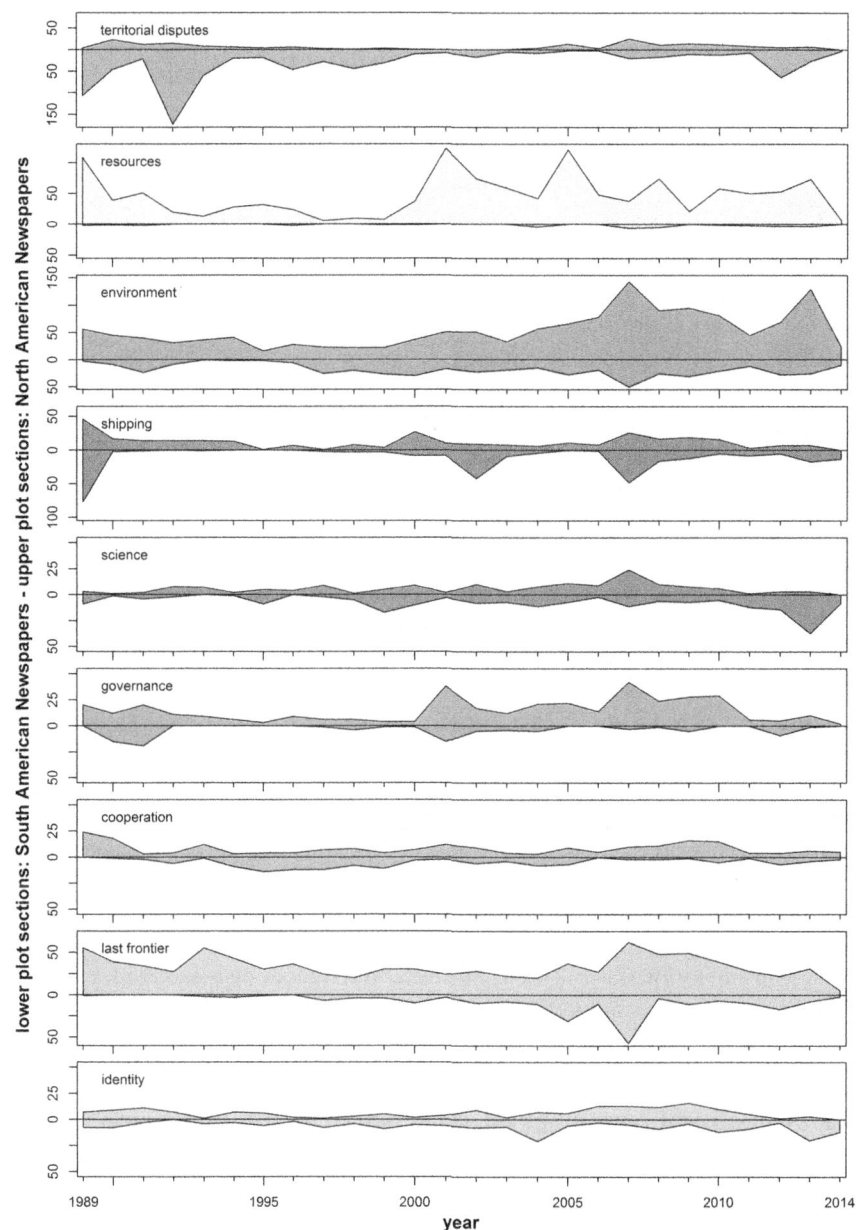

Figure 4.1 Main themes addressed in newspapers

Table 4.1 Actors repeatedly mentioned in newspaper articles

Intergovernmental actors	Industry representatives	State representative	ENGOs	Indigenous peoples
AC	BP	Argentina	ASOC	Gwich'in (International Council)
ATS	EXXON	Alaska	Earthjustice	
IPCC	IAATO	Canada	Greenpeace	Inuit (Circumpolar Council)
OAS	Shell	Chile	Sea Shepherd	Saami (Council)
RAPAL	Statoil	Greenland	Sierra Club	
UN	PetroCanada	Japan	WWF	
UNCLOS	Rosneft	Russia		
UNFCCC		U.S.		

and global warming. These topics were predominantly related to territorial conflicts, resources or environmental concerns, which are also more generally the themes most often addressed in newspaper articles focusing on the Arctic or the Antarctic that again exemplify continuity in the focus of newspaper coverage.

The relevance ascribed to particular actors in newspaper articles also differs depending on which of these different themes is being considered. As Table 4.1 indicates, various actors are repeatedly mentioned across all newspapers, which shows that they are perceived as being of particular significance for the changing Polar Regions and its politics. The ENGO Greenpeace, for example, is referred to many times in connection with both Polar Regions (with reports most often concentrating on Greenpeace's repeated confrontations with Russia, its significance for scientific research in the Polar Regions, its position vis-à-vis the governance of the Polar Regions and its cooperation with ASOC, the Antarctic and Southern Ocean Coalition).

From this, it is already clear that newspaper reporting on the changing Polar Regions and their politics addressed similar events, actors and themes in the four American polar-rim states. Often, however, differences also emerged as some issues (e.g. resources and territorial conflicts) received significantly more attention in one region than another or at different points in time.

4.3 Geopolitical reasoning: the representation of the changing Polar Regions and their politics in newspapers

When considering the representation of views and positions on the two themes that were addressed most often in newspaper reporting in the American polar-rim states ("environmental changes" and "international cooperation"), it is possible to illustrate

1   how specific interpretations and imaginaries were (re)produced in different contexts and at different times,
2   how interpretations have shifted during the period of investigation.

The examples put forward outline not only *region-specific entanglements* involving the Arctic on the one side and the Antarctic on the other, but *interpolar entanglements* as well, and thus provide more answers to the question guiding this book: Why have similar patterns of interpretation become dominant in the politics of the changing Polar Regions?

4.3.1 The (re)production of interpretations and imaginaries

When one looks at *region-specific entanglements involving the Arctic*, it is noticeable that the Arctic has generally been more often covered in Canadian than in U.S. newspapers over the past decades. More specifically, the *Toronto Star* and the *Globe and Mail* have often reported on topics such as the *Exxon Valdez* spill, the establishment of the Arctic Environmental Protection Strategy and the Arctic Council, land agreements in Canada and particularly the "birth of the Inuit homeland" Nunavut (*Toronto Star*, 1999; Walkom, 1999), Canada's sovereignty in the Arctic, the relocation of Inuit and pollution in the Arctic Ocean. U.S. papers, to the extent that they have covered polar issues at all, have focused predominantly on national Arctic concerns, such as the *Exxon Valdez* spill and oil development in the ANWR. They have also discussed Russia's nuclear tests and submarines in the Arctic. But the formation of regional Arctic governance institutions such as the Arctic Council has received little attention. On the rare occasions when Canadian Arctic politics were discussed by U.S. newspapers in the 1990s, the matters most often considered were the negotiation of land agreements and the debate on whether or not Canadians would have to apologise to Inuit families they relocated to Canada's Northern territories in the 1950s. Canadian newspapers, by contrast, more often covered political developments in the U.S. that were of relevance to the Arctic and represented views that supported Canada's aim of intensifying and leading international cooperation in the region. In this regard, their reporting on "Canada as an international leader and its role in the Arctic Council" illustrates how region-specific interpretations and imaginaries were (re)produced at different times to strengthen a particular geopolitical reasoning.

4.3.1.1 Canadian leadership in regional Arctic politics

In contrast to their U.S. counterparts, both the Canadian newspapers selected for analysis were already ascribing great significance to the Arctic region back in the 1990s, and articles often argued for Canadian leadership in regional Arctic politics. In 1998, for instance, Canada was described as "lead[ing] the way in policy for the North" in the Arctic Council (*Toronto Star*, 1998). Even earlier, in 1992 during the negotiation and formation of the Arctic Council, the former principal secretary to Prime Minister Pierre Trudeau, Thomas Sidney Axworthy, described Canada as having "taken the lead in promoting one of the most innovative exercises in international institution-building since the end of the Cold War" (Axworthy, 1992). At a time when Canadian domination of Indigenous peoples in the 1950s was also being much discussed in Canada, Axworthy considered the formation of the Arctic Council to be particularly innovative because of its inclusion of Indigenous peoples' voices, a move that he assesses as supporting a stance directed against their domination:

> Promoting Inuit membership, and encouraging such states as Russia to include aboriginal representatives in their delegations, would help ensure that the forum does not become another government body made up of white men in blue suits making policies for the people who actually live in the region.
>
> (ibid.)

Moreover, he suggested that the inclusion of Russia in the Arctic Council would provide "a modest but real recognition that Russia has joined the democratic community of nations" (ibid.). Like many other authors, Axworthy thus supported Canadian leadership, further arguing that "Canada has a great responsibility to the world to preserve this fragile environmental jewel". He also drew on the imaginary of a "Northern [not Arctic!] nation" to demand the formulation of a "northern foreign policy" (ibid.). These international ambitions were also supported by Canada's issue priorities, as an article published by Kathleen Kenna (1990) in the *Toronto Star* reveals. In it she cited Paul Martin, later Prime Minister of Canada from 2003 to 2006, as encouraging the country to become an international leader "in environmental issues as it has in peacekeeping". The image of Canada as an international leader in peacekeeping was also shared by others. Fourteen years later, for instance, the Chilean newspaper *El Mercurio* would recognise Canada as a "global leader" and as an "ally for peace":

> Canadá es actualmente un líder, en el concierto internacional, en lo económico, político, social y cultural. Un aliado para la paz. En politica exterior, Canadá es un férreo defensor de los valores de la democracia y los derechos humanos y le asigna una importancia fundamental a la preservación de la paz y la seguridad internacionales a través del diálogo y la cooperación. [Canada is currently a leader in the international concert, in matters economic, political, social and cultural. An ally for peace. In its foreign policy, Canada is a fierce defender of democratic values and human rights and ascribes a fundamental significance to the preservation of peace and international security via dialogue and cooperation].
>
> (Mendez Araya, 2004)

In the mid-2000s, Canada's leadership ambitions on issues relating to the environment and peacekeeping in the Arctic were still being widely covered in Canadian newspapers – and, more often than not, Canada's influence in Arctic affairs was related to its global reputation and described as a means of expanding its international significance:

> With two calculated speeches, Stephen Harper has articulated a blessedly coherent vision of Canada's expanding international role. [. . .] Emphasizing

his message, Mr. Harper told the Economic Club: 'Make no mistake, Canada intends to be a player'.

(*Globe and Mail*, 2006a)

Although the public and environmental organisations in Canada were also represented as demanding that Canada's leadership should address environmental concerns, in newspaper articles environmental organisations in particular challenged the perception, promoted by Canadian politicians, that the country had achieved leadership in this field. Jeff Simpson, for instance, described how Canada could influence Arctic politics by offering ideas on "Arctic management, climate change, international institutional reform, oceanic degradation", while at the same time emphasising that it was "certain" that others, like the U.S., would not listen to Canada "when it ha[d] so little to say" (Simpson, 2009b). It is notable too that the U.S. was used as a reference point in this article as it illustrates the significance of the U.S. in Canadian identity-building.

After the first three years of Stephen Harper's prime-ministership, many newspapers highlighted that Canada had neither obtained the "environmental leadership" that had been formerly envisioned, nor acquired a significant international role. Shauna Sylvester (2009) from the Simon Fraser University's Centre for Dialogue, for instance, criticised the fact that, in these two regards at least, the Harper government had not been "think[ing] globally", while Michael Byers (2007b) emphasised "Stephen Harper's antipathy to climate change" as a reason for this and added that Harper had "once dismissed the Kyoto Protocol as 'essentially a socialist scheme to suck money out of wealth-producing nations'". Other articles published by Shauna Sylvester and Martin Mittelstaedt also pointed to a strong entanglement between Canada's reputation in international relations and its position vis-à-vis environmental concerns. Sylvester (2009) opined that: "Conservatives have made the situation worse. On climate change, they withdrew from the Kyoto accord and isolated Canada in international forums by actively campaigning against global carbon reduction targets"; Mittelstaedt (2007) spoke in the *Globe and Mail* of the "government's frayed international reputation on climate change".

In two different articles, Jeff Simpson even pointed to contradictions in the policies pursued by the Harper government, arguing that, although the Conservatives "pushed Arctic policy" (*Globe and Mail*, 2009), and although "[c]limate change has heightened interest about the Arctic, yet climate change remains a secondary, even tertiary issue, for the Harper government" (Simpson, 2009a). In a different article, Rheal Seguin seemed to be aiming to explain this contradiction when he quoted a scientist who considered the absence of leadership by the federal government and, specifically, by the government agency Environment Canada to be caused by the influence of the tar sands lobby:

'Unfortunately, the weight of the tar sands lobby is such that the federal government is not capable at this point to show the leadership that we need,' Dr. McBean said. 'In Environment Canada there are a lot of outstanding people. But I'm not sure that as a department it is functioning in a way that is conducive to providing the kind of leadership that we need'.

(Seguin, 2008)

As a consequence of this lack of environmental leadership "at the top", David Suzuki, a Canadian academic and environmental activist, argued in the *Toronto Star*, this leadership "must build from the ground up" (Suzuki, 2010).

On this basis, it can be argued that, particularly in the late 2000s, the Canadian government was being much criticised for not promoting initiatives to protect the changing Arctic environment. When we consider the leadership role ascribed to Canada in the regional politics of the Arctic in the news coverage from the 1990s and early 2000s as well as the significance of this leadership for Canada's international reputation, the opposition expressed in Canadian newspapers to the lack of focus on the Arctic environment under the Harper government comes as no surprise. Similarly – and again comparing Canada to its neighbour the U.S. – numerous newspaper articles published in Canada demanded greater engagement by the Canadian government in international policy-making concerning the Polar Regions, for instance during the International Polar Year (IPY), which was much covered in Canadian newspapers (and comparatively little covered in U.S. ones). At that time, the *Globe and Mail*, for instance, cited a quote by a scientific adviser to President George W. Bush – originally published in the *New York Times* – who described the interests of the U.S. in the Arctic and applied the intimidating "backyard" imaginary (in)famous from other contexts: "The North Pole is in our backyard. The U.S. has huge geopolitical interests in the Arctic region, and we need to understand the changes that are taking place there" (*Globe and Mail*, 2007a). In this particular article, this position on the part of the U.S. is taken as a stage in the process of building pressure for greater Canadian involvement in the IPY. The statement goes on:

If that is true for the United States, it should also be true for Canada. Yet even though Ottawa hosted Thursday's official Canadian launch of the IPY – a two-year scientific program focused on the Arctic and Antarctica and involving researchers from more than 60 countries – confusion persists over the seriousness of Canada's commitment.

(ibid.)

In direct relation to the U.S., the article again criticised Prime Minister Harper's efforts in the Arctic. He is referred to as having

declared his concern for Arctic sovereignty, emphasizing the need for an enhanced military presence in the North. This is laudable, but it is not nearly enough. He needs to make polar research Canada's answer to the U.S. space program – or even to the U.S. polar research program, which also dwarfs our own.

(ibid.)

Once again, in this article, the Canadian government is indirectly encouraged to exercise leadership in the Arctic and again, through the comparison of Canada's activities in the Arctic to the U.S. space program, the Arctic is depicted as being of global significance and a region through which Canada might be able to increase its international say.

Overall, with regard to international cooperation in the Arctic, the newspapers examined in this chapter recognised that there were tensions between Canada's leadership ambitions in the Arctic region, on the one hand, and a lack of focus on protecting the Arctic environment under the Harper governments, on the other. Consequently, the dominant understanding of Canada as a leader in the Arctic, promoted in the 1990s, changed in the 2000s as a result of the tar sands development and withdrawal from the Kyoto Protocol. Even though – as shown in the previous chapter – the Canadian governments under Prime Minister Harper put particular emphasis on the significance of the North for Canada in their Arctic policies, the significance ascribed to the U.S. as a reference point for Canadian Arctic policy-making mirrored how the understanding of Canada as an Arctic state was perceived as being entangled not only with its international say but also with its positioning towards the only neighbour with whom it shares a national border, the U.S.

4.3.1.2 Entangled concerns: resource development, preserving the environment, and territorial claims

In addition to these region-specific interpretations and imaginaries (re)produced at different times to strengthen the reasoning in favour of Canadian leadership in the Arctic, the representation of resource development, the preservation of the environment, and territorial claims as *entangled concerns* is very noticeable in newspaper reporting on the Arctic. In this regard, articles such as the following – which particularly relate to the representation of Indigenous peoples – illustrate how interest in the Arctic was also spurred by concerns that were generally perceived as being bound up with one another as they evolved.

In Canada, the Indigenous peoples in particular were represented as benefiting from the fact that the opportunities and risks connected with these issues were entangled. In 1989, for instance, the *Globe and Mail* engaged critically with industrial development in the Canadian North. Although the environmental problems that Arctic oil development might cause were seldom emphasised, articles did more often discuss the impacts on the people living in the North. There was criticism, for instance, of the fact that although the "arrival of petroleum company workers [. . .] has meant work for local people", most people from the North work in "positions as unskilled laborers" (Fagan, 1989). Industrial activities were accordingly perceived as neither providing long-term employment opportunities for the people from the North nor as supporting the improvement of their skills. The negotiation of settlement rights between Indigenous peoples and the Canadian government, however, was represented as a means of counteracting domination and the exploitation of Northern territories by industry. In 1989, the *Toronto Star*, for instance, reported that "Native groups oppose export of Arctic gas" as long as the land claims of the Dene and Metis in the Mackenzie Valley remain unsettled (*Toronto Star*, 1989). Moreover, representatives of the Indigenous peoples, such as the president of the Inuit Tapirisat of Canada, used the growing interest in Arctic territories to emphasise: "We are users of the land and the stewards for the land. If you force the stewards and the original land-users off the land, you make room for industry to move in" (*Toronto Star*, 1991a), which could ultimately lead to "a cultural genocide" (ibid.). Despite the creation of Nunavut in the 1990s, by which the Canadian government gave legal title and administration rights to the Inuit who inhabited Canada's most northern territory (which is "a third larger than the state of Alaska", Farnsworth, 1992), domination by others who might settle on "their" land was presented in Canadian newspapers as a continuous source of fear for Indigenous peoples. When the Nunavut agreement was signed, for instance, an Inuk is referred to who described this moment as "both exciting and frightening at the same time", stating

> [w]ith the threat of global warming and climatic change upon us, it is conceivable that one day, more and more southerners will settle in Nunavut. Higher salaries and opportunity have always led non-Natives to the North. In Yukon, nearly a century after the Klondike Gold Rush, Native people are still a minority in their own land. In the Western NWT, non-Native people now make up a majority – though not by nearly the same margin as in Yukon. In Nunavut, most residents are Inuit, but one large diamond mine or industrial megaproject would forever alter the demographics and make the Inuit a minority in their own land.
>
> (Zellen, 1993)

In general, the negotiations leading to the creation of Nunavut did receive attention in Canadian newspapers (see e.g. Lewis, 1989; Delacourt, 1991). At the same time, however, other land agreements were being negotiated between Indigenous peoples and the federal government in Canada. These concerned, for instance, areas located along the boundaries of the Yukon and the Northwest Territories, which were also disputed among different Indigenous peoples, such as the Dene and the Inuit (see Roberts, 1998). Some of these articles supposed that the Inuit were not only aiming to protect "their" land and culture but were also pursuing their own economic interests in view of the growing development of oil, gas and tourism in the North. In this regard, it was argued that

> [t]hey will be selecting the most valuable land, based on the potential for tourism, minerals, oil and gas or revenue potential. The agreement provides royalties from oil and mineral extraction in the claim area.
>
> (McInnes, 1989)

Similarly, representatives of both Indigenous groups were quoted as not being "against the idea of development at all" but as being "concerned about what benefits will flow to the Dene/Metis" (*Toronto Star*, 1989).

Despite these accusations, however, the Nunavut Land Claims Agreement was not generally opposed in Canadian newspapers and was also perceived as a means "to strengthen the country's sovereignty there" (Byers, 2009a). This understanding was represented as being shared by the Canadian government and the Indigenous peoples (despite the forced relocation of Indigenous peoples in the 1950s). The phrase "Canada's sovereignty over the waters of the Arctic archipelago is supported by Inuit use and occupancy", for example, was included in the agreement and said to be based on a suggestion to that effect made by the Inuit negotiators (ibid.). Again, Indigenous peoples were represented as instrumentalising the interests of others (in this case the Canadian government) to achieve their goals. Although these representations challenged the understanding of Indigenous peoples as being dominated, other articles clearly enforced the perception that Indigenous peoples were still discriminated against and perceived as weak. Ten years after the "birth of Nunavut" (Mahoney, 1999), for instance, the living conditions of its Inuit inhabitants were again receiving attention in Canadian newspapers, which represented voices demanding leadership by the federal government to fight high rates of unemployment and suicide in the territory. Authors such as Colin Alexander (2009) went so far as to deny that the Inuit had the capacity to continue administering Nunavut on their own.

In addition to these region-specific entanglements spurred by the *(re)production of interpretations and imaginaries* and the representation of different issues as *entangled concerns*, recurring positions on resource development in the Arctic illustrate particularly how newspaper reporting on the Arctic in Canada and the U.S. has not only been subject-driven but has also become *entangled over time*.

Concerning resource development more explicitly, in the 1990s and early 2000s U.S. and Canadian newspapers focused particularly on the controversy about the opening of the Arctic National Wildlife Refuge (ANWR). This controversy encouraged numerous ENGOs in the U.S. to focus on the ANWR in their campaigns – one interviewee even stated: "The refuge really is a huge environmental cash cow [. . .] so, you know, people fundraise on it all the time" (Interview, April 2014). The ANWR is mainly located in the U.S. state of Alaska but also extends into Canada's Yukon territory. Its resource potential and the proposed opening of the protected area encouraged the so-called "Great Alaska Debate" (Egan, 1991), "[o]ne of the hottest environmental battles" (Kristof, 2006), over "Another Attack on the Arctic" (Babbitt, 2004) that was often depicted as a "Moral Divide" (*New York Times*, 2006). Many of the arguments and the related reasoning introduced in articles on the opening of the ANWR were later reproduced in the context of Arctic offshore oil and gas development. Examples of the different positions on the opening of the ANWR illustrate the broad panorama of recurring and entangled positions in both contexts at different times that related to

1 the significance of the territory in the Arctic for Indigenous peoples and for subnational governments,
2 benefits that would arise from oil and gas development, particularly in regard to infrastructure, employment and energy security, as well as drawbacks related to climate change and environmental risks,

3   reasons promoted by politicians and political parties, by scholars and environ-
    mental representatives in favour of and against oil and gas development in the
    Arctic,
4   references made to events such as the *Exxon Valdez* spill,
5   the reproduction of arguments discussed in the U.S. and Canada.

In U.S. newspapers, representatives of the Gwich'in in particular, who (in maintaining their so-called "traditional lifestyle") depend on the Porcupine caribou
herd that settles in the ANWR, were quoted as criticising the proposed opening at
various points in time by drawing attention to their cultural heritage:

> Much of the public has been led to believe that the Arctic is a barren white
> nothingness. This couldn't be further from the truth. The people of the Gwich'in
> Nation have lived south of the Brooks Range in this region for 20,000 years.
>
> (Beach, 2010, see also Wald, 1989)

Aside from the historical significance ascribed to the ANWR by the Gwich'in,
this commentary further exemplified that the Gwich'in regard the Arctic as being
misrepresented in dominant discourses – as a place where no one lives (as its
description as a "white nothingness" indicates), as a place where the settlements
of the Gwich'in are not recognised. In addition, in this commentary a clear stance
was taken against "development on lands or in waters that sustain so many of
us" (ibid.). Instead, "[c]onservation, improved technology and alternative energy"
are depicted as being "long overdue" (ibid). A similar position was also promoted by other Indigenous peoples, such as the Inupiat who "have a resolution
opposing development" in the ANWR, even though the government representative "H. Sterling Burnett said that the Inupiat favour development in the Arctic
National Wildlife Refuge" (Beach, 2005). In other articles, too, the Inupiat were
represented as supporting the opening of the ANWR. It was, for instance, argued
that it would be "only fair to give weight to the views of the only people who
live in the coastal plain: the Inupiat Eskimos who overwhelmingly favor drilling
(they are poor now, and oil could make them millionaires)" (Kristof, 2003). These
articles (and the language used in them) illustrate why Indigenous peoples feel
dominated and perceive their views to be misrepresented and ignored even though
they concern their "Alaskan homeland" (De La Torre, 2005).

An interviewee explained these contradictory representations by accusing
corporations interested in off- and onshore drilling of wanting to "feed that mis-
conception" on purpose, arguing that this positioning of Indigenous peoples with
respect to oil development "was intentionally mischaracterized" (Interview, April
2014). Very notably, however, it was not only Indigenous peoples such as the
Inupiat who were represented in the newspapers as being in favour of the opening
of the ANWR because of the prospect of financial gain, citizens of Alaska were
also described as supporting oil development because living (particularly energy)
costs were higher in Alaska than in other U.S. states. The higher cost of living also
served as a justification for the oil dividends (ranging between $1,000 and $3,000)

that each citizen of Alaska receives every year (Interview, March 2014). In some articles, like the Indigenous people of Canada, the people of Alaska were depicted as needy and dependent on oil revenues:

> Behind that support is another uncomfortable reality. As much as Alaskans like to cultivate their rugged, self-reliant image, many need financial help to get along here, and they get it: from oil money and from the federal government, which gives more money per capita to Alaska than to any other state, according to census figures.
>
> (Verhovek, 2001)

Just as the Indigenous peoples were represented as feeling dominated when they demanded territorial rights, domination was an argument used by politicians from Alaska, who criticised the fact that the federal government in Washington D.C. was able to decide about the opening of the ANWR, not the Alaskans themselves. Although President Bush was quoted as explaining the opening of the ANWR for drilling on grounds of "first things first, are the people who live in America" (even if this disappoints "other nations seeking a commitment to reduce emissions", Seelye, 2001), the Republican Senator Lisa Murkowski of Alaska, for instance, was cited as having said:

> We in Alaska are starting to feel cut off from the rest of the world, [. . .]. The rest of the country would just as soon lock us up and say, Nothing, nada, zip, you cannot do anything. You are not responsible enough to carry on development because we are concerned about the environment.
>
> (Firestone, 2003)

Thus, the opening of the ANWR was highly loaded in political terms. In the same article, reference was also made to another Republican Senator from Alaska, Ted Stevens, who was quoted as threatening: "People who vote against this are voting against me. [. . .] I will not forget it" (Firestone, 2003). Lisa Murkowski's father, Frank H. Murkowski, a former Senator and Governor of Alaska, had already promoted the opening of the ANWR arguing in the *New York Times* that "the coastal plain could supply a significant portion of U.S. needs over 20 to 30 years" (Murkowski, 1991).

On the federal level too, the opening of the ANWR was often represented as being necessary to meet the growing energy demand in the U.S.: "as temperatures have risen, so has the demand for oil from the Slope, which holds the country's last big known domestic reserves" (Revkin, 2004a, see also *USA Today*, 1990). Similarly, President Bush was cited as having "declared that, once again, the nation has an acute shortage of energy" (Kahn, 2001). Moreover, on the state level, the people in Alaska were represented as suffering from "skyrocketing heating bills" (Salzman, 2006), which is why they were described as demanding "more tax money to pay for oil" (ibid.) – revenues that they hoped to receive from the opening of the ANWR and the development of a "potential million barrels a day from the wildlife preserve" (ibid.). Although this estimated figure was reckoned as likely "boost[ing] pressure to tap [the] Alaska Refuge" (Warrick, 1998), it was also highly controversial as Jeff Gerth (2005) pointed out: "Even the most optimistic backers of the drilling plan agree that any oil there would meet only a small fraction of the nation's energy needs". Moreover, the author argued that President George W. Bush first proposed the opening of the ANWR in 2000 "after oil industry experts helped his presidential campaign develop an energy plan" and thus – as with the accusations brought against Prime Minister Harper and the tar sands lobby in Canada – criticised the influence of the oil lobby in Washington D.C. The "first steps to open the reserve", however, were – according to a different article published in the *New York Times* – ascribed to the preceding Clinton administration, which had supported a "two-year study involving hundreds of scientists and representatives of Inupiat communities" that "returned with a recommendation to begin oil leasing" (Babbitt, 2004). Even the outcome of this study, however, was represented differently in different newspapers. An article published in the *Washington Post*, for instance, stated the contrary:

> A new study by the Clinton administration concluded that drilling for oil in the Arctic National Wildlife Refuge in northern Alaska would do significantly more damage to the region's fragile ecology and wilderness character than previously thought.
>
> (Kennworthy, 1995)

Thus, with regard to the ANWR the circulation of contradictory information was not the exception and clearly contributed to heating up the controversy. A further article published in the *New York Times*, for instance, referred directly to another article that it represented as being "so rife with errors and outdated information, they demand correction" (Alexander, 1995). Politicians too, like the former Senator from Alaska, often spoke about "myths" when referring to the arguments of opponents of the opening of the ANWR (e.g. the *New York Times*, 2006).

Another argument often presented to justify the opening of the ANWR also related to the growing energy demand but particularly concerned the U.S.'s dependency on foreign oil. Articles emphasised, for instance, that the opening of the ANWR would contribute to "reduce the country's dependence on imported oil" (Banerjee, 2004) – a position that was shared on the federal and state levels, being, for instance, also supported by the former Alaskan Republican Senator Ted Stevens. Similarly, the former president of the American Petroleum Institute wrote in the *New York Times*:

> The potential oil from the refuge cannot be dismissed. Projected production equal to 20 years of current imports from Saudi Arabia is significant, particularly when we are reminded daily of the need for greater energy security.
>
> (Cavaney, 2006)

Four years earlier, Cavaney had argued in the *New York Times*: "The refuge holds the potential to provide half the oil we now import from the Persian Gulf and to enhance our national energy security for the next quarter century" (Cavaney, 2002). Whether or not oil development in the ANWR would actually provide "greater energy security", as Cavaney argued, is much disputed, not only in terms of the expected quantities of oil located in the ANWR. The Democratic Senator for Illinois, for instance, was quoted in a different article as arguing: "We are going to defile this wildlife refuge to export oil that won't even benefit the United States. Drilling for oil in the Arctic so we can export it to China is no answer to America's energy security challenge" (Pear, 2005). Cavaney, however, opined that the opening of the ANWR would also create jobs and further accused his opponents of showing "[e]motionalism", which – in contrast – "won't power economic growth or create jobs" (Cavaney, 2006). Aside from additional financial benefits and new jobs, it was also often argued that "oil would mean roads" (Kristof, 2003), that oil development in the ANWR would encourage the improvement of Alaska's infrastructure – as discussed earlier, all these pro-arguments were often expressed but were represented as being unaddressed by the oil industry in the 1990s. However, an interviewee argued in the same vein:

> You know, frankly, that's what makes all that work. There is a production of oil and gas [. . .] if you want to have schools and internet and running water and electricity and all those things that the people want especially in the Arctic where it is outrageously expensive to do all of that . . . the thing that underpins it all – and I think that is true largely for the whole state of Alaska – is oil and gas production. So, you know that is part of the reason that Senator Begich and Senator Murkowski and everyone is kind of on the same page [. . .] . Cause we can't run the state without this income. [. . .] You know that is what pays for the schools and the police and pretty much everything.
>
> (Interview, April 2014)

Also in *USA Today* the opening of the ANWR for drilling was depicted as "probably inevitable" to meet the nation's energy needs that "with adequate environmental safe-guards, [. . .] need not be the disaster opponents fear" (*USA Today*, 2005). The definition of "adequate safeguards" and of risks, however, was also a recurring and controversial concern in the newspapers under analysis. Andrew C. Revkin (2003a), for instance, wrote in the *New York Times*:

> Even though oil companies have greatly improved practices in the Arctic, three decades of drilling along Alaska's North Slope have produced a steady accumulation of harmful environmental and social effects that will probably grow as exploration expands, a panel of experts has concluded.

In a different article, Eric Pianin (2001) cited the (at that time) designated Interior Secretary Gale A. Norton as having argued "that oil and gas exploration in Alaska's Arctic National Wildlife Refuge could be done with minimal threat to the environment". To challenge this view, the conservation organisation the Audubon

Society recalled the 1989 *Exxon Valdez* oil spill, asking "Do we really believe there won't be any more spills?" (Alvarez, 2001).

The pro-development positions presented previously are in sharp contrast to views more prominently represented in the 2000s. Instead of dependency on foreign oil, for instance, it was argued that dependency on oil and gas resources in general and also "inaction" needed to be overcome to meet the "single biggest challenge to our planet" (Kristof, 2006) – global warming: "[T]he sad reality is that much of the Arctic plain will probably be lost anyway in this century to rising sea levels. That should be our paramount struggle: to stop global warming" (ibid.). In this regard and when promoting "conservation and the development of alternate energy sources" (Katcoff, 2001, see also Benedetto, 1990), as in the practical geopolitical discourses discussed before, reference was also made to "future generations": "That way, our children can inherit an environment marked by political stability and economic prosperity, as well as a clean arctic wilderness" (ibid.).

In a similar vein, the environmental community in the U.S. was represented as a "unified opposition to drilling in the Arctic" (Alvarez, 2001), whose representatives often referred to the ANWR as a "national treasure" (Kolton, 2000, see also Kanamine, 1991) or "a unique environmental cathedral" (Friedman, 2001) but who did not necessarily exhibit the "emotionalism" often ascribed to them. Instead, they were represented as arguing that

> a decision to place the refuge off-limits to oil companies should be based on a thoughtful balancing of complex economic, national security, environmental and aesthetic values. Demonization of the entire oil industry is not justified. For you to say that the companies are 'hatching an assault' on the refuge's coastal plain is overwrought.
>
> (Carr, 1999)

Moreover, ENGOs were depicted as encouraging dialogue with the industry. There were, for instance, references to the director of the World Wildlife Fund who "meets regularly with BP executives" (Banerjee, 2000). Similarly, Indigenous groups and environmentalists were stated to "have met with BP executives and board members to try to persuade them to promise not to drill in the refuge, just as the company pledged a decade ago to stay out of Antarctica" (ibid.).

In Canadian newspapers, on the other hand, the concerns expressed by the Canadian government towards the U.S. government over the opening of the refuge were also presented as relating particularly to the potential "threats to the way of life of the Gwich'in First Nations of northern Yukon" (Harper, 2005). In this regard, the former Prime Minister and board member of the Canadian World Wildlife Fund John Turner also referred to the Gwich'in and argued:

> The stakes for both wildlife and the people who depend on it couldn't be higher. Unfortunately, this issue doesn't lend itself to a compromise where industry can drill so as not to affect the caribou; we must make an 'either/or' choice.
>
> (Turner, 2005)

Moreover, like environmental groups quoted in the U.S. newspapers, Turner emphasised the need to act "responsibly" also in view of "future generations":

> The Arctic is one of the last places on Earth where you can still be alone with the land, with wildlife and with your own heartbeat. I can't imagine that future generations will blame any-one for having protected too much of the planet. They may well hold us responsible for having protected too little.
>
> (ibid.)

Turner also outlined how U.S. and Canadian conservation groups shared the position that the opening of the ANWR for drilling "would contravene three long-standing international agreements between Canada and the U.S. to conserve caribou, migratory birds and polar bears" (ibid.). Turner thus also hinted at the shared self-understanding of environmental groups in both American Arctic-rim states when putting emphasis on their monitoring role.

In line with the perceived intimidation of Canadian interests by the U.S. discussed before, and also in view of the opening of the ANWR, the *Globe and Mail* wrote that "Canadian diplomats have quietly told Capitol Hill officials that U.S. law requires Washington to consult with Ottawa before taking significant action in the coastal plain" (*Globe and Mail*, 1995a). More generally, throughout the period of investigation the Canadian government was represented as opposing the opening of the "ecologically fragile Arctic National Wildlife Refuge to drilling" and as "remain[ing] confident [that] the U.S. Senate will block the bill" (Koring, 2001). In his *Globe and Mail* article, Paul Koring (2001) also recited popular pro and con arguments that were circulating in the U.S. and reproduced the accusation that the U.S. government was influenced by the oil industry when he highlighted the fact that the oil industry was "the biggest contributor to the Bush–Cheney campaign, want[ed] the refuge opened to drilling and [said] new technology would limit environmental damage" and that "the United States, which imports more than 50 per cent of its oil, [was] vulnerable to foreign producers and must increase domestic production" (ibid.). In this regard, Mike Allen (2000) even emphasised that George W. Bush and Richard B. Cheney were often referred to as "the Big Oil ticket". Similarly, Alanna Sullivan (1995) reported that in "the bitter debate over development of the wildlife refuge" references were often made to the *Exxon Valdez* disaster to underline concerns over potential oil spills. Drew Fagan also referred to the spill when quoting the former Valdez mayor as having said "The reality is that Alaskans are small cogs in this wheel. We can alter the course a little, and the *Exxon Valdez* spill may have done that, but we can't change it" (Fagan, 1990). Fagan further summarised the clash of interests between those opposing the opening on the federal and those supporting it on the state level when referring to Alaska's former governor Jay Hammond Fod as having said "We are asked to be the oil barrel for America and the national park for the world" (ibid.).

Although the possible opening of the ANWR received attention over decades in newspapers published in both North American countries, in quantitative terms much more attention was paid to offshore resource development in the Arctic in the late 2000s and early 2010s. In Canada and the U.S., news coverage focused particularly on the lease-sale processes and on environmental risks and often reproduced views and positions expressed in articles covering the disputed opening of the ANWR.

As in the debate on the ANWR, environmental groups were represented as expressing particularly strong opposition to offshore drilling in the Arctic and emphasising environmental security (see e.g. the Sierra Club in Mufson, 2009). Also, like the *Exxon Valdez* spill in discussions on the opening of the ANWR, the BP *Deepwater Horizon* spill in the Gulf of Mexico contributed to a growing discussion of security measures in the context of Arctic offshore oil development. It was described as "the worst [spill] in U.S. history" (Schmit, 2010) and referred to as the "BP tragedy" (Nick, 2010). Particularly in the aftermath of the incident, critical voices in Canada received a great deal of attention (see e.g. Butts, 2010; Galloway, 2010; Harden-Donahue, 2010; McCarthy, 2010; Winson, 2010). Similarly, U.S. and also Chilean newspapers reporting on the *Deepwater Horizon* accident often used the oil spill to also emphasise the "incalculable" environmental risks caused by drilling in the Arctic – "una de las áreas más sensibles de la Tierra" [one of the most sensitive areas on Earth] (*La Nación*, 2013a). Even though "Shell said its emergency response plan was far more robust than the one BP had in the gulf" (Krauss, 2010), it was also reported that "[a] number of investigations into the 2010 BP *Deepwater Horizon* explosion and oil spill in the Gulf of Mexico identified poor oversight of contractors as a central contributing factor to the accident" (Broder, 2013). In this regard, the moratorium on all new offshore drilling plans announced by the Obama administration after the spill was depicted as "a courageous and correct move in the face of an ongoing environmental crisis caused by repeated, institutional failures of both government and industry" (Nick, 2010). When Shell announced that it would pause exploratory drilling off the coast of Alaska in 2013 (see e.g. Eilperin and Mufson, 2013b), environmental groups such as Oceana were also cited as taking this as an admission: "Shell is finally recognizing what we've been saying all along: that offshore drilling in the Arctic is risky, costly and simply not a good bet from a business perspective" (Mufson, 2014). With respect to other oil leases in Alaska's Chukchi and Beaufort Seas, it was also reported that ConocoPhillips had later suspended its plans "because of uncertainties over federal regulatory and permitting standards" (Krauss, 2014), after the Department of the Interior had obliged the Shell Oil Company to address safety issues before resuming its drilling operations. In this article, Krauss referred both to the environmental group Oceana as having regarded this as a "good choice" arguing that drilling "cannot be done safely" in Arctic waters and to Alaska's Senator Lisa Murkowski, who "expressed disappointment" and furthermore reminded readers that

> Shell was forced to remove its two drilling rigs from the area and send them to Asia for repairs after a series of ship groundings, weather delays and environmental and safety violations during the 2012 drilling season.
>
> (ibid.)

With regard to the latter, "the runaway oil rig Kulluk" (Fountain, 2013) received particular attention in U.S. newspapers (cf. Broder, 2012a; Mufson, 2013). Senator Murkowski kept supporting Shell's operations in the Arctic and was quoted, for instance, as arguing "the most important thing is for Shell to continue to make progress and demonstrate once again that Arctic drilling can be done safely" (Broder, 2012b, see also Mufson, 2012), while President Obama was also represented as wanting "to go forward" (Broder and Krauss, 2012).

The increased coverage of the opposition provided by environmental advocates at those times can be at least partly explained by the events that empowered their arguments (see e.g. McCarthy and Vanderklippe, 2010) and by the measures taken by an alliance of ten environmental advocacy groups who "filed suit in federal court in Alaska" (Broder, 2012c). They criticised, for instance, the industry and also the government for not being "prepared to handle the risks of drilling in the fragile and unforgiving region" and compared allowing drilling in the Arctic with

> a building inspector letting a developer start construction on a skyscraper on shaky grounds before the safety plans are even complete. It's premature, it's unwarranted and it's wrong – especially when it's happening in one of the most pristine places on earth.
>
> (Broder, 2012b)

Like the references to environmental risks, the representation of different positions in regard to offshore development in Alaska evoked the arguments and imaginaries drawn upon in the debate on the ANWR. For instance, domestic offshore drilling was represented as a requirement "if our country is to increase its stable and secure energy" (Baker, 2011). In a similar vein, President Bush continued to be cited as arguing that the ban on offshore drilling must be lifted "to help reduce oil prices and America's dependence on foreign oil" (Boyer, 2008), but it was still suggested that "offshore drilling wouldn't do [any] such thing. No law says that oil found there must stay in America" (ibid.), an argument also put forward by Democrats who said that a ban on offshore drilling "would help companies more than customers" (Jackson, 2008). Moreover, the Republicans and many oil companies continued to be represented as sharing the same view that through "technological advances" it would be possible to "explore oil offshore in ways that protect the environment" (Mufson, 2008), and as justifying the opening of new offshore exploration areas by saying "that it is hypocritical to pressure other countries to open new exploration areas, especially offshore, when the United States refuses to drill in its own waters" (ibid.). Also the *Exxon Valdez* disaster was still depicted as encouraging fears, although residents in Alaska were described as depending on the revenues generated by the oil production there over "the past 50 years", and Indigenous peoples in particular were represented not only as "rely[ing] on oil for their livelihoods" (Mouawad, 2007) but as also depending on fishing in their "garden" – the Arctic ocean – as a result of which they "were [passively!/DW] put in the agonizing position of choosing between their two worlds" (Broder and Krauss, 2012). The wording used in the newspapers was often very similar. Indigenous peoples, for instance, were still often generalised as composing one homogeneous group and called "Eskimos" (ibid.). Alaskan politicians were depicted as not being able to "survive as an opponent of any oil development" and opposition by conservationists was discredited and related to a "philosophical bent" (Broder and Krauss, 2012). Like those who opposed the opening of the ANWR, environmental groups were said not to have formed an alliance with "North Slope residents", who described their food security as being "threatened, not just by climate change, but by offshore development" (Eilperin, 2012) but were also referred to as "resent[ing] intervention by organizations based outside of Alaska" (Broder and Krauss, 2012), although environmental groups such as Pacific Environment are represented as being particularly concerned with "human rights issues, access for subsistence users to resources, and the protection of endangered species" (Mouawad, 2007).

4.3.1.3 Environmental concerns: dominated by outsiders and irrelevant

Turning now to *region-specific entanglements that concern the Antarctic*, fear of being dominated or misrepresented by "outsiders" was an interpretation reproduced particularly often in the South American press. Newspaper articles repeatedly pushed the idea of Chilean conservationists being influenced and dominated by outsiders. In *El Mercurio*, for instance, Agüero Garcés (1996) argued that Chilean ecologists "al querer imponer una escala de prioridades ajena a nuestra realidad" [who wish to impose an order of priorities alien to our reality] were influenced by industrialised countries and thereby illustrated "nuestra dependencia cultural de naciones que ya son ricas" [our cultural dependency on nations that are already rich]. To make his case, the author evoked the fear of being dominated by a reasoning developed and promoted outside of Chile (in more wealthy countries) in order to criticise the views put forward by representatives of environmental organisations – six years after the end of the Chilean dictatorship under General Augusto Pinochet who came to power after a military coup supported by the Central Intelligence Agency under the U.S. federal government (Kornbluh, 2000). Agüero Garcés outlines very explicitly that it was not the "exigencia de protección medioambiental en si" [the necessity of protecting the environment in itself] that he regarded as a problem. Instead, he stated that the priorities and models "that are not ours" promoted by these organisations contradicted the use of resources that Chilean society wished for and needed: "Y es ahí donde se demuestra la indudable 'dependencia cultural' de estas organizaciones respecto de modelos de sociedad que no son los nuestros" – [And it is here that the undoubtable cultural dependency of these organisations is shown with respect to models of society that are not ours] (Agüero Garcés, 1996).

The critical attitude towards environmental organisations operating in Chile expressed by Agüero Garcés in the preceding article was not an exception. Two years later, a representative of the Chilean mining industry also described environmentalists as "populists" whose reasoning "no tiene razón al lógica" [is not right nor logical] (Urquidi Fell, 1998). He criticised environmental organisations for demanding that the private sector cover environmental damage financially and spoke of injustice when arguing that those wishing to initiate industrial projects needed to assume responsibility for the total risk and were, moreover, held responsible for problems that arose from "la iniciativa de terceros o del Estado" [the initiative of third parties or the state]. Despite the scepticism towards environmental organisations evident here, when the representation of environmental problems in Chilean newspapers over time is considered, it is noticeable that awareness of environmental problems grew in the 1990s and was being particularly supported by the government of the day. Back in 1991, for example, Patricio Aylwin, the first president of Chile after the Pinochet dictatorship, announced the development of the first long-term policy "to protect the environment for future generations" (*El Mercurio*, 1991a). Moreover, while until the 1990s the guiding principle in Chile was "crecer primero, limpiar después" [grow first, clean up later] (O'Bryan, 1997), this policy was replaced when the government started to assume an active role in the protection of the environment, established a National Commission for the Environment, adopted a plan directed against pollution and a law in 1994 relating to this issue.

In contrast, in neighbouring Argentina, environmental concerns were increasingly discussed from the mid-2000s and related to pollution (Ramos, 2008), the depletion of the ozone layer (e.g. *La Nación*, 2009) and ice losses in the Polar Regions caused by climate change (e.g. Lipcovich, 2009; *La Nación*, 2012a). As Carmen María Ramos pointed out, however, environmental concerns were never a priority in Argentinian politics. Ramos explained this by the series of crises Argentina had been through in the previous decades, which had resulted in questions relating to the environment being perceived as irrelevant (Ramos, 2008).

On the regional level, Chile's efforts in particular during the negotiation of the Madrid Protocol were the object of great attention and were often covered on the front pages of Chilean newspapers. The country's engagement in this regard was reviewed in exclusively positive terms (e.g. "Chile is among the group of nations that wish to create a protocol that modernises the original treaty", Valle, 1990) and its actions were described as legitimate, given Chile's and Argentina's geographical proximity to Antarctica: "Chile y la Argentina serían, además, los primeros en ser perjudicados por una contaminación ambiental" [Chile and Argentina, moreover, will be the first to suffer from environmental pollution] (*La Tercera*, 1990a). Even the Chilean stance against an indefinite moratorium (promoted by ENGOs, France, Belgium and Australia) was perceived as being shaped by farsightedness as it was presented as an important decision that would – again, like the reasoning promoted in practical geopolitical discourses and in newspaper articles published in the U.S. and Canada – principally affect future generations:

> La posición chilena es que no se pueden tomar tan importantes decisions que afectarán principalmente a las futuras generaciones [the Chilean position is that we cannot take such important decisions, which will principally affect future generations].
>
> (*El Mercurio*, 1991b)

The international negotiations over the Kyoto Protocol were also widely covered in Chilean newspapers. Christopher Zegras, for instance, wrote on behalf of the International Institute for the Conservation of Energy in *El Mercurio* that the prospect of the negotiations resulting in an agreement "es casi milagroso" [is almost a miracle] but that it would be no surprise if such an agreement fell short of what was really needed to confront the problem ("enfrentar el problema") of climate change because of the variety of diplomatic, economic and private interests involved (Zegras, 1998). Zegras further identified the Global Climate Coalition in particular as a powerful player ("una de las fuerras más poderosas") under which the liquid fuel industry was organised (ibid.).

Rolando Stein, on the other hand, weakened any criticism and clarified that, in signing the Kyoto Protocol, Chile – as a developing state that contributed comparatively little to global pollution – would not be obliged to reduce its emissions (Stein, 1998). Argentine newspapers too often differentiated between the responsibilities of "rich countries" and "developing countries" (cf. e.g. Lean, 2007) when discussing climate change. In this regard, the failure of the U.S. to ratify the Protocol was severely criticised, for instance, by Juan Pablo Montera, affiliated with the Universidad Católica and the Massachusetts Institute of Technology, who wrote that President Bush could not pretend not to know about constantly increasing emissions and the associated problems of climate change and who referred in this regard particularly to the findings of the IPCC (Montera, 2001). *La Tercera* (2004) also supported criticism of the position of the U.S. when quoting a Greenpeace member arguing that the ratification of the Kyoto Protocol by Russia put pressure on the U.S. More drastically, Oscar Spinelli opined in an article published in 2003 by the Argentine *Clarín* that, if "Russia and the U.S. don't support the Kyoto Protocol the South will one day turn into hell" ("EE.UU. y Rusia no lo apoyan, y así el Sur un día se volverá un infierno," Spinelli, 2003). In a similar vein, the Argentine *Página 12* described the government under President Bush as a government that "ha buscado sabotear sistemáticamente los esfuerzos internacionales para luchar contra el cambio climático" [has systematically tried to sabotage international efforts to fight against climate change] (Lean, 2007).

During the negotiations for the Kyoto Protocol, Chilean newspapers such as *La Tercera* were already identifying the U.S. as the "primer país contaminante del planeta" [the most contaminating country on the planet] (*La Tercera*, 1997a) that needed to reduce its emissions but, on the other hand, was also demanding the reduction of emissions by "las naciones en vías del desarrollo" [developing nations] (*El Mercurio*, 1997). In this regard, it was reported that during the negotiations "los países ricos" [the rich countries] were accused by the "países del tercer mundo" [third-world countries] of preventing the latter from achieving a similar degree of economic development to their own through their calls for environmental protection (*La Tercera*, 1997b) and it was this perception that was regarded as strongly affecting the negotiations (*La Tercera*, 1997c). Despite this criticism, the occurrence of climate change and global warming and its anthropogenic causes were generally not disputed in Chilean and Argentine newspapers (cf. e.g. *La Nación*, 2003a). However, although notable changes in the climate were being reported on as early as 1989, these were not then linked with climate change or global warming (cf. e.g. *La Nación*, 1989). The collapse of Larsen B and further ice melting were, however, directly said to have been caused by the warming atmosphere (Cecchi, 2002; *La Nación*, 2005). After the collapses, references were increasingly made to findings published by scientists in *Nature* who argued that global warming did exist despite the growth of ice in some parts of Antarctica (Chang, 2002; *La Nación*, 2002a). With the particular attention paid to climate change effects in the Polar Regions during the International Polar Year and in the IPCC report of 2007, the Polar Regions became more often represented as the most sensitive regions of the planet (cf. *La Nación*, 2011a) and the extensive news coverage of both (cf. Bocchicchio, 2006; *La Nación*, 2006a, 2006b; Moreno, 2007) seems to have inspired a greater focus on climate change as well.

Like the articles published in Chile, newspaper reporting in Argentina represented climate change and the related environmental changes in the region as a major challenge (cf. Bär, 2003), arguing, for instance, that climate change provokes catastrophes (*Clarín*, 2007) caused by rising sea levels (*Clarín*, 2012a) and floating icebergs (*Clarín*, 2012b). The newspapers also criticised the environmental movement for not having succeeded in drawing sufficient attention to the protection of the environment (Ramos, 2008). Other positions were repeated in both countries and over time illustrating discursive entanglements across national boundaries, such as the shared significance ascribed to Argentina's and Chile's geographic proximity to Antarctica and the shared fear of being dominated by others, particularly by conservationists and "rich countries". Although in both countries, climate change was not prioritised in newspaper reporting, it was presented as a problem at a time when it also received more attention at the international level (e.g. during the negotiations for the Kyoto Protocol). In both countries, climate change was further represented as resulting from anthropogenic causes. Instead of arguing, however, in favour of a general limitations of emissions, newspapers in both countries differentiated between obligations of "los países ricos" [rich countries] and "países del tercer mundo" [third-world countries].

4.3.1.4 Entangled concerns: multilateral cooperation, territorial disputes, and geographic proximity

In addition to these region-specific entanglements generated by the *reproduction of interpretations*, multilateral cooperation, territorial disputes and geographic proximity were often represented as *entangled concerns* in news reporting in Argentina and Chile.

Although issues specifically related to the Antarctic were more often addressed in Chilean than in Argentinian newspapers, the *Bahía Paraíso* spill and the territorial disputes relating to the Southern Patagonian Ice Field and the Islas Malvinas/ Falkland Islands were continuously recalled in both South American countries when their (shared) geographical proximity to Antarctica was in question. In a similar way to articles in the Canadian press, for instance, articles in the *Clarín* present Argentina and Chile as the "vecinos muy próximos" [nearest neighbours] to Antarctica and the "primeras víctimas" [main victims] if conflict in the Antarctic region were to escalate (Becerra, 1989). Particularly in the 1990s, this proximity was also emphasised in articles that pointed out shared interests including sovereignty ("intereses comunes sobre el cual se reconocen mutuamente soberanía", Leal, 1994), often framed under the expression "Antártida Sudamericana" [the South American Antarctic] (cf. Becerra, 1989; Leal, 1994).

Similarly, the United Kingdom's claim to the Islas Malvinas/Falkland Islands has many times been represented as fundamental to its geopolitical interests in Antarctica, because the Falkland Islands are said to serve as the reference point for its territorial triangle (sector claim) in the Antarctic and thus to provide it with access to Antarctica and the Southern Ocean (*La Nación*, 2010; Gamba, 2012; Télam, 2013). Also, thanks to this understanding, Argentina's claim to the Islas Malvinas is represented as being supported by the 33 Latin American countries (with the exception of Paraguay) that form part of the Economic Commission for Latin America and the Caribbean (CEPAL, *Página 12*, 2013) particularly by Chile. It is even emphasised that Argentina and Chile are united by a "hermanidad" (sisterhood) that is "inconmovible e inalterable" [implacable and unalterable] (*Clarín*, 2012d; see also Niebieskikwiat, 2007; Tokatlian, 2013). By some, this special partnership is, however, perceived as based on their shared dismissal of the United Kingdom's claim (cf. Niebieskikwiat, 2009), implying that otherwise the relationship between Chile and Argentina would be different. More often, however, Chile and Argentina are both represented as supporting the strategy of governing Antarctica as a place of peace and a heritage of mankind (Tokatlian, 2013).

Shared sovereignty interests are also described as having encouraged the Argentine initiative to establish the Reunión de Administradores de Programas Antárticos Latinoamericanos (RAPAL), with the aim of promoting cooperation among Latin American countries in Antarctica (ibid.). More specifically, regional cooperation among Latin American states is often portrayed as particularly necessary with reference to cooperation among scientific organisations (e.g. Guzmán, 2008: "América se une en la Antártica" [America unites in Antarctica]), among environmental organisations (Mansilla, 1994: "Ecologistas piden integrar el Mercosur" [ecologists ask Mercosur to integrate]) and among military forces in Antarctica (Leal, 1994: "el territorio debe ser un espacio integración regional. Antártida Sudamericana" [the territory must be a space of regional integration. The South American Antarctic]). However, geographical proximity to Antarctica is also often depicted in Chilean and in Argentine newspapers as being of particular significance for international activities there more generally, when, for instance, Ushuaia is represented as a "gateway to Antarctica" (cf. e.g. Skliarevsky, 1997; *La Nación*, 2003b). Similarly, the director of the Instituto Antártico Chileno, José Retamales, has pointed out that the increasing number of parties to the Antarctic Treaty can be useful in strengthening Chile's sovereignty in Antarctica as their national campaigns and research in Antarctica are mostly conducted via Punta Arenas ("A través de nuestro país el vínculo aéreo es más sencillo" [the simplest air link is through our country], García, 2013). Retamales is also cited as opining that Chile needs to demonstrate that "a Chile le importa la Antártica" [for Chile, Antarctica matters], which is why the INACH reports regularly on all Chilean activities in Antarctica.

In news reporting more generally, Argentina and Chile are understood as close allies that share a similar burden and responsibility with respect to the Antarctic because of their geographic proximity, which is also the basis for their territorial claims there – a basis that is not challenged in the articles investigated but represented as legitimate. Other reasons for Argentina and Chile being represented as important allies – "sisters" – are their successful negotiations concerning their overlapping territorial claims in Antarctica and their shared opposition to the United Kingdom. To exemplify multilateral cooperation, the newspaper articles considered all refer to the participation of Argentina and Chile in inter-American organisations and forums (such as the Comisión Económica para América Latina y el Caribe-CEPAL, Mercosur and RAPAL), which are also seen as means to support their sovereignty claims in Antarctica. Unlike the Canadian and U.S. news coverage examined, it is notable that in Chile and Argentina regional cooperation is also represented as a strategy to defend interests against "powerful outsiders" and thus mirrors a shared fear of domination.

4.3.1.5 Interpolar entanglements: the resource race in the Polar Regions

Specific interpretations and imaginaries have also been (re)produced in newspapers published in countries bordering the opposite Polar Region to the one immediately in question. Such cases exemplify *interpolar entanglements* as they illustrate how specific representations and a related reasoning have been adopted, empowered or challenged outside the region in which they originated. One such case is the representation of the Polar Regions as being affected by a "resource race".

In the U.S. and Canada, for example, news reporting on the Arctic significantly increased from 2004 to 2008 and, at that time, the topics of concern were often related to global warming and climate change, disputed territory in the Arctic and the "Arctic resource race". Items included, for instance, the growing interest of countries in the Arctic, the publication of the U.S. Geological Survey and the seabed mapping for UNCLOS applications. In 2005 much news reporting in Canada concerned the disputed ownership of Hans Island (see e.g. Gilbert, 2005; O'Neill, 2009, on the symbolic importance ascribed to Hans Island) and, in particular, the Russian flag planting at the North Pole.

The "Arctic resource race" ("la carrera por riqueza del Artico"/"Una carrera geopolítica") a topic often linked to the Russian North Pole flag planting, however, was also covered in Chilean and Argentine newspapers (particularly the Chilean ones, cf. *La Tercera*, 2007a; Bermúdez Liévano, 2012). This illustrates the widespread attention that the latter event received. Numerous articles published in the *Clarín* in Argentina, for instance, focused on international reactions to the Russian flag planting and emphasised that the U.S. rejected Russia's ambitions with regard to North Pole resources. Often cited quotes from the Canadian politician Peter Mac Kay and the Russian minister Sergei Lavrov were also reproduced. In an article published in *La Tercera* (2007a), however, in contrast to most of the newspaper coverage on the flag planting in the North American newspapers, the

Russian action was not predominantly depicted as a "land grab" (see e.g. Borgerson and Antrim, 2009 in the *New York Times*). Instead, Russia's expedition was compared to a joint North American scientific expedition that had similarly aimed to provide scientific evidence for submissions to the Commission on the Limits of the Continental Shelf (CLCS). Thus, the article did not exclusively support the notion of an "Arctic resource race" although it argued that estimates of resources had inspired growing interest in the Arctic and depicted both the U.S. and Russia as competing over the "riqueza del Ártico" [the wealth of the Arctic] (*La Tercera*, 2007a).

The growing attention paid to a possible "Arctic resource race" further encouraged a spillover of representations of the resource development also occurring in Antarctica. Despite the regulations in force, for instance, potential resource development in Antarctica was touched upon in Canadian newspapers (e.g. the *Toronto Star*, see Comier, 2007). More explicitly, *El Mercurio* in Chile claimed that the majority of consultative parties to the Antarctic Treaty were developing geological programmes just to investigate the existence of resources in Antarctica (*El Mercurio*, 2008) – even though the Madrid Protocol prohibits the exploitation of mineral resources. The Argentine newspaper *Clarín* (Sierra, 2008) even used the imaginary of "la conquista de la Antártida" [the conquest of the Antarctic] and referred particularly to Russia and China as wanting to play an important role there. The Russian flag planted at the geomagnetic South Pole on February 14, 2008 and Russia's expansion of its scientific bases in Antarctica are also used as references (ibid.). These representations are particularly notable if one recalls that all the original signatories to the Antarctic Treaty installed their flags at the South Pole (Interview, January 2015). Another *Clarín* article described activities in the Arctic and Antarctic as "las luchas por los recursos naturales" [the struggles for natural resources] (*Clarín*, 2008a) encouraged by rising oil prices, which provided an incentive to open up the large reserves estimated to exist in the Arctic and Antarctic (again, figures from the U.S. Geological Survey were cited). A *Clarín* article further emphasised that, apart from its influence on the ecological balance in the Polar Regions, the possible development of hydrocarbons would have geopolitical consequences as numerous states would likely redefine their influence in these areas (*Clarín*, 2008a). Rosendo Fraga too referred to disputes over natural resources in both Polar Regions as in both resources are seen as being of economic importance "en el largo plazo" [in the long term] (Fraga, 2012). Illustrating Cold War thinking par excellence, Fraga's article also represents the Arctic Council as a forum whose primary purpose is to enable the U.S. and Russia to divide resources in the Arctic. The author thus challenges the perceptions of others who are depicted as preferring to see the Arctic Council as an international organisation whose purpose is to protect the Arctic environment (ibid.). Twenty-three years after the end of the Cold War, Fraga describes the U.S. and Russia as the two most important military powers by which the Arctic is surrounded. It is notable that no reference is made to Canada.

In regard to the Antarctic, Fraga (2012) concluded that sovereignty claims were going to intensify and also forecast more conflicts between states over this issue (in both Polar Regions). As a means to substantiate the presence – and sovereignty – of Chile and Argentina in Antarctica, all the newspapers considered encourage cooperation, particularly scientific cooperation, in Antarctica (cf. *Clarín*, 2012c; *La Nación*, 2011b; *El Mercurio*, 2012). Against this backdrop, Chile's investments in Antarctic activities – such as the establishment of a new base at the Polar Circle as well as investments in the infrastructure (and scientific institutions) of Punta Arenas and Puerto Williams as main ports of entry for Antarctic tourism – are reviewed positively. In Argentine newspapers the Chilean strategy of increasing its presence in Antarctica via investments in scientific endeavours is, moreover, regarded as a good example that Argentina should follow ("siguendo el ejemplo de Nuestro vecino" [following our neighbour's example] (*La Nación*, 2011a). Similarly, INACH's efforts to improve Chileans' knowledge of Antarctica through education is put forward as a model for Argentina and said to be important in "aclarar a la opinión pública" [clarifying public opinion] (*Clarín*, 2012c).

It is notable that, when comparing Argentina's Antarctic strategy to Chile's activities in Antarctica, the Argentine newspapers under analysis often adopt a critical stance towards the former (cf. e.g. Tokatlian, 2013). The Argentine Antarctic Institute, for instance, is criticised for not investing enough in science (*La Nación*, 2004), for not having a long-term strategy or not having any Antarctic strategy at all – criticism that the director of the Institute, Mariano Memolli, frequently dismisses. *La Nación* also represents Argentina's diplomats and scientific and military personnel in Antarctica as neither professional nor efficient – a judgement that Memolli rejects as "absurdo" (*La Nación*, 2011c), recalling Argentina's success in negotiating the establishment of the Antarctic Treaty Secretariat in Buenos Aires ("un importante triunfo de la diplomacia de nuestro país" after "una década de conflicto diplomático" [a significant triumph for our country's diplomacy after a decade of diplomatic conflict] (*La Nación*, 2001a).

These examples illustrate how the increasing importance given to the representation of a "race for resources" taking place in the Arctic in the late 2000s encouraged a similar debate about the Antarctic. It is notable, however, that despite representations of clashing geopolitical interests among different states – non-state actors again are mostly left out of the picture – what the newspapers under analysis generally envision is not a stronger (unilateral) military presence from their respective countries but more multilateral cooperation. Investments in scientific activities and joint scientific endeavours are reviewed particularly positively in this regard.

4.3.1.6 *Interpolar entanglements: defending sovereignty in the Arctic and Antarctic*

Interpolar entanglements are also observable when we compare representations in newspaper reports of national positions with respect to the Arctic and Antarctic. It is notable, for example, that in Argentina and Canada similar actions are discussed as means of defending their polar sovereignty (e.g. with regard to the Northwest Passage and the Islas Malvinas/Falkland Islands). The Islas

Malvinas, for instance, are regarded as Argentine territory for historical and geographical reasons (cf. e.g. *Clarín*, 2013). It is often pointed out that Argentina inherited the islands from Spain (ibid.). It is even questioned whether the "Argentine nation" is complete without the Malvinas ("Sin las Malvinas, ¿es la nuestra una nación inconclusa?" [Without the Malvinas, is ours a nation left incomplete?] (Grondona, 2012). Some articles challenge this understanding of the role of the Islas Malvinas/Falkland Islands. Mariano Grondona, for instance, describes this kind of reasoning as immature (ibid.). Most often, however, that reasoning is reproduced and supported. In an article published by *La Nación*, Pino Solanas (2012) even condemned the re-establishment of bilateral relations with the United Kingdom after the dictatorship and during the negotiations for the Madrid Protocol as "un acto de capitulación" [an act of capitulation]. Solanas also reviews the British submission to the CLCS as an attempt to "mantener su enclave colonial en las islas" [retain its colonial enclave on the islands] (ibid.), which is why he urges the Argentine government to include them in its submission as well. Martín Dinatale (2012) called attention to the fact that the Islas Malvinas/Falkland Islands are listed as forming part of the European Union according to the Treaty of Lisbon of 2007 and described this as a strategy to advertise the United Kingdom's claims. Gustavo Barra judged the inclusion of the islands in the Lisbon Treaty to be against "lo que dice las Naciones Unidos" [what the United Nations says] (Barra, 2007) namely that the United Kingdom and Argentina should find a peaceful solution. Similarly, the Secretary General of the Organization of American States is quoted as criticising the United Kingdom for stationing a destroyer and a nuclear submarine in the Falklands, which, he claims, "se trata de una militarización del Atlántico Sur" [amounts to a militarisation of the South Atlantic] (*La Nación*, 2012b).

Canada's claim that the Northwest Passage is an internal strait has been similarly described as being of great significance for Canadians for historic, geographical and cultural reasons. It is, for instance, argued that the very name Northwest Passage "is an intrinsic part of Canadian culture" that was used "throughout the three centuries of British exploration" (followed by the formal transfer of titles over the Arctic islands to Canada by the British government, Byers, 2009b). In a similar vein, James Delgado (2000) writes in the *Globe and Mail* "[o]ur northern frontier is ours thanks to the blood sacrifice the Arctic exacted from our explorers. We have a special affinity with what is up there; it's like the roof of our own house". Reference is also made to the settlements of Indigenous peoples when pointing out, "Canada's case is unique, if only because it claims jurisdiction over that melting Arctic ice on which the Inuit have lived for generations" (*Globe and Mail*, 2007b). Against this backdrop, the U.S. is often criticised for not recognising "Canada's sovereignty over the Arctic waters" (see e.g. *Toronto Star*, 1990a; Byers, 2007a) even though – as Mark Collins (2012) emphasises – others too, "notably members of the European Union and Japan", dispute Canada's claim to the Northwest Passage. He further regards the claim as "actually quite shaky in terms of international maritime law" and declares this to be "one reason we don't take it to the International Court of Justice" (Collins, 2012) – a position shared by

Franklyn Griffiths (2006) of the University of Toronto in the *Globe and Mail*. In his article Griffiths, moreover, encourages the Canadian government to build on existing bilateral agreements and intensify cooperation with the U.S., which he regards, in contrast to growing militarisation, as a "practical and inexpensive way of exercising Canadian jurisdiction in the Arctic waters" (ibid.). Cooperation with the U.S. – particularly the NORAD (North American Aerospace Defense Command) agreement – was, however, not always reviewed positively in Canadian newspapers – for instance, when Canada was depicted as being "too reliant on the U.S. for security" (Lowman, 1989).

In both Argentina and Canada, election campaigns and visits to the Arctic and Antarctic have received a great deal of attention (*Globe and Mail*, 2008b; Grondona, 2012; *La Nación*, 2002b, 2011c; Pizzi, 2013; Skliarevsky, 1997; Woods, 2011). After being criticised for his "High Arctic rendezvous and the meticulously staged photo-op with our military" (Roberts, 2010) – viewed as a strategy on the part of "the Conservatives" for "fluffing up Mr. Harper's image" instead of "dealing with the real issues facing Canadians" (ibid.) – Stephen Harper is quoted as explicitly defending his annual trips as necessary means for nation-building: "We're doing it because this is about nation building. This is the frontier. This is the place that defines our country" (Buzza Smith, 2010) – an understanding that also circulated in the 1990s when Canada was often represented as "an Arctic nation" (cf. *Toronto Star*, 1993a). However, in the *Toronto Star* as well, Harper's Arctic rhetoric and his military orchestration were taken as a strategy pursued by the Conservatives to win the next election:

> Fear is always a powerful political force. In times as nervous as these it becomes the next best thing to a doomsday weapon. Conservatives want Canadians to be very afraid as the next election creeps closer.
>
> (Travers, 2010)

Numerous newspaper articles on these visits, moreover, refer to Canada's most Northern and Argentina's most Southern and far distant territories with pride when highlighting, for instance, Ushuaia as the principal entrance to Antarctica and depicting the Argentine bases in Antarctica as being of great significance for the penguin colonies. In Argentina, the "Día de la Antártida" (Antarctic Day) is covered in newspaper articles on an annual basis and it is repeatedly stated that this day commemorates the planting of the first Argentine flag on Antarctic territory on Laurie Island, which forms part of the South Orkneys (cf. *La Nación*, 2002b; *Clarín*, 2004; *Página 12*, 2014). Moreover, in Canada numerous articles seem to relate to the Arctic also in order to raise awareness of a region not much considered by the public. The *Toronto Star*, for instance, published an article stating:

> Many maps of Canada, including the ones typically found in schools, textbooks, and on classroom walls, ignore the Canadian Arctic. The map I just posted on my classroom wall cut off everything north of the 70th parallel. Canada's territory stretches much further north. Ellesmere Island reaches past

the 83rd parallel! We have to get the Canadian Arctic on the Canadian policy map, but first, let's try to get the entire Canadian Arctic on our regular maps.
(Ahmad and Jaya, 2005)

In Argentina, on the other hand, such a demand is unnecessary. As the Argentine *Clarín* (2010) outlined in an article, according to law no. 26.651 it is obligatory for all public institutions and school books to depict Argentina's territorial claim in Antarctica as forming part of Argentinian national territory.

Overall, in both Argentina and in Canada territory (claimed) in the Arctic and Antarctic is discussed in newspaper reporting as being of great significance for the "nation" due to its geographic location and proximity, and historic and cultural entanglements. While some relate to these territories with pride and emphasise the need to defend Argentina's and Canada's sovereignty there, others consider the arguments used to justify a feeling of "belonging" as rather weak and accordingly argue against an overvaluation of the significance of these territories in regard to other strategic interests. In both countries, therefore, similar patterns of interpretation are used to emphasise geopolitical relationships to the Polar Regions.

A similar *interpolar entanglement* is also observable in the *recurring representation* of the Arctic and Antarctic as colonies. The colonisation of the Arctic and the Antarctic is a widespread point of reference and fear in all polar-rim countries. The relocation of Inuit to Canada's Northern territories in the 1950s, for instance, was much discussed in Canadian newspapers in the 1990s and 2000s (Platiel, 1993; *Toronto Star*, 1993b; Ferguson, 1994; *Globe and Mail*, 1995b; Aubry, 1996; Mason, 2008, 2009). This practice was criticised as an act of colonisation until the Indian Affairs Minister officially apologised in 2010 in the name of the Canadian government (*Globe and Mail*, 1990; *Toronto Star*, 1990b, 1990c; Curry, 2010). But even after the official apology by the Canadian government, the imaginary of an "Arctic colony", of people from the North being dominated by "outsiders" persisted in the newspapers under analysis. This imaginary supports particularly an understanding of the Canadian Arctic territory as being different but dominated by central Canada and Canadian decision-makers located below the Arctic Circle (cf. Chapter 3). In contrast to the former understanding, however, in more recently published articles this domination is no longer exclusively related to people but also to the territories in general, which mirrors the more widespread representation of the Arctic as no longer solely a border area (to be defended and populated by inhabitants) but also as a place with potential "riches" – an understanding spurred amongst other things by the frequent citing of the U.S. Geological Survey in Canadian newspapers (McCarthy, 2009; McKenna, 2008; *Toronto Star*, 2008). Similarly, U.S. newspapers have often made references to Alaska being perceived as an "Arctic colony" of the U.S., dominated in its policy making by the federal administration because of its resource potentials and pristine environment (e.g. Egan, 1991).

In Argentine and Chilean newspapers, references to colonised territory in the Antarctic were mostly made when reporting on the Islas Malvinas/Falkland Islands – for instance, in 2003, when the Argentine president Néstor Carlos

Kirchner spoke at the United Nations Special Committee on Decolonization against the "colonisation" of the Islas Malvinas by the United Kingdom. At the same time, however, he emphasised that Argentina feels obliged to protect the interests of the international community in Antarctica and intends to find a peaceful solution to ensure that all activities pursued by Argentina are in line with the Antarctic Treaty and the Madrid Protocol (*Clarín*, 2003). Like other politicians before him, he did not prioritise Argentina's interests in the Islas Malvinas/ Falkland Islands over its interest in forming part of multilateral cooperation in Antarctica. Similarly, Foreign Minister Jorge Taiana emphasised in an interview on the CLCS submission five years later that Argentina's sovereignty claim did not conflict with the regulations provided for in the Antarctic Treaty even though Argentina maintains its understanding that the Islas Malvinas/Falkland Islands belong to it (Curia, 2008).

In more general terms, until the United Kingdom made its submission to the CLCS in 2009, the regulations provided by the Law of the Sea remained largely unaddressed in Chilean and Argentine newspapers. Many newspaper articles focused exclusively on the joint rejection of the United Kingdom's submission by Chile and Argentina (e.g. *Clarín*, 2009), which was represented as intending to "reclamar en la ONU un millon de kilometros cuadrados mas del continente blanco" [to claim one million square kilometres more of the white continent at the UN] (ibid.) even though the British did not include their claims in Antarctica in the submission but reserved the right to do so in the future (it was also represented as such earlier, e.g. in *Clarín*, 2008b). As an illustration of the general scepticism directed against the United Kingdom, it is worth noting that this misrepresentation (the submission included the Islas Malvinas/Falkland Islands but not Antarctica) and the rather alarmist news coverage took place in 2007 although the British ambassador to Argentina, John Hughes, had, for instance, published a commentary in *La Nación* in which he explained the reasons for this submission (to clarify borders in the Antarctic with regard to scientific information). He emphasised

> El Reino Unido y la Argentina han formado parte de todos estos procesos constructivos de cooperación e información compartida. Esperamos que continúe así durante largo tiempo. [The United Kingdom and Argentina have both participated in all related processes constructively and cooperatively and shared information. We expect this to continue for a long time].
>
> (Hughes, 2007)

Although some articles published in Argentina criticised the inclusion of any territories located in the Antarctic in the United Kingdom's submission to the CLCS (e.g. Oliva, 2007), like Argentina's presentation of its scientific documentation at the CLCS, the submissions as such are most often described as being justified (e.g. Polack, 2009). The director of the Argentine Antarctic Institute, for instance, is quoted as defending the submissions as being a necessary preparation in case of renewed conflicts over territory in Antarctica (Oliva, 2007). Although he states that he does not believe this will happen, he refers to the "puja por recursos energéticos" (dispute over energy resources) and particularly to the global *interest in the Arctic* at times of high oil prices and to Exxon who presented the development of Arctic oil as a good way of helping to satisfy growing energy needs (ibid.). Memolli points out that this understanding is also shared by the Argentine Energy Institute (ibid.). Linking again to the CLCS process, the article also refers to the U.S. Center for Strategic and International Studies (CSIS) according to which submissions to the CLCS are said to be encouraged by the Arctic 5's interests in petroleum ("la revitalization de los reclamos por soberánia tiene el color del petróleo" [the revitalisation of sovereignty claims is coloured by petroleum] (Oliva, 2007).

These examples illustrate that the understanding of territories located in the Arctic and in the Antarctic as colonies has been recurrently reproduced with respect to both Polar Regions. It is important to note, however, that the strategic relevance ascribed to these representations has changed. In Canada the understanding of the Arctic as a colony used to relate predominantly to the people living in the North, who were represented as defending the Canadian border in the Arctic. With the growing significance ascribed to the resources estimated to be located in the Arctic and its pristine environment, however, the current understanding of domination no longer relates exclusively to the people but also to the territory. With respect to the Antarctic as well, the representation of the Islas Malvinas/Falkland Islands as being colonised by the British is no longer exclusively related to the strategic significance of the islands for the United Kingdom's territorial claim in Antarctica. It is increasingly described as being encouraged by the estimated resource potentials in Antarctica and the rising global demand for energy. These patterns of interpretation consequently reinforce the understanding of a resource race taking place in the Arctic and Antarctic. With this in mind, it comes as no surprise that, particularly in consideration of their estimated resource potentials, the predominant representation of the Arctic and Antarctic in all the newspapers under analysis is as regions possibly attracting future conflicts. At the same time, however, both regions are considered to be characterised by international cooperation, which is said to be particularly inspired by environmental challenges:

4.3.1.7 Disentanglements: the internationalisation of the Polar Regions

The *internationalisation of both Polar Regions* is generally represented differently in the newspapers under analysis as they consider different states to be "powerful players". However, they all subscribe to the notion of "polar orientalism" coined by Klaus Dodds and Mark Nuttall (see Chapter 3) when questioning the interests of far distant countries in the Polar Regions. U.S. newspapers have more often discussed Russia's role in the Arctic, while Canadian newspapers have focused particularly on China's interests in the region itself and in the Arctic Council (cf. *Globe and Mail*, 2007a, 2008a; Grant, 2010; Schiller, 2010). Newspaper articles published in Argentina also reflect more often on China's role in the Arctic, pointing out, for instance, that China's investments in Arctic research have exceeded those of the U.S. (Bermúdez Liévano, 2012). Chinese investments are thus viewed as being particularly driven by the opening up of the Arctic region "to commercial and international exploitation" even though China justifies its interest as being based on the global environmental significance of the region (ibid.). Similarly, newspapers published in Canada and the U.S. have commented critically on the inclusion of other non-Arctic states such as India, Italy, Japan, Singapore and South Korea as observers in the Arctic Council. These states are described as having "sought economic opportunities in the region and viewed participation in the Arctic Council as means of influencing the decisions of its permanent members" (Lee Myers, 2013).

China's role in the Antarctic, its expeditions and construction of new research bases, has likewise been particularly touched on in Argentine newspapers. In *Clarín*, Gustavo Sierra regards China as having the potential to replace the unipolarity said to have been obtained by the U.S., which is why he speaks of a possible "enorme confrontación" between the two (Sierra, 2012). In regard to the Antarctic, Sierra (2008) depicts "los países ex comunistas" [the ex-communist countries], i.e. Russia and China, as countries aiming to play an important role in the future of the Antarctic while pointing out at the same time that Argentina is in comparison "un páis pequeño y solo podemos tener presencia" [a small country able only to maintain a presence] in Antarctica. Against this backdrop, in both articles Sierra (2008, 2012) argues in favour of strengthening the cooperation in Antarctica among Latin American countries to keep a strong say in the governance of the continent. Similarly, in the Arctic, despite the presumed strategic positioning of "outside countries" and in contrast to the depiction of the "resources races" and "international conflicts" that are about to take place in both regions, actors promoting international cooperation are also often cited in the newspapers (increasingly so when climate change is being referred to). Paul Arthur Berkman, for instance, emphasises the need for this cooperation in the *New York Times* asking "if these nations are still too timid to discuss peace in the region when tensions are low, how will they possibly cooperate to ease conflicts if they arise?" (Berkman, 2013).

In 2010 and after the widely covered Russian North Pole flag planting, the so-called "militarisation of the Arctic" was still being discussed in all the newspapers under analysis. At the same time, however, articles were published that questioned the depiction of an emerging international conflict in the Arctic and instead pointed to cooperative actions among the Arctic states. In an article published by the Canadian historian Whitney Lackenbauer, for instance, the Canadian Operation Nanook, which is often represented as reflecting the militarisation of the Arctic, was reviewed very differently, as military activity that was "an exercise to hold up Canadian military capabilities" but not aimed at intimidating "our closest neighbours, the Danes and the Americans" who are – quite the contrary – said to have been invited to participate (Lackenbauer, 2010). In this article, Lackenbauer also challenges the imaginary of an Arctic "military theatre", often used to relate to potential conflicts in the Arctic, and shifts its meaning to emphasise cooperative actions, entitling his article "High Arctic theatre for all audiences" (ibid.). In 2011 as well, after the successful negotiation of a boundary agreement with Norway, then Prime Minister Vladimir Putin is quoted as having said "If you stand alone in the Arctic you do not survive" and emphasising that "Canada has a good working relationship with Russia with respect to the Arctic. There is no likelihood of Arctic states going to war" (Byers, 2011).

Similarly, in regard to the Antarctic, *La Nación* (2001b) comments that cooperation among states is crucial on the "continente blanco" [white continent] because of the challenging environment. Among scientists from different countries of origin, it is said that the difficult life on the bases "se genera una convivencia compleja en lo que hace a la solidaridad" [generates a complex cohabitation that encourages solidarity]. In the Arctic, environmental changes, such as the opening of formerly ice-covered water in the Arctic Ocean, are also more often related to the need for greater international cooperation – for example in efforts "to prevent and mitigate the effects of climate change", which are not doubted to be global in scope (Berkman, 2013). More explicitly, the *New York Times*, for instance, speaks about "the urgency of establishing common standards for protecting the Arctic environment and patrolling shipping lanes" (Rekin, 2006a). It is notable, however, that only rarely are references made to the Arctic Council in this regard, which – although often described by politicians and scientists as the main intergovernmental forum in the Arctic (cf. Chapter 3) is not represented as such in Canadian and U.S. newspaper reporting.

In the U.S. newspapers a great deal of attention was paid to the negotiation of marine protected areas in the Antarctic – another concern closely related to the internationalisation of both Polar Regions that depends on successful cooperation. In 2013, for example, Dan Bilefsky reports for the *New York Times* that China, Russia and Ukraine have rejected the proposals introduced by the U.S. and New Zealand. In a different article published by the *New York Times* earlier, both proposals were described as "excellent proposals to create two major marine reserves" located close to Antarctica, "this planet's only continent wholly protected from mining and other economic activity, save tourism – a place where nature, not commerce, rules" (*New York Times*, 2013). In his report on the rejection of the marine protected areas, the author refers to experts from the WWF and quotes a climate adviser for the ENGO Friends of the Earth as having said that "economic and political interests had trumped global environmental imperatives" (Bilefsky, 2013). In 2007, however, Mariano Memolli also introduced a different reasoning directed against the implementation of marine protected areas in the Antarctic, when calling attention to the question of whether, under the umbrella of preservation, countries with the resources to administer these regions were not, in fact, creating new provinces:

> Desde hace tiempo se viene hablando de generar regiones protegidas en aguas internacionales. Habría que preguntarse si, bajo el paraguas de la preservación, los países con recursos para administrar esas regiones no estarían creando nuevas provincias. [For some time there has been talk about the creation of protected areas in international waters. One has to ask if, under the umbrella of preservation, the countries with the resources to administer these regions are not creating new provinces.]
>
> (Oliva, 2007)

In a similar vein, recalling the negotiations regarding the whaling sanctuary in the 1990s, an interviewee from Chile mentioned that critics had emphasised the understanding that such a sanctuary would constitute "una limitación a nuestra soberanía" [a limitation to our sovereignty] (Interview, February 2015). In contrast, and unsurprisingly, in all of the articles considered, environmental organisations were represented as being clearly in favour of the establishment of marine protected areas in both Polar Regions (*Toronto Star*, 1991a; Goar, 1995; Mittelstaedt, 2001; Aguas, 2008; Bilefsky, 2013).

Overall, newspapers in Argentina, Chile, Canada and the U.S. perceive the internationalisation of the Polar Regions positively when arguing that shared challenges can only be addressed through multilateral cooperation. Particularly those articles that relate to the potential resources thought to exist in the Arctic and Antarctic, however, question the motives of "outside" actors in cooperating, and represent fears of domination and escalating conflicts. Similarly, even though the negotiation of marine protected areas and the agreement on common shipping standards are reviewed as positive outcomes of international cooperation in the Polar Regions, as, most notably, the director of the Argentine Antarctic Institute stated, all these standards are also perceived as limiting the scope of actions and affecting matters formerly considered to fall under the sovereignty of individual states. Interestingly, in articles stressing the latter no references are made to the consensus principle that applies in both regional settings – the Arctic Council and the Antarctic Treaty System – an instrument aimed at preventing the domination of smaller states by more powerful ones. Accordingly, it can be observed that, either on purpose or through lack of knowledge, well-known fears of domination are reinforced in regard to the internationalisation of the Polar Regions.

4.3.1.8 The positioning of non-state actors

Even though most articles concentrate on other actors, newspaper reporting in all American polar-rim states has increasingly covered the positions of non-state bodies, mirroring their growing say in practical geopolitics. Particularly the positions of ENGOs – and among this group especially the positions and actions pursued by Greenpeace ("the global environmental watchdog", Comier, 2007) – have been much discussed. In the 1990s, for example, Greenpeace's protests in the Arctic against Russia's nuclear missile tests and its opposition to whaling in the Southern Ocean received particular attention. In the 2000s and 2010s, newspapers published in the American polar-rim states focused more on Greenpeace's stand against whaling carried out by Japan and on its protests against offshore drilling in the Arctic.

More specifically, with regard to Arctic missile tests the *Toronto Star*, for instance, focused on Greenpeace's strategies to influence governmental policies and observed a change in the organisation, which was represented as putting greater emphasis on lobbying instead of organising peace marches and demonstrations (Dexter, 1989). In 1990, the *Globe and Mail* pointed out that Greenpeace's nuclear disarmament campaign in the Arctic aimed to influence government policies as Greenpeace was depicted as trying to persuade "the superpowers to sign a comprehensive treaty" (McLarien, 1990). Greenpeace's reports on the "large-scale dumping of nuclear waste" were likewise considered to be "the first to estimate the total amount of radiation in the dumped material" (Wilson, 1993), thus attributing a monitoring role to Greenpeace. These examples show that in the 1990s both Canadian newspapers were already depicting Greenpeace as a non-state actor increasingly intending to influence domestic and international politics through intensive lobbying and scientific reports (reports delivered by the World Wildlife Fund were depicted in the same way, e.g. *Globe and Mail*, 2008c).

Newspaper articles published by the *New York Times*, the *Washington Post* and *USA Today* reported less on Greenpeace's political strategies and more on its confrontations with Russia. In this regard, the year 2013 marked a peak in U.S. newspaper reporting on Greenpeace. When the "Arctic 30" were arrested in Russia, numerous articles focused on the conditions under which the activists were held and the diplomatic efforts to free them. At the same time, Greenpeace's "Save the Arctic" campaign was often covered in a way that supported its "enemisation" of the oil industry. For instance, a Greenpeace activist is quoted in the *New York Times* article as saying: "The Arctic is melting before our eyes, and yet the oil companies are lining up to profit from its destruction" (Kramer, 2013). This article does not take a critical stance towards the attempt by Greenpeace activists to scale the Prirazlomnaya drilling platform operated by Gazprom in the Pechora Sea. Instead, the author takes an affirmative position and explains "Greenpeace International sent the ship to the Pechora Sea to draw attention to the potential environmental threats caused by a rush to exploit natural resources in the Arctic" (ibid.). The article further promotes an understanding of Greenpeace's actions as being legitimised in political terms by stating that "there was little doubt that Western governments and rights groups would have used the Olympics as an opportunity to draw attention to the plight of such high-profile prisoners" (ibid.). Moreover, the author refers to the 1985 attack on the *Rainbow Warrior* in a New Zealand harbour in which one activist was killed to illustrate that, like the arrested "Arctic 30", Greenpeace activists had defended their views against strong opposition before. This reference supports the author's depiction of Greenpeace as a brave and authentic actor. If compared with earlier news reporting on Greenpeace activities that were not of international significance but directed against the U.S., this article can also be read as exemplifying how public opinion draws on an understanding that "our enemy's enemy is our friend". In a different newspaper article, for instance, also published by the *New York Times* in the year 2000, Greenpeace's opposition to the building of the Trans-Alaska oil pipeline and another pipeline along the Alaska–Canada Highway is depicted more critically and – particularly in comparison to the positions of other ENGOs – as rather extreme: The author James Brooke states:

> Greenpeace opposes them, saying Americans should wean themselves off fossil fuels and invest the billions of dollars it would take to build pipelines in alternative energy sources like solar and wind. Other groups, recognizing

that gas pollutes less than oil, plan to focus on reducing the environmental impact of the lines by pushing to have them follow the existing Alaska oil pipeline and the Alaska – Canada Highway.

(Brooke, 2000)

In a similar vein, Greenpeace and the World Wildlife Fund are also represented as the environmental groups that most prominently "have been lobbying global corporations for years to cut their carbon output" (Eilperin, 2010).

In Chilean and Argentine newspapers too, reports on environmental organisations have most often referred to Greenpeace (even though "Greenpeace has not been active in Antarctica for many, many years", Interview, June 2014). For instance, during the imprisonment of the 30 Greenpeace activists from the *Arctic Sunrise* who protested against Gazprom's oil rig Prirazlomnaya in the Pechora Sea, Emiliano Ezcurra, former director of Greenpeace, pointed out in Argentina's *Página 12* that this was not the first time that Greenpeace activists had been arrested by "los poderosos" [the powerful] and released not only "por presión pública" [because of public pressure] but also "porque la ley así lo indicaba" [because the law prescribed as much]. He thus depicts the activists as the less powerful party but the one that is acting in accordance with the law. Ezcurra further legitimises the actions of the Arctic 30 by stating that this is also not the first time that Greenpeace activists have put a banner on an oil rig and arguing that there is no other way to draw attention to what is happening in the Arctic "un ecosistema clave para regular el clima en la tierra" [a key ecosystem for the regulation of the climate of the planet] (ibid.). Moreover, Ezcurra refers to the "petroleras" [oil companies] as actors who only care for their own benefits ("vamos a sacar el petróleo del Ártico y no nos importa nada más" [we're going to get the oil out of the Arctic and nothing else matters to us]) and who threaten that other protesters "van a pasar muy mal como estos 30 chicos" [are going to get themselves into a lot of trouble like those 30 guys, ibid.]. In the end, Ezcurra uses this representation of the oil and gas industry as "peligroso" [dangerous] to call for the establishment of a "parque mundial, al igual que la Antártida, un área consagrada a la ciencia y la paz" [a global park similar to the Antarctic, an area devoted to science and peace] (Ezcurra, 2013) in the Arctic and, in this regard, he also refers to Greenpeace's successful "fuerte campaña global" [strong global campaign] in the 1980s that he depicts as having resulted in the negotiation of the Madrid Protocol (ibid.).

Also in *La Nación* Natalia Oreiro, another member of Greenpeace, emphasises the good intentions behind the actions of the Arctic 30 when stating "saben lo importante que es salvar el Ártico" [they know that it is important to save the Arctic] (*La Nación*, 2013c). Another article published in the same paper invokes the IPCC reports and the Greenpeace campaign as means to remind us that it is humanity's right that activities should not cause the planet to deteriorate ("el derecho de la humanidad a que las actividades no deterioren el planeta") and warns against any possible contamination: "[N]o somos conscientes de nuestra fragilidad" [we are not aware of our own fragility] (ibid.). The article thus takes a clear stance against oil and gas development in the Arctic and very notably also argues against any other activities that may have a negative impact on the planet. However, even before the Arctic 30 were arrested, Argentinian newspapers were referring to Greenpeace's Arctic campaign. Often, these articles referred in particular to the low extent of the summer sea ice in the Arctic and reproduced the reasoning promoted by Greenpeace activists who argued that the world is turning into a boiler ("una caldera") that humans have switched on (Gaffoglio, 2012).

In comparison to the Argentine newspapers under analysis, the newspapers from Chile paid less attention to the protest of the Arctic 30. There was a difference in this respect between the late 1980s and the early 1990s, however, when Chilean newspapers referred to Greenpeace and its activities in Antarctica to support the positions taken by Chilean politicians who argued for the adoption of the Madrid Protocol. In this regard and choosing similar wording to that used by the Argentine newspapers mentioned before with respect to the Arctic, *El Mercurio* (1990) also referred to the "frágil ambiente" [fragile environment] of Antarctica and cited Greenpeace (and the WWF) who similarly called Antarctica "la última gran reserve ecológica del mundo y asi debe permanecer" [the last great ecological reserve in the world that needs to stay as it is] and demanded international cooperation to protect it. *La Tercera* (1990b, 1990c) also reported on the "Antarctic World Park Campaign" by Greenpeace and cited the Chilean Chancellor as arguing in favour of the protection of the pristine environment in Antarctica, which he described as the "unica seguridad de nuestro future" [the only security for our future] (*La Tercera*, 1989) when emphasising Chile's long coastline shortly after the *Bahía Paraíso* spill.

During the mid-1990s and mid-2000s, articles published in Argentine and Chilean newspapers most often referred to Greenpeace when reporting on its actions against Japanese whaling in Antarctic waters (*Clarín*, 1998, 1999, 2001; *La Tercera*, 2007a). The newspapers all supported Greenpeace's positions, described its agenda-setting success when promoting the establishment of the Antarctic Whaling Sanctuary (*Clarín*, 1998) and also referred to its monitoring activities, for instance, when reporting on how the Greenpeace ship *Arctic Sunrise* followed the Japanese ship *Nisshin Maru* in the Antarctic to observe whether or not the Japanese did any whaling (see e.g. *La Tercera*, 2007a). In Argentina, Greenpeace was also depicted as the ENGO whose scientific explorations (in collaboration with the Instituto Antártico Argentina) indicated that the crack in the Larsen Shelf was caused by greenhouse gases (Santana, 1996), and that drew attention to the significance of the ozone hole and criticised the Argentine government for not having formulated an environmental policy (Santana, 1997a). Similarly (though without direct reference to the Polar Regions), environmental organisations were also described as demanding a greater focus on the sustainable development of the environment on the part of Mercosur, "transparencia y permita la participación civil en la construcción de una integración verdadera" [transparency and permission for civilians to participate in the construction of a green integration] (Mansilla, 1994).

Besides Greenpeace, ASOC and the WWF are two other ENGOs prominently mentioned in the five newspapers from Chile and Argentina – but no Chilean or Argentinian ENGOs receive coverage. In all the newspapers under analysis, reference was made to ASOC as an alliance of ENGOs demanding greater awareness of the growth of tourism and the implementation of regulations to control it (Aguas, 2008; Goycoolea and Luco, 2007). Newspapers often refer to ASOC's position when shipping accidents take place, for instance, after the grounding in the Antarctic of the cruise ships *Explorer* in 2007 (*La Nación*, 2007) and *Ocean Nova* in 2009 (Higgins, 2009). These articles supported ASOC's lobbying for stronger regulation of tourism, for example, to be provided by the Antarctic Treaty System. In contrast, in U.S. and Canadian newspapers, for instance in *USA Today*, environmental groups were represented as the main drivers of discourses on climate change and global warming (Mastio, 1999).

Overall, in the newspapers from Argentina, Canada, Chile and the U.S. Greenpeace's activities in the Arctic and in the Antarctic received a great deal of attention and were constantly reviewed positively. In Canadian newspapers especially, Greenpeace was most often represented as a "watchdog" and an important actor influencing domestic and international politics. In Argentina and Chile, Greenpeace was usually perceived as an important actor providing opposition to powerful industries. This focus on Greenpeace is remarkable, as the organisation's formal say, for instance, in regional politics is very limited: Greenpeace does not participate in the Arctic Council nor in the Antarctic Treaty System, whereas other ENGOs such as WWF and ASOC do. Although numerous newspaper articles also relate to WWF and ASOC, their positions and say in Arctic and Antarctic politics receive significantly less attention. It is also notable that ENGOs with their origins in Argentina and Chile that focus on the Antarctic go almost unrepresented. As explained in more details in the next chapter, among the many reasons for this is their lack of resources to conduct or participate in Antarctic campaigns that are often very expensive.

4.3.2 Shifting interpretations

Besides revealing the region-specific and interpolar entanglements illustrated by the interpretations and imaginaries dealt with earlier, newspaper reporting in the American polar-rim states also provides a record of how interpretations, particularly the significance ascribed to environmental changes and to international cooperation, have changed over time.

Articles in newspapers from all four American polar-rim states have repeatedly discussed the ozone layer, global warming and climate change (the phenomena and their implications). As far as climate change is concerned, both Canadian and U.S. newspapers have increasingly referred to UN climate talks since the mid-2000s. They have also paid growing attention to the melting ice in both Polar Regions (the fourth IPCC report that outlined "strong evidence of climate change" received a great deal of attention in this regard). Many articles have focused on the partly growing and partly decreasing ice in Antarctica and particularly on the record low minimum extent of Arctic summer sea ice observed in the 2010s. However, even before that, in the 1990s, U.S. papers reported on the melting ice shelf in Greenland and considered climate change and global warming in this context. In 1997 the *New York Times*, for instance, reported that the Arctic sea ice "shrank at a rate of 2.9 percent per decade" between 1978 and 1996 (Browne, 1997). Climate change and global warming were also repeatedly addressed in Canadian papers and the melting permafrost and its implications for Indigenous peoples had already been addressed in 1991 (*Toronto Star*, 1991b).

It is noticeable though that in articles published in the 2000s, the wording is generally stronger since climate change is depicted as a more certain phenomenon. Earlier, for instance in 1997, when Canadian and U.S. newspapers reported on the ice shelf collapse in Antarctica, it was still much discussed whether or not the causes of global warming and climate change were anthropogenic (e.g. *Globe and Mail*, 1997). The article in question, for instance, asked: "What about the climate-change story? Although science cannot be certain it will happen until it has happened, there are many suspicious fingerprints showing up on the globe. The world's surface temperature is rising and many glaciers are retreating" (*Globe and Mail*, 1997). In contrast, in 2007 the severity of climate change was compared to an Al Qaeda terrorist technique:

> If we learned that Al Qaeda was secretly developing a new terrorist technique that could disrupt water supplies around the globe, force tens of millions from their homes and potentially endanger our entire planet, we would be aroused into a frenzy and deploy every possible asset to neutralize the threat. Yet that is precisely the threat that we're creating ourselves, with our greenhouse gases.
>
> (Kristof, 2007)

When climate change is being written about, Canadian newspapers often cite ENGOs. In an article published in the *Globe and Mail* in 2002, for instance, a representative of International Union for the Conservation of Nature and Natural Resources (IUCN) is quoted as having stated: "Hardly a week goes by when we don't read new evidence of climate change in the news" (Mitchell, 2002). In this article, the author also supports an acknowledgement of shared responsibility of a kind often demanded, particularly by ENGOs, as the article also provides space for the personal perspective of the representative in question:

> I feel helpless as far as reversing climate change goes. We can offer advice but the ultimate test is whether humans on a global scale are willing to take action and even if they do, whether the ecological momentum of climate change can be turned around.
>
> (ibid.)

The other side of this "shared responsibility" coin is often depicted as the side of shame and ignorance, represented as being "morally inexcusable" as inaction is very often seen as leading to catastrophes (see e.g. Chivian, 2007). It is notable that this imaginary of a "shared responsibility", in the context of growing human activities in the Arctic, had already been picked up in news reporting on the *Exxon Valdez* disaster in 1989 by the *Globe and Mail*. In his article David Suzuki, the Canadian academic and environmental activist, reacted to a report in which the columnist Terence Corcoran blamed environmental activists for feeling "no shame in resorting to distortion and exaggeration" and asserted that "Governments are going to have to assume more responsibility" (Suzuki, 1989). In his reply, Suzuki argued that "[t]his uncritical defence of business-as-usual by giant multinational corporations is not acceptable", thus ascribing responsibility not simply to governments but also to industry and – like Mitchell 13 years later – to all humans:

> The sudden and massive increase in human numbers and technological muscle-power in this century is without precedent. Armed with the notion that economic imperatives must subordinate all else, we are as a feral species running amok. Unless we learn the deep lessons from incidents like the Valdez spill, more of the planet will pay the price for our inaction.
>
> (Mitchell, 2002)

Although ice losses caused by climate change are depicted as being greater in the Arctic than the Antarctic (*La Nación*, 2012a), much attention is similarly paid to climate change in Chilean and Argentinian newspapers. In the case of Chile, an interviewee stated that, due to the significance ascribed to climate change, Antarctica was also the object of greater interest among the public. At the "nivel de gobierno" [government level], for instance, "yo diría que hay una consciencia mayor y un ámbito de cooperación" – [I would say that there is a greater consciousness and an atmosphere of cooperation] (Interview, February 2015). Similarly in Argentina, the collapse of the Larsen A Shelf in particular – often explained as being caused by global warming (see e.g. Santana, 1997b) – has been related to the growth of an environmental consciousness ("hay mayor consciencia ambiental" [there is greater environmental awareness] Santana, 1997a).

As in the U.S. and Canada, newspaper articles in Argentina and Chile that relate to the IPCC reports use stronger wording when reporting on climate change. The 2007 IPCC report, for example, is cited as illustrating a global consensus that the very complex phenomenon of climate change is a reality (*La Tercera*, 2007b). Rapidly melting glaciers in countries such as Chile are represented as indicating the future impacts of global warming, which is why preparations are said to be needed for a world without glaciers ("[a] prepararse para un mundo sin glaciares", Rojas, 2009). Also in a 2008 *New York Times* article, climate change is depicted as "changing all the rules" but here reference is made particularly to the Arctic (*New York Times*, 2008). The article states that with melting sea ice, the "accessible" Arctic is "leading to fierce disputes over territory and natural resources" (ibid.). James Gustave Speth (2006) likewise describes climate change as "the biggest thing to happen here on earth in thousands of years, with incalculable environmental, social and economic costs" and asks "Why Aren't We Marching?" The Premier of Quebec, Jean Charest, and the Premier of Manitoba, Gary Doer, had already answered this question in 2005 when they declared it to be a "key step" to shift "the public discourse on climate change from being a matter of environmental concern to one that includes the social and economic merits of taking concrete action" (Charest and Doer, 2005). In contrast to previous representations of climate change, the greater certainty (cf. Simpson, 2009c: "It gets harder to ignore the signs of climate change") and priority ascribed to the phenomenon are also reflected in the wording used in articles published by *USA Today* that consider climate change, for example, a "defining issue of our time but [one that] can be difficult to understand" (Koch, 2014). Others describe it even more drastically: McKibben (2007), for instance, writes: "We're in a desperate race. Politics is chasing reality, and the gap between them isn't closing nearly fast enough. [. . .] Global warming has a huge start; the sprint to catch up is the story of our time". And the *Washington Post* (2013a) speaks of "the real horror show, the true existential threat, is yet another crisis of our own making: the catastrophic effects of climate change".

Since the 2000s, significantly more attention has also been paid to the discussion of multiple climate change impacts. Newspapers in the American polar-rim states emphasise, for example, the "alarming rate" (O'Neill, 2008) of collapsing ice shelves (see e.g. Campbell, 2008, on the Markham shelf; Calamai, 2003a; Strauss, 2003; Revkin, 2003b; Gugliotta, 2003; Weber, 2008; Leeder, 2008, on the Ward Hunt Ice Shelf; Revkin, 2006b; Gillies, 2006; Lillebuen, 2006, on the Ayles Ice Shelf; Rother, 2005, on the Larsen A Shelf; Grant, 2002; Revkin, 2002, on the Larsen B Shelf). Resulting side-effects are more often covered, such as health threats (Suzuki, 2006) and rising sea levels that – "if a major portion of Antarctica" melts – are represented as "flood[ing] not only many islands but also inundat[ing] China's industrial region and threaten[ing] coastal cities in North America and western Europe" (Calamai, 2003b). In this regard, references are also made to the Arctic Climate Impact Assessment published in 2004 by the Arctic Monitoring and Assessment Programme (AMAP) – a working group operating under the auspices of the Arctic Council. When reporting on the assessment, however, newspapers often make no reference at all to the Arctic Council – so that they neither promote international cooperation in the Council nor critically engage with the production of this report (cf. *Globe and Mail*, 2004; Revkin, 2004b, 2004c; *Toronto Star*, 2004; Watt-Cloutier, 2006).

While in the 1990s, most newspaper articles did not refer to the anthropogenic causes of climate change, in the 2000s this changed with the growing news coverage of the IPCC reports. In this regard, the Polar Regions often served as a reference point, empowering their image as climate change barometers (Weber, 2010). Newspapers in Argentina, Canada, Chile and the U.S. alike often represented ENGOs as raising awareness and demanding (political) action to combat climate change. Often, though, despite the strong wording used, the reporting of their demands remained rather unspecific.

The significance ascribed to scientists and scientific findings also changed in this context and, as with the shifting interpretation of environmental changes, it is notable that, particularly in U.S. newspapers, articles to do with the validity of scientific findings on environmental changes often involved controversy (cf. Revkin,

2000: "When will we be sure?"). The repeated "nonsense about the uncertainty of global warming" as Naomi Oreskes calls it in the *Washington Post* has also been seen as contributing to the absence of discussion concerning the measures needed to deal with climate change, which – in retrospect – is also regarded as having led to climate change and global warming becoming "politically polarizing" issues (Bostrom, 2010). In 1997, for instance, an article published in the *New York Times* stressed that "[f]orecasting the regional and local impact of climate change is even harder because the global models do not deal with climatic change on a regional and local scale very well". The author then went on: "But nothing prevents scientists from zeroing in on areas and activities that could be vulnerable to the predicted alterations in climate, and they are doing so more intensively" (Stevens, 1997). In 2007, on the other hand, the "2,000 scientists who are recognized as experts in their respective fields" (Gorrie, 2007) and are "involved in writing or reviewing" the IPCC report are cited as being "nearly certain to conclude that there is at least a 90 percent chance that human-caused emissions are the main factor in warming since 1950" (Kanter and Revkin, 2007) and that there "is no doubt that one of the regions that is affected by climate change is the Arctic" (Scott, 2007), a finding that had still been disputed 10 years earlier (see e.g. Warrick, 1997). Also, in an article published by the *New York Times* in 2010 entitled "Who Cooked the Planet?" (that drew on the imaginary inherent in the accusation "we're cooking our favorite planet" published in a 2007 article also in the *New York Times*, cf. Kristof, 2007), the author Paul Krugman stressed the view that the U.S. Senate "didn't fail to act because of legitimate doubts about the science. Every piece of valid evidence [. . .] points to a continuing, and quite possibly accelerating, rise in global temperatures. Nor is evidence tainted by scientific misbehavior" (Krugman, 2010). Instead, he accuses politicians, industry and think tanks of "greed and cowardice" because of their investments in disinformation campaigns and in scientists who question climate change and support understandings promoted by industry (e.g. "any effort to limit emissions would cripple the economy", ibid.). Emphasising the injustice of mostly "non-causers" of climate change suffering most from the related environmental changes (as already addressed above), Krugman concludes: "Greed, aided by cowardice, has triumphed. And the whole world will pay the price".

In contrast, Canadian newspapers often promoted an affirmative position in regard to scientific findings on climate change, for instance, in relation to the IPCC. In an article published by the *Globe and Mail*, Alanna Mitchel was already writing in 2001 that the IPCC's third assessment report "is by far the strongest warning ever issued on climate change by an international body". The author also uncritically quotes scientists involved in the IPCC as regarding science as "the primary driver for action on climate change" and recommending the use of "alternative power sources and less energy [. . .] to reverse the warming trend" (Mitchell, 2001). Generally speaking, in fact, scientific research and findings are most often uncritically presented in Canadian newspapers, as scientists (including those from private think tanks) are often represented as being highly esteemed, for instance, through being described as "distinguished" (cf. *Toronto Star*, 1993a).

In the Argentine and Chilean newspapers under analysis – and despite the fears expressed for instance by Agüero Garcés in 1996 about being dominated by foreign ideologies and organisations – scientific findings in regard to climate change remain unquestioned and the understanding that climate change is human-caused is not challenged. Criticism is, in fact, levelled at those who challenged these findings (the U.S. and Canadian governments under President George W. Bush and Stephen Harper) and who did not support international agreements such as the Kyoto Protocol. Moreover, in Chilean and Argentine newspapers scientific research on climate change is often represented as being supported by national governments. In 2010, the Chilean president Michelle Bachelet was depicted as encouraging particularly "buena ciencia rigurosa" [good and rigorous science] to counteract climate change (*El Mercurio*, 2010). She urged the "política mundial" [world politics] to support the development of better knowledge of climate change instead of neglecting it ("No se saca nada con negar el cambio climático" [No one gains anything by denying climate change], ibid.).

It is notable that, in newspapers from Chile and Argentina, the information used in news reporting on the changing environment in Antarctica, climate change and global warming is most often identified as being obtained from foreign press agencies. This type of information, in other words, was not developed in Chile and Argentina. At least in Argentina in the mid-2000s, this seems to have been related to the dominant perception, also reflected in *La Nación* (cf. Ramos, 2008), that environmental problems are mostly the concern of experts (Ramos, 2008) whose knowledge is of crucial significance in counteracting climate change (*El Mercurio*, 2010) but who are most often seen as not being located in the "países en vías de desarrollo" [developing countries] (Montera, 2001). Particularly in regard to the discussion of environmental changes in the Polar Regions, this representation in newspapers supports the often criticised and rather technocratic representation of environmental concerns in discourses (see e.g. Kothari, 2013).

Overall, shifting interpretations in the newspaper coverage in the American polar-rim states bring not only trans-polar but also, and particularly, region-specific entanglements to light. In regard to both the Arctic and the Antarctic, climate change and global warming were addressed in the 1990s but the significance ascribed to both phenomena increased significantly in the 2000s. However, Canadian and U.S. newspapers most often addressed melting ice shelves and the extent of sea ice while giving less consideration to the "human dimension". This human dimension – emphasised by the use of the imaginary of a "shared responsibility" and references to "future generations" – was increasingly drawn upon in the 2000s but had already been addressed before in regard to the Antarctic. Newspapers published in Chile and Argentina, for example, underlined this human dimension by applying the same imaginaries to the *Exxon Valdez* accident in 1989 and the negotiation of the Madrid Protocol in the early 1990s. In Chile and Argentina, furthermore, newspaper reporting did not question climate change and global warming's being responsible for environmental changes in the Antarctic, though it was not until the 2000s that U.S. newspapers depicted climate change as a rule-changer for human activities with regard to the Arctic as well. In contrast to the scepticism shown to climate change in newspaper reporting before, at that time the wording used to describe climate change impacts also changed, for instance, in newspaper articles that spoke about "alarming rates" and designated the changing climate as something that "is killing us". Likewise, in all American polar-rim states the understanding of the Polar Regions as "climate change barometers" was often applied (directly and indirectly) to promote an urgent need to respond to climate change and was increasingly used from the mid-2000s on.

Like the representation of environmental concerns, the representation of scientists and of scientific findings also changed. Moving on from the uncertainty often emphasised when relating to scientific assessments of climate change in the 1990s, the mid-2000s and early 2010s saw politicians, industries and think tanks that expressed doubts about climate change being more often represented as driven by their own greed. In both the U.S. and Canada, governments that denied climate change and refused to act against environmental changes were accused of corruption, as being influenced by either the tar sands or the oil industry. The newspapers under analysis thus clearly challenged practical geopolitical reasoning in this regard.

To sum up: As in practical geopolitical discourses, in popular geopolitical discourses too environmental concerns have been increasingly considered in interpretations of political processes in the Polar Regions. Through shedding light on the representation of environmental concerns and international cooperation in regard to the Arctic and Antarctic, numerous entanglements between issue-specific interpretations and the representations of actors have been brought to light, which have all contributed to a recurring understanding of the Polar Regions as being of global significance, as attracting political conflict and cooperation, as being characterised by a "race for resources" and/or as spaces of particular (e.g. cultural and national) significance to some. Some of these entanglements are region-specific but many of them are also interpolar in scope. Some of them enforce a particular understanding, while others challenge such understandings (most notably, territorial claims and resource development have attracted conflictive understandings).

In regard to the representation of environmental changes in the Polar Regions, the interpolar imaginary of a "barometer" has been increasingly applied in newspaper reporting in all four American polar-rim states. This imaginary was most often used to underpin the significance of climate change impacts. It was also often employed to demand political action. Moreover, in the newspapers under analysis the representation of the "barometer" imaginary enforced a geopolitical interpretation of the Polar Regions as spaces of global significance and spaces increasingly structured by the idea of a "global responsibility". The danger inherent in the influence that this imaginary might exert in the future lies in the possible marginalisation of the alternative, local and region-specific interpretations outlined previously. Or, to put it differently, the growing dominance of the interpretation of the Polar Regions as spaces of global significance certainly encourages the inclusion of new actors and perspectives from outside the regions who are now more often regarded as needed to meet the challenges in and beyond the regions that arise from these environmental changes. Such an interpretation, however, does not support the exchange and discussion of conflictive views in newspaper reporting in regard to resource development in the Arctic and to territorial claims outlined before. As a consequence, the solutions developed to deal with current prominent environmental challenges in the Arctic and Antarctic are likely to remain unsupported by those who still fear domination or feel dominated by those who shape these solutions. These fears may thus encourage political disentanglements in the Polar Regions, which were also exemplified in the previous chapter most notably with regard to Alaska's Arctic Policy. Particularly in newspaper reporting, however, the dominance of this "global reasoning" applied to the changing Polar Regions might also result in diminishing interest in the local perspectives that are needed to encourage a critical consideration of concerns that are not exclusively regarded as affecting the entire planet.[1]

The next chapter pays particular attention to the role of non-governmental theorists and strategists in order to analyse the reference points considered in formal geopolitics and to assess whether or not the dominant reasoning promoted by them can be regarded as either pushing or challenging a particular strand of reasoning in the practical and popular geopolitical discourses explored so far.

Note

1 Local concerns, however, are most often discussed in local newspapers, which were not considered in this analysis. This points to a significant limitation. In order to determine whether or not the interpretation of the Polar Regions as spaces of global significance has also been increasingly drawn upon in the local press and whether local concerns have received critical consideration in local newspapers in addition to the growing significance ascribed to the interpretation of the Polar Regions as "barometers", additional research that takes into account local newspapers in the American polar-rim states will be needed.

5   Formal geopolitics

The changing Polar Regions in
assessments by non-governmental
theorists and strategists

The entanglements outlined in practical and popular geopolitical discourses contribute to a better understanding of how representations of environmental concerns and international cooperation in the Arctic and Antarctic evolved in positions adopted by political practitioners and in newspaper reporting. But how do these representations in practical and popular geopolitical discourses resonate with academic discourses and representations? To what extent have the latter, the so-called formal geopolitical discourses, challenged or encouraged governance actions (or failures to act) in the politics of the Polar Regions as outlined in the previous chapters? Moreover, are these academic assessments entangled and do they thus contribute to transnational and interpolar dynamics in the regions' politics?

Set out in a quotation from Castree (2003, p. 429), "discourses and practices are both cause and effect of practical and formal geopolitics today", this chapter is based on the understanding that formal geopolitics significantly impacts and is impacted by discourses and thus strengthens the prioritisation and interpretation of practices. It is, of course, difficult if not impossible for anyone other than the political practitioners and journalists themselves to identify the knowledge (discourses and practices) they considered. Still, investigating the views that evolved in American polar-rim states and are referred to in formal geopolitics offers insights into the scientific discourses conducted in the countries in question. These scientific discourses matter, given that both political practitioners and journalists often cite researchers and their findings to underpin their aims and points of view – though often selectively (see also Postigo et al., 2013).

Despite the different perspectives of non-governmental theorists reflected in their interpretations of the political processes of the changing Polar Regions, their interpretations and representations often correspond and contribute to the dominance of understandings previously identified in practical and popular geopolitical discourses. Representations of the following topics in particular illustrate entanglements among practical, popular and formal geopolitical discourses:

1   The "Race for Resources" in the Arctic,
2   The Arctic and Antarctic as regions of future conflict,
3   Climate change as a trigger for the internationalisation of the Polar Regions,
4   Canada's nation-building and leadership in the Arctic,
5   Competition between the U.S. and Alaska as polar players.

5.1 Non-governmental theorists and strategists in the politics of the Polar Regions

> In a discourse approach, knowledge is political and related to power as well as to situation, and the structuring metaphors (the frame) are what guide people towards or away from cooperation. Researchers are no exception: they choose topics, among other things, according to research interests formed in and through discourse, and are guided towards or away from cooperation in relation to its framing [. . .] their positions are constructed in and through discourse in the interlinkage between power and knowledge.
>
> (Keskitalo, 2004, p. 181)

The Critical Geopolitics approach understands non-governmental theorists and strategists (academics, commentators, intellectuals of statecraft) as actors who contribute to the shaping of geopolitical ideas insofar as they provide "policy-oriented geographical templates" (Dodds, 2014, p. 41).[1] They conduct research in strategic studies institutes, policy institutions or privately and provide recommendations for political practitioners. Like Castree, as quoted at the very beginning of the chapter, Carina Keskitalo and Laura Nader emphasise that researchers have a "highly subjective position" (Nader, 2011, p. 217), at the same time as being influenced by the discourses that they are surrounded by and contribute to shaping. As other Critical Geopolitics scholars have highlighted, research programmes are additionally entangled with political agendas and the funding provided for them (see e.g. Ó Tuathail, 2006).

In the context of polar politics and on a regional level – in the Arctic Council (AC) and the Antarctic Treaty System (ATS) – researchers have contributed to the publication of scientific reports and participated in national delegations as representatives of non-state actor groups (particularly ENGOs). Specifically, in the ATS, scholars have had a very important role as – so far – only scientific associations have obtained observer status. The inclusion of scholars in these regional governance settings is also closely connected to the general significance ascribed to science in the Arctic and Antarctic and in the understanding of environmental changes in the Polar Regions. In both regions, science is further regarded as having "drive[n] the initial cooperation" (Nicol, 2015, p. 54) that led to the establishment of the AC and encouraged the negotiation of the Antarctic Treaty.

Domestically in Canada and the U.S., researchers particularly affiliated with think tanks such as the Centre for International Governance Innovation (CIGI), the Center for Strategic and International Studies (CSIS), and the Council on Foreign Relations (CFR), have published on the politics of the Arctic in recent years. Their interpretations have also been considered in governmental policy reports and newspaper articles.[2] Mirroring the significance ascribed to science in the politics of the Polar Regions, ENGOs too have increased their research activities. An interviewee highlighted, for example, that ENGOs like Pew work with "scientists around the world, to bring a more scientific basis to our work", which also facilitates Pew's being "seen more as an independent, non-partisan nor lobbying ENGO", as "a think tank" (Interview, April 2014). Similarly, the Alaska-based Institute of the North says it considers itself "a principal investigator just like others in academia would be or a government agency would be" (Interview, March 2014). Also, more generally, it is notable that, in regard to the Arctic, the growing acknowledgement "of local knowledge [has] created new partners in scientific inquiry and new publics to which science has become accountable" (Mason, 2015, p. 142).

In Chile, by contrast, scientists focusing on environmental concerns are mainly affiliated with universities in metropolises and not based in the areas affected (Blanco Wells and Fuenzalida, 2013). Moreover, researchers exploring environmental concerns in Chile are most often educated in research fields that are of particular significance for the economy, such as agricultural science, forest management and biology (ibid.). As with research conducted on Antarctica and Argentina, few studies derive from the social sciences (Interview, February 2015). Interviewees moreover stated that, in Chile, science is not a prioritised field of work, which is reflected in the comparatively small "número de científicos que nosotros tenemos por cada mil habitantes" [number of scientists that we have per thousand inhabitants] (Interview, February 2015) and the overall number of think tanks based in Argentina and Chile (139) compared to the equivalent number (447) based in Canada and the U.S. (On Think Tanks, 2018). The same discrepancy is to be found in the substantially smaller number of publications on the changing Polar Regions that derive from Argentina and Chile. Interviewees also highlighted limited access to quality education as a significant hindrance for those who wish to participate in discourses on Antarctica because it is "un tema de la élite [. . .] una discusión técnica [. . .] no solo en Chile pero en muchos países" [an elite subject [. . .] a technical discussion [. . .] not only in Chile but in many countries] (Interviews, February 2015). Overall, interviewees identified only four institutes that work on topics relating to the Antarctic in Chile: the INACH, the military, the Universidad de Magallanes, and the government of the region Magallanes and Chilean Antarctica. Of these, only researchers affiliated with the university conduct independent research. An interviewee described this as "muy pobre" [very poor] in view of "la diversidad del pensamiento" [the diversity of thinking] (Interview, February 2015). Sometimes the institutes mentioned are said to form alliances with individuals interested in Antarctica, but in general, all interviewees agreed that Chilean society does not seem to be much interested in Chile's activities in Antarctica. Likewise, with respect to ENGOs, a government official stated in an interview "nunca, nunca ha conversado yo con una ONG que quiera trabajar en la Antártica" [I have never, ever spoken with an NGO that would like to work in Antarctica] (Interview, February 2015). Against this backdrop it comes as no surprise that the information provided in newspaper articles in Argentina and in Chile on the Larsen Shelf collapses, for instance, came from foreign sources.

Similarly, in Argentina, national research activities in and on Antarctica are orchestrated not by private research institutions but by the government-run Argentine Antarctic Institute (Interview, February 2015). Unlike Chilean Antarctic researchers, many of whom are based in its "ciudad puerta de entrada" [gateway city], Punta Arenas, however, most Argentinian researchers who focus on the Antarctic are not to be found in the Argentine counterpart Ushuaia, which "no tiene una comunidad científica" [does not have a scientific community] (Interview, February 2015), but in the capital, Buenos Aires.

5.2 Geopolitical reasoning: the representation of the changing Polar Regions and their politics by non-governmental theorists and strategists

The rationale of this section, unlike that of equivalent sections in previous chapters, is not to trace how different perspectives evolved, because the accessible data do not allow the potential political impact that a specific scientific perspective (and its evolution) has had over time to be classified (political practitioners, for example, do not reveal which scientific publications they take into account). So, with the aim instead of outlining entanglements and disentanglements in the discourses under consideration, this section focuses on the question of how the positions identified in the previous chapters are reproduced or challenged by scholars from the American polar-rim states.

In dealing with the changing Polar Regions, specifically with environmental concerns and international cooperation in the Arctic and Antarctic, researchers in the polar-rim states have provided different interpretations. In particular, the *race for resources* in the Arctic was repeatedly and controversially addressed in formal geopolitical discourses, especially during the late 2000s and early 2010s. Like newspaper reporters, non-governmental theorists from Canada and the U.S. often invoked the "resource race" when referring to the potential "Arctic riches" identified by the U.S. Geological Survey, and used imaginaries such as the "last frontier" to underpin their understanding of the "tremendous economic development" (Higginbotham et al., 2012, p. 1) and the "gold rush" that they forecast would take place in the Arctic. In both countries, non-governmental theorists who challenged the understanding of the Arctic as a "hydrocarbon Eldorado" questioned the figures provided by the U.S. Geological Survey and criticised colleagues and the media for reproducing alarmist interpretations, arguing that most of the estimated resources are located in zones of countries bordering the Arctic (and that their ownership is accordingly regulated under international law, see e.g. Pelletier and Lasserre, 2012).

Particularly after the Russian flag planting at the North Pole and the Ilulissat Declaration of the Arctic 5, scholars began to use an alarmist tone in their assessments. In a much-cited 2008 Foreign Affairs article, for example, Scott Borgerson argued: "Global warming has given birth to a new scramble for territory and resources among the five Arctic powers" (Borgerson, 2008, p. 63). In a similarly dramatic way, he forecast "[t]he coming anarchy" in the Arctic, arguing that "[t]he combination of new shipping routes, trillions of dollars in possible oil and gas resources, and a poorly defined picture of state ownership makes for a toxic brew" (ibid., 2008, p. 71). John Higginbotham, affiliated with the Canadian Centre for International Governance Innovation, referred to "strengthened governance [a]s the key to containing chaos and achieving order in the New Arctic" that is affected by "the 'great melt'" (Higginbotham, 2013a, p. 2) and the analyst Roger Howard (2009), who also works as a journalist, spoke of "the Arctic gold rush" in his book of the same name.

In order to enforce the representation of a race for resources taking place in the Arctic, some academics seem to deliberately ignore facts that challenge these alarmist interpretations. For example, even though the Arctic 5 had prominently stated their intention to stay committed to the United Nations Convention on the Law of the Sea a year earlier, Howard speaks about "disputes as to whom the resources belong because the ownership of the Arctic region is unclear" (2009, p. 45). This interpretation has been continuously challenged, for instance by Tavis Potts and Clive Schofield (2008) or by Jean-François Pelletier and Frédéric Lasserre, who all question the most often cited estimates provided by the 2008 U.S. Geological Survey. They explicitly warn that "[o]verestimated reserves might also mislead governments" and criticise "[p]oliticians and the press [who] often forget to point out that almost all of these reserves (95%) would find themselves inside non-contestable exclusive economic zones of countries bordering the Arctic" (Pelletier and Lasserre, 2012, p. 554f.).

Although in the more recent past, fewer scientific articles have been published that use such alarmist forecasts, as an article published in 2012 by (Captain) Melissa Bert, who is affiliated with the Council on Foreign Relations, exemplified, even controversial and often challenged understandings of the Arctic continue to be reproduced. Bert (2012, p. 6) states that "[t]he race for resources is on" and refers to the estimates of resources "in the last frontier" provided by the U.S. Geological Survey when urging the U.S. to "[j]oin [. . .] Arctic neighbors in the exploration for and extraction of oil, gas, and rare earth minerals in the U.S. portion [that] could provide an economic boom to the failing economy of the United States" (Bert, 2012, p. 6). In their book *The Scramble for the Arctic. Ownership, Exploitation and Conflict in the Far North* (2010) Eugene Potapov and Richard Sale (2010) forecast that the world energy demand and expanding economies of "developing nations [such as] China, India, Brazil" will "[i]nevitably [turn] the eyes of the developed world [. . .] towards their northern boundaries" (Sale and Potapov, 2010, p. 179).

5.2.1 Discursive entanglements: the resource race and regions of conflict

As in practical and popular geopolitical discourses, in formal geopolitics too, the much-cited imaginary of the "Arctic resource race" is closely entangled with the representation of the Arctic as *a region prone to international conflicts* or a region "in crisis" (Byers, 2010, p. 137). Even though no interstate conflict has arisen in the Arctic since the end of the Cold War, non-governmental theorists increasingly referred to an "Arctic conflict" after the Russian flag planting and submissions to the CLCS. They describe these practices as "some kind of land grab" (Potts and Schofield, 2008, p. 158) and have often used the terms "dispute" and "conflict" interchangeably to empower their description of the Arctic as a region of future hostilities. Rob Huebert, for instance, refers to the status of the Northwest Passage and the boundary delimitation of the Beaufort Sea as "two current disputes" and then identifies "three other potential conflicts involving the determination of the outer limits of the polar continental shelf that are still percolating, and they may or may not develop into diplomatic disputes" (Huebert, 2007b, p. 10).

As they previously criticised the "Arctic Resource Race", scholars in the 2010s have often attacked this "overstated" potential for conflict and the reproduction of this imaginary in "media narratives" that they characterise as being based on "flawed knowledge and the persistence of bipolar narratives" (Scassola, 2013) and "alarmist accounts [and] implausibility" (Welch, 2013, p. 3). Even Huebert (2013, p. 195) castigated the depiction of the Arctic as a place of either cooperation or conflict for being "too simple a dichotomy". Even so Huebert, who is known for "consistently fram[ing] twenty-first century Arctic dynamics through a threat narrative" (Lackenbauer and Manicom, 2013, p. 4), illustrates the fact that, despite a growing sensitivity with regard to wording, the understanding of the Arctic as a region of potential future conflict has continued to circulate in formal geopolitical discourses. Robert W. Murray too (2012, p. 18) emphasised that, although even during the Cold War no "hard power conflict" arose in the Arctic, he still regards the region as being shaped by "increased security competition and [being] a potentially conflictual region in the future" (p. 18).

As also cited in the 2013 Arctic Strategy released by the U.S. Department of Defense (cf. Chapter 3), other non-governmental theorists such as Whitney Lackenbauer and James Manicom (2013, p. 7) have emphasised the power of this "conflict imaginary" when arguing that the "vigorous academic and media debates, and hyperbolic rhetoric over boundary disputes like Hans Island and the status of the NWP" have given rise to "acute concerns about Canadian sovereignty". In collaboration with their colleague Phil Steinberg from the United Kingdom, Jeremy Tasch and Hannes Gerhardt (2015, p. 17) have also argued that it is not the interactions between states bordering the Arctic Ocean that shape the representation of a contested Arctic but that "contestation is occurring within and between imaginaries". Many academics such as Tavis Potts and Clive Schofield (2013, p. 437) have accused the media of promoting a "rather misleading vision of the Arctic as a potential or indeed likely zone of conflict". Likewise, Oran Young's assessment (2011, p. 193) was that "resource wars and even armed clashes in the Arctic during the foreseeable future may make headlines, but they are more alarmist than alarming".

Instead of representing the Arctic as a region of future conflict, scholars writing during the 2010s began to emphasise more often that policy making was encouraging cooperation there. Heather Conley and Jamie Kraut (2010, p. 18), for example, considered "Canada's dual-track strategy of diplomacy and defense" as a policy that "has eased concerns of armed conflict and contributed to constructive engagement within the framework of international governing institutions". Lackenbauer and Manicom (2013, p. 18), meanwhile, have recommended more transparency to challenge the imagery of rising conflicts in the Arctic,

> Canada should develop a clear message that clarifies its Arctic agenda, indicates opportunities for cooperation and collaboration in science and economic development, and corrects misconceptions about Canada's position on sovereignty and sovereign rights in the Arctic.

Brigham (2010) pointed out that shared governance and scientific cooperation are "at record levels in the Arctic today" and more intense than "at any other period in history" and thus contribute to diminishing the potential for conflict. From a state-centric perspective, he also viewed the development status of the countries involved in the Arctic (leaving it unclear how development was to be defined) as another factor weakening Borgerson's understanding of the Arctic as a region characterised by likely future conflict.

Notably, with regard to the Antarctic too, academics such as Jeanette Irigoin Barrene (2011, p. 2) have promoted the understanding that shared governance and scientific cooperation (as exemplified by the signatories of the Antarctic Treaty) reduce the likelihood of an escalating conflict in Antarctica, a prospect that was – as in the Arctic – being increasingly addressed by scholars in the late 2000s and early 2010s. Twenty years earlier, in line with scholarly assessments of the Arctic, non-government theorists had discussed the potential for future conflicts in Antarctica controversially: Jorge Alberto Fraga (1992, p. 118), for example, used the understanding that "[e]l futuro de la Antártida es incierto" [the future of Antarctica is uncertain] as a stepping stone to recommend support for national programmes undertaken by South American countries. Calixto Armas Barea and Juan Carlos Beltramino (1992, p. 47), on the other hand, emphasised the contrary, observing that "mucho hielo se ha esparcido sobre las reclamas territoriales" [a lot of ice has grown over the territorial claims]. They explained this shift by pointing to the less state-centric positioning by "los Estados territorialistas" [the territorialist states] such as Chile and Argentina, encouraged by their shared aim of protecting Antarctica as "una reserva natural" [a nature reserve].

Thus, although the representation of future conflict in the Polar Regions illustrates an *interpolar entanglement* in formal geopolitics, one needs to be aware that these representations relate to different issues of concern. Frank Klotz (1990, p. 116) regarded "the sovereignty time bomb" to be "the ultimate challenge" for the Antarctic Treaty System. David Welch (2013, p. 5) used a similar wording 23 years later when speaking about the "potential climate change time bomb" in the Arctic but emphasised that "[s]ecurity is not at stake in any meaningful sense in the Arctic, but is very much at stake because of it" (Welch, 2013, p. 2). Unlike Klotz, who was referring to political control over territory, Welch considered the Arctic to be a region of geopolitical interest because of "the threat it poses to ecospheric security" (ibid.) and used the imaginaries of the "potential climate change time bomb" and the "proverbial canary in a coal mine" (2013, p. 5f.) to encourage cooperation with respect to environmental changes.

5.2.2 Discursive entanglements: climate change and the global dimension

"The canary in a coal mine" metaphor has also been used by non-governmental theorists to promote different strategies. Scott Borgerson (2008, p. 77), for example, places the canary imaginary in the context of "planetary health", acknowledging a global dimension. In contrast to Welch, however, he uses it to demand unilateral actions. He describes, for instance, the "melting Arctic [as] a harbinger of how the warming planet would profoundly affect U.S. national security" and demands "[s]elf-preservation in the face of massive climatic change requires an enlightened, humble, and strategic response". Instead of promoting awareness of the far-reaching significance and causes of climate change, as most scholars do when comparing the Arctic with the canary (e.g. Martello, 2008), Borgerson (2008, p. 77) goes in the opposite direction: "Both liberals and conservatives in the United States must move beyond the tired debate over causation and get on with the important work of mitigation and adaptation and managing the consequences of the great melt".

Generally speaking, the "threat of climate change" has often been seen as having had "quite different political consequences in the two Polar Regions" (Howkins, 2016, p. 155): in the Arctic it has encouraged representations of the region as the site of a "resource race" and a "region of future conflict"; in Antarctica, according to Howkins, climate change has "functioned to support a political status quo that is based on the idea that it is doing 'science for the good of humanity'" (ibid.). In a similar vein, non-governmental theorists have recommended different actions. However, scholars seem to share an understanding of both Polar Regions as "barometers of global climate change" (Broadhead, 2010; Joyner, 2011; Kao et al., 2012; Lackenbauer and Manicom, 2013; Young, 2011) and agree on the significance of the polar environment for the planet when stating, for instance: "Lo que allí occura afecta a todos los países y no solo a los mienbros del Tratado" [What occurs there affects all countries, not only the members of the Treaty] (Roy, 1992, p. 102); or "la Antártica juega hoy día en el problema del cambio climático un papel semejante al que los agujeros negros juegan en comprender el destino final del universe" [at present, Antarctica is of similar significance for the problem of climate change as black holes are for an understanding of the final destiny of the universe] (Weitzman, 2008, p. 314); or "If the Arctic is the barometer by which to measure the earth's health, these symptoms point to a very sick planet indeed" (Borgerson, 2008, p. 67).

As in practical and popular geopolitical discourses, this reasoning supports those who argue in favour of the *internationalisation of the Polar Regions* – the inclusion of "new" actors, not only in scientific endeavours, but also in the governance of the regions, ranging from environmental organisations, "newly recognized indigenous experts and political voices" (Martello, 2008, p. 369), to actors from outside the affected areas. Non-governmental theorists consequently often understand the growing global significance ascribed to environmental changes in the Polar Regions as a legitimate reason for distant countries having an interest in contributing to their governance. However, non-governmental theorists have also been aware of "polar orientalists" whose emphasis on environmental concerns they regard as a cloak to hide geopolitical intentions. Lackenbauer and Manicom (2013, p. 3) remain cautious:

> The rising interest of so-called 'new actors' in circumpolar affairs, particularly China and other East Asian states, offers renewed uncertainty and the possibility of a new threat narrative. Canadian commentators have been

accordingly suspicious of East Asian intentions, despite Canada's positive bilateral relations with all three Northeast Asian states.

Most often, however, scholars have emphasised that the opening up of regional governance settings such as the AC is necessary to encourage an exchange of knowledge on global challenges, such as climate change, and increase the visibility of the Council and people's awareness of it, and – as with the ATS in the late 1980s and early 1990s – to gain greater acceptance for it. The controversial question that scholars have frequently discussed in this regard is how best to strengthen the AC and include non-Arctic states. Academics such as Andrea Charron argued (2012, p. 777), for example, that AC member states such as the "U.S. and Canada want to avoid [. . .] the Arctic Council [becoming] a UN-like body with so many priorities, working groups, observers etc." and supported this position by arguing that such an expansion of the AC might cause "the modest, but important advancements made to date [to be] lost or overshadowed". Douglas C. Nord (2016) also focused on structural and operational features of the Council "black box", the primary functions of the polity, the activities of the different subsidiary bodies, and specifically considered the roles of Canada and the U.S. in shaping its structure – though not in a systematic manner. In general, the different steps taken towards greater formalisation of the AC have appeased the numerous voices calling for an Arctic Treaty, as many non-governmental theorists consistently did during the period of investigation.

Scholars from the Canadian Centre for International Governance Innovation, for instance, claimed that a treaty would improve environmental stewardship in the Arctic and recommended that it ought particularly to "deal [. . .] with liability and compensation for pollution caused by offshore oil infrastructure" (McCallum et al., 2013, p. 2). Often, however, proponents of "an Arctic-wide treaty" (McCallum et al., 2013) neglected to define its potential geographic scope (which is also determined by other existing international instruments) and the explicit issue areas such an instrument might regulate. Likewise, the purposes attributed to such treaty differed. The U.S. Council on Foreign Relations, for example, published Scott Borgerson's article, which focuses less on environmental risks and rather promotes the idea of "an overarching treaty that guarantees an orderly and collective approach to extracting the region's wealth" (Borgerson, 2008, p. 75). The general format of a treaty is controversial in formal geopolitical discourses too. A number of academics in Canada and the U.S. assessed that an Arctic Treaty would be neither "feasible nor desirable" as the regulatory instrument for the Arctic and argued that such a treaty would conflict with the interests of Arctic states and likely remain shallow (see e.g. Young, 2012b, p. 175).

With regard to the Antarctic Treaty System, on the other hand, scholars have more often criticised the fact that the norms and measures agreed on in this framework have not been reflected and implemented at the national level. In Chile and Argentina, non-governmental theorists regarded this as being partly caused by incapacity of the part of responsible actors (see e.g. Waghorn Gallegos, 2007). To remedy this situation in Chile, Infante Caffi (2006, p. 48) recommended, for example, that national activities in Antarctica should be monitored annually from the perspective of the National Antarctic Policy and demanded the distribution of the necessary resources to institutions that maintain scientific bases in Antarctica and conduct scientific activities there (see also Bombin Sanhueza, 2009, p. 454). Her assessments thus challenged positions expressed in practical geopolitical discourses. Along with the Chilean National Antarctic Policy, Infante Caffi, however, also emphasised the significance of "los temas globales y el aporte de la Antártica al conocimiento de los fenómenos que afectan o impactan al planeta" [global topics and Antarctica's contribution to the knowledge of phenomena that affect or impact the planet] (Infante Caffi, 2006, p. 48). Others such as Juani Soledad Bombin Sanhueza (2009, p. 448) also stressed the importance of "trabajo conjunto a nivel internacional" [joint work at an international level] when promoting a multilateral project named "Clima de Antártica y Sudamérica" [the Climate of Antarctica and South America] initiated by Brazil, Chile and the U.S. during the third International Polar Year, which he evaluated as a good example of this kind of joint endeavour. In Argentina, scholars have often highlighted multilateral scientific cooperation as a benefit of the ATS (Colacrai de Trevisan, 1994). Argentina's participation in the System and the location of the Antarctic Treaty Secretariat in Buenos Aires are further seen as contributing to the country's international reputation: "Sería la primera vez que Argentina se convierte en sede de un organism internacional de estas características" [It will be the first time that Argentina has become the headquarters of an international institution of this type] (Colacrai de Trevisan, 1994, p. 344).

In line with the formal geopolitics of the Arctic, environmental concerns in the Antarctic have also been depicted as being closely entangled with international cooperation, not solely among states but also among non-state actors (even though most studies focus on state actors). Although scholars also relate this collaboration in the ATS to national sovereignty interests, they usually regard collaboration among a variety of actors positively, as encouraging a multilateral response to global problems and contributing to the diffusion of norms. In the case of Chile, however, despite the consensus principle, according to which all consultative states have an equal say in the ATS, scholars also ascribe different responsibilities to different actors. José Javier Gorostegui and Rodrigo Waghorn (2012, p. 320), for example, speak of Chile's special responsibility as an "estado de partida" (rim state) to Antarctica in terms of ensuring the protection of the environment and marine living resources, and the compliance of activities with the UNCLOS. Javier Paredes, on the other hand, addresses these points without calling them responsibilities. Instead, he describes them as Chile's "interests" that are manifested in its status as a "'bridge' country" (Paredes, 2009, pp. 138 and 146), a status that, in addition to the perception of Chile being "a key actor in the Antarctic and in the system of the Antarctic Treaty", is seen as providing "potentialities [. . .] for the development of the Antarctic activities in general" (Paredes, 2009, p. 138) and determining that Chile "es un país antártico" [is an Antarctic country] (ibid., p. 147). Paredes does also speak of responsibilities, however, as there are "responsabilidades, funciones y derechos" [responsibilities, functions and rights] that go along with Chile's status as an Antarctic rim state with competence over the "vía de acceso a la Antártica" [access route to Antarctica] (Paredes, 2009, p. 147). In this regard, and corresponding to more recent practical geopolitical reasoning, Bombin Sanhueza identifies lack of adequate equipment and the necessary number of experts to realize scientific activities and investigations in "el territorio antártico" [the Antarctic territory] as "[u]no de los puntos más débiles de Chile" [one of Chile's most significant weaknesses] (2009, p. 454) with respect to the "muchos desafíos que [quedan] superar por nuestro país" [many challenges that our country needs to overcome] (ibid.), without clarifying explicitly whether or not these challenges relate to scientific endeavours in Antarctica or beyond.

5.2.3 Discursive disentanglements: identity constructions

When formal geopolitical discourses on the changing Arctic and Antarctic are compared, the scholarly assessments considered underline the perception that the Antarctic is addressed considerably less by non-governmental theorists from Argentina and Chile. This is remarkable, as the scant attention paid to the Antarctic in formal geopolitical discourses in Chile and in Argentina conflicts with the emphasis placed on the "national Antarctic identity" in the respective countries' practical geopolitical discourses.

Non-governmental theorists in Canada, in contrast to their South American counterparts, have particularly enforced *Canada's Arctic identity* in their assessments. They often allude, for example, to Canada's national anthem, that speaks of the "True North strong and free" (Emmerson, 2011, p. 73), in order to emphasise that:

> We are a northern country. The vast expanse of our Arctic – indeed 40% of our landmass – is an integral part of Canada. [. . .] Sovereignty is Canada's number one Arctic Foreign Policy priority – an issue that is closely connected with governance in the Arctic.
>
> (Wright, 2013, p. 105)

Academics too have often stressed this particular "Canadian identity" when demanding federal investment in Canada's northern territories. Although some scholars acknowledge that this understanding of the Arctic is "associated with Canadian national identity" that is said to have "ebbed and flowed since an independent, sovereign Canada was born in 1867 [. . .] by Canadians who reside south of sixty" (Zellen, 2009b, p. 1), at present, according to Huebert (2011a, p. 809) for example, "[t]here is little doubt that Canadians see themselves as a northern people – even if the vast majority live along a narrow band along its southernmost border".

When discussing economic development, however, scholars evoke the objective of nation building to enforce a demand that is also supported by newspaper reporting, namely that "hiring policies for new industries in the Arctic [should] include Northern peoples" and not only for unskilled positions that "do not allow for advancement or growth" (McCallum et al., 2013, p. 4). Like practical and popular geopolitical discourses, non-governmental theorists often emphasise the "human dimension" in the context of the changing Arctic (e.g. Martello, 2008, p. 352). The imaginary of the "Arctic home" is also frequently used in scholarly assessments to demand the consideration and general empowerment of Indigenous peoples who "have called the Arctic home for a very long time" (Huebert, 2009b, p. viii). Again, as with practical and popular geopolitical discourses, scholars refer particularly to the Inuit when drawing attention to "the main problems of environmental management in the Inuit homeland of arctic Canada" (Jull, 1990, p. 139) and the effects of the changing Arctic on the "culture and identity in the Inuit homeland" (Alia, 2007; also Broadhead, 2010, p. 913; Shadian, 2014). Other Indigenous peoples living in the Arctic and non-Indigenous citizens, on the other hand, have been significantly underrepresented in scholarly assessments. Some academics such as John Higginbotham, moreover contribute to the romanticisation of Indigenous peoples by highlighting the fact that their "isolated communities face a range of modern social and economic pressures, while striving to maintain traditional values and occupations" (Higginbotham, 2013b, p. 1), without qualifying this contrast more explicitly nor considering that not all Indigenous peoples regard these changes as "pressures", but instead welcome changes in their lifestyle.

Similarly, some emphasise Nunavut's "geography and special identity" (Higginbotham, 2013b, p. 9) to argue in favour of "[f]ederal investment in Nunavut [which] should be seen as an effort of nation building, like the Trans-Canada Highway and the Confederation Bridge in southern Canada" (ibid., p. 10). Here, an identity derived from a place that differs in many ways from the rest of Canada is put forward as something that can be created and that should remain connected to Canada on the grounds of "the North American continent's long-term economic and security interests" (Higginbotham et al., 2012, p. 3). Others too see such interests as significant on a regional ("North American") level and similarly assess that "infrastructure development and nation building is needed in the Canadian Arctic to meet Canada's objectives" (McCallum et al., 2013, p. 1).

Canadian nation-building interests are, however, also described as being entangled with the perception of a sovereignty (and to a considerably smaller degree, an environmental) "threat to the North" (Coates et al., 2008, p. 201). The former often serves as a stepping stone for those demanding the acquisition of military equipment, such as "new ships that can operate in at least some ice conditions" to protect Canadian Arctic waters (Huebert, 2007a, p. 9). In tune with Canada's Arctic policies, scholars refer to the Northwest Passage as "a part of the Canadian identity – a nationalistic issue that has always evoked a strong emotional, even chauvinistic, response from a Canadian population not normally prone to such reactions" (Lajeunesse, 2008, p. 1040), that is "[i]ntimately linked to Canadian nationalism" (Elliot-Meisel, 2009, p. 204) and "has long been the subject of sagas and epic journeys and is part of the Canadian identity" (Zellen, 2009a, p. 91). Overall, and in contrast to the U.S., where "the Arctic is not in the minds of the US public" (Bergh, 2012, p. 2), it is often argued that "[t]he Arctic is a highly politicized issue in Canada that is closely tied to national identity" (ibid., p. 15).

Although scholars both promote and challenge Canada's identification with the Arctic, they generally support Canada's regional leadership aspirations there. Jérémie Cornut (2010, p. 948), for example, asserts that "Canada has been at the forefront in promoting environmental protection of the Arctic. Canadian policy had a great influence on other countries, and finally led to significant changes in international environmental laws". Andrea Charron (2012, p. 776), however, challenged the perception that Canadians identify and are interested in the politics of the Arctic, arguing that, despite this alleged leadership, Canadians do not know much about Arctic governance: "there is confusion among the domestic public for a variety of reasons [. . .] about the UN process and the role of the Arctic Council".

More recent formal geopolitical reasoning often builds on the perception that Canada is not fulfilling its leadership role in crucial areas of concern. Michael Byers (2010, p. 1), for example, speaks about "a major opportunity – which has not yet been seized – for Canada to lead in the creation of a truly cooperative, permanently peaceful North". Whitney Lackenbauer and James Manicom (2013, p. 15), on the other hand, reproduce the understanding of Canada's being "a leader in Arctic science", which was also drawn upon in practical and popular geopolitics. More explicitly than Byers, Lackenbauer and Manicom build on this understanding to promote international scientific cooperation, particularly with "East Asian scientists" as Korea and China have made "heavy investments in icebreakers and research stations" and science is promoted as an area that "can serve as a conduit for international collaboration, influence and confidence building" (Lackenbauer and Manicom, 2013, p. 15). Instead of reproducing a "polar orientalist" understanding of East Asia's interest in the Arctic driven by possible "resource gains", Lackenbauer and Manicom expand the line of reasoning well-known from the Antarctic region, in which experiences in scientific collaboration are means to overcome sovereignty disputes and the "enemisation" of the other.

Particularly before and during Canada's second AC chairmanship, non-governmental theorists took different positions on whether the country's regional leadership agenda should focus on domestic or regional issues of concern. The CIGI research fellow James Manicom (2013) published a noteworthy theory-driven article in which he focused on Canada as providing a model for the analysis of the relationship between domestic and international policy making in respect of the Arctic. He considers interests formulated by domestic stakeholders with regard to disputed territory in the Arctic, examines the impact of these negotiations on negotiations taking place internationally, particularly with Russia (p. 68ff.), and thereby addresses a significant gap in research on the changing Arctic, which often fails to include domestic and subnational perspectives in the analysis of international politics. Most often, however, researchers have not applied a theory-driven approach of this kind when arguing either in favour or against domestic or regional priorities in Canada's Arctic policies.

In line with the subnational Arctic policies discussed in Chapter 3, John Higginbotham (2013a), for example, emphasised that "[t]he challenge to Canada starts at home" and demanded that the Canadian government focus on "better national planning, smart new federal infrastructure investment, and improved private – public partnerships". Andrea Charron also drew particular attention to the living conditions in Canada's northern territories and called for a focus on "sustainable communities" (2012, p. 776). Monique McCallum, Nabeel Sheiban and Simone Stawicki, on the other hand, promoted the expansion of "P3's mandate to include community development programs where large infrastructure development will occur that focusses on skills-based education programs" (2013, p. 6). In contrast to Higginbotham, however, they not only understand "a mantra of nation building" as a means to meet domestic challenges, but also consider "the ambitious pursuit of Canada's domestic priorities" as an opportunity to "increase [Canada's] legitimacy and authority over Arctic matters, thereby enhancing its international reputation and improving its stature in the Arctic Council" (ibid., p. 7). Andrea Charron (2012, p. 776) also emphasised this entanglement between Canada's regional and domestic priorities in regard particularly to environmental issues and Canada's withdrawal from the Kyoto Protocol: "Canada will have far less credibility pushing for an ambitious Arctic Council agenda that focuses on the environment and northerners if it has not made any progress at home". Similarly, in view of its tenure of the AC chairmanship, Higginbotham (2013b, p. 1) urged the government to establish a "maritime and economic leadership in the 'New Arctic'" that builds on an "Arctic Maritime Corridors and Gateways Initiative" and thus addresses an issue that is closely related to Canada's identification with the Arctic, namely its sovereignty over the Northwest Passage. In view of growing human activities in the Arctic, McCallum, Sheiban and Stawicki suggested that the Canadian government should promote both domestic and regional objectives during its chairmanship, such as "the adoption of a standardized liability cap by all Arctic States" (2013, p. 5) and effective cooperation among "businesses and governments" on both levels, for instance, "infrastructure partnerships with mining companies" on a domestic level and support for "the application of the Association of Oil and Gas Producers for Observer status in the Arctic Council" (McCallum et al., 2013, p. 6).

Lackenbauer and Manicom relate the growing significance of Canada's role in the politics of the Arctic to "the idea of Canada as an 'Arctic superpower'" (2013, p. 2). They regard this leadership idea, though, as encouraged less by the AC chairmanship, and more by a domestic driver in the shape of Stephen Harper who, when he became prime minister in 2006 "initially trumpeted the idea that 'use it or lose it is the first principle of sovereignty'", trying to reach "voters with deep-seated anxieties about Canada's potential loss of sovereignty" (ibid.). This reasoning and the specific significance ascribed to Stephen Harper reappeared in practical and popular geopolitical discourses. Jérémie Cornut (2010) also identified this correlation in outlining the special interest in Canada's politics in the Arctic shown under prime ministers Pierre Trudeau (1968–1977, 1980–1984), Brian Mulroney (1983–1993) and Stephen Harper (2006–2015), highlighting that the study of Canadian foreign policy indicates that interest in the Arctic "waxes and wanes" (Cornut, 2010, p. 943). Rob Huebert, on the other hand, regarded public knowledge "about the challenges that are now emerging" (2011b, p. 824) as the determining factor in the prioritisation of Arctic issues by the Canadian government. His perception, however, contradicts the recognition of Arctic environmental concerns in popular geopolitics in the 2000s and their subordination and absence in practical geopolitics.

Particularly in view of the lack of priority given to environmental concerns, Heather Smith (2010) and Lee-Anne Broadhead (2010) criticised the combination of "two dangerously outdated [mental] maps" promoted in Canadian politics, "one affirming state sovereignty as paramount, the other identifying permanent growth as indispensable", as a result of which "politically and ecologically absurd arguments [are] presented as commonsense facts" (Broadhead, 2010, p. 924). Broadhead strongly argued against the depiction of Canada as facing military threats in the north. Instead, she emphasised the danger and scope of the environmental changes taking place in the Arctic, arguing that Canadians were on a level with "all fellow human beings, [who] are threatened by the dramatic changes taking place as a result of climate change" (ibid.). Smith similarly stressed that "[a]ll of the Arctic, indeed all of the world, will be affected by climate change" (Smith, 2010, p. 941). Confronting the entangled objectives of defending sovereignty and promoting economic growth in the Arctic, Smith explicitly criticised the "Conservative government of Stephen Harper" for "consider[ing] the melting Arctic as an opportunity for economic development" (Smith, 2010, p. 931) and "minimiz[ing] the depth and breadth of climate change impacts" (ibid., p. 937). Both Broadhead and Smith thus challenged Huebert's understanding of Canadian policy making in the Arctic as inspired by issues of public interest. They further protested against the depiction of Canada as an environmental leader in the Arctic (as popularised in the 1990s) under the Harper government and beyond, arguing that, like previous U.S. administrations, "previous Canadian governments were not great climate change leaders either" (Smith, 2010, p. 939).

Although climate change was increasingly considered in Canada's national and subnational Arctic policies in the late 2000s and early 2010s, in practical geopolitics it remained subordinate to the potential and interest ascribed to economic development in the Arctic. Academics such as Smith assessed this prioritisation in "the Conservative discourse on the Arctic" as "naïve and dangerous", arguing that "[c]limate change impacts will not be linear and are not limited to melting sea ice" thus restricting any attempts to "control the environment" (Smith, 2010, p. 941). Smith also underlined this understanding, by using the imaginary of the "Arctic home" ("[t]he Arctic is more than a map of potential resources; it is someone's home", ibid., p. 942) that has often been applied in popular and practical geopolitical discourses too. Smith, however, uses this imaginary to criticise the positively connoted self-understanding of Canadians:

> So when the Conservative government tells us we will care for the Arctic for all of humanity, perhaps we should question the credibility of this statement. We are not taking care of the Arctic for all humanity; we are taking care of the Arctic for ourselves. We are protecting our consumptive lifestyles and turning a blind eye to the juggling game in which we are involved.
>
> (Smith, 2010, p. 942)

In contrast to the formal geopolitical reasoning promoted with regard to regional cooperation, which is said to be encouraged particularly by scientific cooperation and shared environmental concerns, the bilateral cooperation between the U.S. and Canada (the "natural partners in the Arctic", Higginbotham, 2013a) is said to be based on "a unique framework of bilateral treaties and pragmatic arrangements" as well as "the foundation of a shared northern border and unique historic, economic, military and people-to-people ties" (Higginbotham et al., 2012, p. 8). In both cases, differences in the respective domestic Arctic policies are perceived as influencing multilateral cooperation in the AC. Spence, for instance, argued that, while Canada focused on economic development for the people of the North during its chairmanship, the U.S. "demonstrated a similar interest in its domestic audience by putting climate change front and centre" (Spence, 2015, p. 4). This prioritisation of domestic concerns has been criticised by scholars who highlight the "erratic shifting of the council's priorities" while demanding that "serious effort must be placed in advancing a strategic discussion about a vision for the Arctic region and the role that the council can play to achieve it" (ibid.).

In line with practical geopolitical discourses in other American polar-rim states, in formal geopolitical discourses in the U.S. non-governmental theorists often represented the U.S. as an Arctic nation "by way of its Alaskan coastline" (Conley and Kraut, 2010, p. 7). Particularly in the 2010s, with the growing significance ascribed to the Arctic in U.S. foreign policy (until the 2010s, the U.S. was often perceived as "ha[ving] forgotten in the past that it is an Arctic nation"; Charron, 2012, p. 789), assessments from non-government theorists increasingly addressed the *relationship between the U.S. and Alaska in the politics of the Arctic*, a relationship that has been shaped by competition.

Like their counterparts in Canada, for example, U.S. non-governmental theorists demanded investment in infrastructure and development projects to increase the "ability of Alaska and US interests to navigate areas with ice" (McBeath, 2013, p. 6), emphasising that Alaska is "the only Arctic state of the United States" (ibid., p. 1). Scholars also frequently reinforced Alaska's "nostalgic, romantic imagery" and the "images of expansive, undeveloped land and abundant wilderness and wildlife" (Hogan and Pursell, 2007, p. 63) by referring to it as the "last frontier" (Coates, 1989, p. 1). Domestically, this representation was often entangled with economic development, particularly since "the region's lack of development" provided "an opportunity for investment" (ibid.). In this regard, some also claimed that the Arctic had "come to be an accepted part of national identity" (Emmerson, 2011, p. 74). Others, however, stated that Alaska remains "remote and exotic for Americans" as "a place – if thought about at all – to go on a once-in-a-lifetime cruise vacation that will allow them to witness both the American symbol (the bald eagle) and America's wilderness" (Hitchins, 2011, p. 974).

As in practical and popular geopolitical discourses concerning the Arctic, academics often focus on the benefits and challenges of economic development in Alaska. Andrey Petrov and Philip Cavon (2013, p. 348), for example, highlighted the fact that the "state of Alaska is an example of northern economy that is largely dependent upon the petroleum industry", that is shaped by "the weakness of its internal economic capacities and institutions that indicate potential long-term economic difficulties" and, similar to "many other frontier economies, is under perpetual threat of cataclysm associated with resource bust or federal budget cuts". In this regard, Alaska-based scholars, such as Diddy Hitchins (2011, p. 976), have highlighted "the impotence of Alaskan interests", as in the failure to open the Alaska National Wildlife Refuge (ANWR) for oil drilling by the federal government, when arguing that "despite more than a decade with Republican majorities in both house and senate and a pro-development administration, Alaska was quite unable to obtain the necessary majorities to gain passage". The perceived domination by "the lower-48" is also reflected in the "often-hostile rhetoric from leaders in the United States' farthest north frontier" (McBeath, 2013, p. 1). Opponents of the ANWR opening repeatedly stressed the imaginary of Alaska's wilderness and emphasised its pristine environment – an imaginary that non-governmental theorists applied to the Arctic when describing it as a region characterised by "vast, unchanging and timeless wilderness, undisturbed by humans, in which harsh environments challenge the ability of plants, animals and people to survive and flourish" (Nuttall and Callaghan, 2000, p. xxv). Proponents of the ANWR opening, on the other hand, urged people not to forget that "Alaska has long been inhabited" (Haycox and Mangusso, 1996, p. xvii; also Huebert, 2009b, p. viiii), and non-governmental theorists frequently asserted that "Aboriginal communities need to be engaged and consulted in economic development that directly affects and could benefit them" (Higginbotham et al., 2012, p. 6).

While in the 2010s Alaska's say in domestic politics of the Arctic is more often perceived as being dominated by national interests, scholars have often pointed out that, at the regional level and particularly in the 1990s, "Alaskans had participated eagerly in the opening of relations around the Arctic after the Cold War" (Hitchins, 2011, p. 975). Alaska, therefore, is described as not only having "sought trade and investment ties with East Asian nations" (McBeath, 2013, p. 1), but also engaging with non-state actors such as the Northern Forum to "bring [. . .] about economic development and the improvement of communications infrastructure" (Hitchins, 2011, p. 975). These representations strengthened local positions outlining how Alaska contributed successfully to transnational cooperations at a time when new forms of cooperation, such as the creation of "effective public – private partnerships" (McCallum et al., 2013, p. 1) were often being recommended as solutions for the challenges that the U.S. was facing in the Arctic, such as "the cost of securing the Arctic. At a time when the American military budget is facing large cuts, finding room for polar resources will be difficult" (Pincus, 2015, p. 167).

Also, more generally, climate change effects were often regarded as having "dramatically altered the landscape of the Arctic and heightened concerns about the region's future", which have been seen as providing "an urgent imperative for the United States to adapt its maritime strategy and posture" (Conley and Kraut, 2010, p. 7). Though in the 2000s "the administration continued to resist the idea of human responsibility" for climate change (Hitchins, 2011, p. 975), in the 2010s non-governmental theorists recognised a change that also strengthened "calls for cooperation" between the U.S. and Canada (Charron, 2012, p. 780). Scholars, for example, more often recommended the ratification of UNCLOS to avoid the U.S. finding itself "in a growing strategic disadvantage in shaping future policy vis-à-vis the Arctic" (Conley and Kraut, 2010, p. 26). They also suggested cooperation particularly with the Arctic 5 "in the fields of environmental clean-up and remediation, as well as search and rescue" (see also Bert, 2012, p. 9) and with "other nations interested in the region as a whole" based on an institutional framework (Conley and Kraut, 2010, p. 26).

Despite these fairly specific suggestions made by academics, however, Jerry McBeath (2013, p. 2) assessed "[t]he strategy announced by President Obama in May 2013 [to be] predictably general and vague". Scholars were critical of the fact that the U.S. "remains in a holding pattern" and "lacks a focused strategy, forestalling its own progress in the region" (Bert, 2012, pp. 9 and 16). Moreover, despite the greater focus on climate change responsibilities claimed by the U.S. government, in 2013 scholars still found that "[t]oday, U.S. Arctic policy is increasingly shaped by economic factors, primarily concerning oil, gas, and mineral resource development" (Conley et al., 2013, p. 1). Instead of actively contributing to changes affecting priorities set in the politics of the Arctic, the U.S. has thus been seen as aiming to "keep things largely as they are" (Hough, 2013, p. 25). In this regard, scholars also criticised the "lack of public discussion" (Pincus, 2015, p. 167).

To sum up: Following the approach of Critical Geopolitics, the interpretations provided in formal geopolitics contribute to considerations promoted by political practitioners in policy making and by journalists in newspaper articles. The reasoning, ideas and recommendations summarised under formal geopolitics do have an impact on policy making, political attitudes and public opinion. Science, however, is not a neutral distributor or corrector of information but is also formed by "the social", which shapes scientific processes through interaction and negotiation.

The controversial assessments provided by academics from the American polar-rim states in formal geopolitical discourses on the changing Polar Regions mirror this understanding of science. Particularly the examples discussed earlier illustrate how some scholars seem to deliberately ignore facts to promote a particular line of reasoning and how challenged (and in some cases alarmist) interpretations of the Polar Regions continue to be reproduced. Although much criticism has also been expressed in this regard by academics themselves when emphasising that the task of scholarly assessments is to reflect and not to reproduce (and in some cases even denigrate) recurring patterns of interpretation known from practical and popular geopolitics, a lack of systematic and theory-driven research, in particular, often seems to have facilitated entanglements among practical, popular and formal geopolitical reasoning of this kind.

As scientific exchanges do not take place in "national containers" but are more often international in scope, a circulation of such interpretations in formal geopolitical discourses has further been facilitated. In this regard, the dominance of assessments from Canada and the U.S. may help to explain interpolar and inter-American entanglements. In the examples discussed earlier, even those scholars who address different issues in the Arctic and the Antarctic use a similar wording (e.g. in relation to climate change- and sovereignty challenges) or similar patterns of interpretation to emphasise national identities.

Notes

1 Theorists and strategists at governmental agencies such as Environment Canada, the National Oceanic and Atmospheric Administration in the U.S., the Chilean Antarctic Institute (INACH) and the Argentine Antarctic Institute (IAA) provide knowledge that is in different ways considered in negotiations on governance decisions to act or not to act. Their interpretations are, however, classified as practical geopolitical reasoning when following the approach of Critical Geopolitics and are accordingly not considered in this chapter.
2 The 2009 Canadian Coast Guard Report "Rising to the Arctic Challenge", for instance, includes explicit references to non-government theorists such as Rob Huebert from the University of Calgary and Michael Byers from the University of British Columbia (Canadian Standing Senate Committee on Fisheries and Oceans, 2009, p. 3). The report also cites Scott Borgerson from the Council on Foreign Relations, which is headquartered in the U.S. (ibid, p. 25) and thus illustrates the consideration of knowledge "across national borders".

6 Conclusion

Entangled/disentangled actors and discourses in the politics of the Polar Regions

This book has revolved around the question: "Why have certain patterns of interpretation become dominant in the politics of the Polar Regions?" Based on the understanding that policy making that concerns the Polar Regions is embedded in political contexts that influence the negotiation of policies and actions to manage the changing Arctic and Antarctic, this book has focused on the analysis of these political contexts by investigating entanglements in

1 the representation of different issue areas in practical, popular and formal geopolitics,
2 interpretations introduced in national and regional contexts in and across both Polar Regions and their recurrence over time.

This entanglements-based perspective opens up the different contexts in which policy making on the Polar Regions is embedded and allows the ordering of different interpretations in the politics of the Polar Regions to be investigated.

In order to better understand that ordering, this book has examined

1 the inclusion and political say of state and non-state actors in policy making that concerns the Polar Regions,
2 the representation in newspaper reporting and scientific studies of environmental concerns and international cooperation in national and regional policy making, together with the overall significance ascribed to them, and
3 the extent to which these representations correspond, conflict and have changed in the American polar-rim states since 1989.

The book has further indicated how these interpretations are bonded to specific geopolitical reasonings (such as an understanding of Antarctica as common heritage, of the Arctic as a global region or of either as a place to which sovereignty rights apply). Despite the systematic consideration of numerous factors and different dimensions that the entanglement perspective brings to light, however, the central findings discussed next cannot, of course, provide a "complete" answer to the question under analysis, as even if general positions guide actions, actors do not always follow a utilitarian principle in policy making and discourses are also influenced by the occurrence of sudden and unexpected events (as references to the *Exxon Valdez*, *Bahía Paraíso* and *Deepwater Horizon* spills in the discourses under analysis have exemplified). In that sense, like the Critical Geopolitics approach, the entanglements perspective is limited in its power to explain the overall contexts that guide policy making. It allows one to illustrate, however, how the ordering principles evolve and operate that ultimately affect and guide policy making and governance structures. The polar entanglements identified in this book (among actors, discourses and representations) and the analysis of the contexts in which these entanglements are embedded thus help to explain past and recent dynamics in political processes that concern the Polar Regions and contribute to a more comprehensive understanding of the processes that shape their politics.

6.1 Changing political influence? State and non-state actors in the politics of the Polar Regions

The Arctic Council (AC) and the Antarctic Treaty System (ATS) evolved at different points of time, which impacted the institutional structures of both, most notably their formal implementation as an intergovernmental forum and a treaty system respectively. They are the most important regional institutions shaping the politics of the Polar Regions and, despite their differences, over the course of time they have developed similar strategies to cope with challenges. Growing scepticism concerning their legitimacy, for instance, contributed to an "opening up" of both. Nowadays, despite the "polar orientalist" understanding of non-polar-rim states and non-consultative parties (that has recurred particularly in countries with polar sovereignty claims such as the American polar-rim states), "outside" states are more often included as observers and parties in both settings.

The rapid and complex environmental changes that have been increasingly ascribed to climate change since the mid-2000s (previously, environmental concerns related primarily to the ozone hole and pollution in the Arctic and Southern Oceans) as well as uncertainty as to the scope of its effects have also contributed to this "opening up" and to the diversification of subsidiary bodies. In both cases, the understanding of environmental concerns as being of transnational significance encouraged the inclusion of non-state actors. While scientific cooperation has been the foundation on which the ATS was built, in the AC too scientific cooperation is being increasingly recognised as a means to guide policy making on the rapidly changing Arctic environment. Consideration of the "traditional knowledge" promoted by Indigenous peoples in the AC as well as the enhanced scientific dialogue among "recognised" researchers independent of their institutional and national backgrounds (as promoted by the newest AC agreement) and the knowledge introduced by ENGOs underpins the perception that, over the course of time, the political influence of non-state actors has grown in the (regional) politics of the Polar Regions. Decision-making rights and agenda-setting power, however, have remained in the hands of member states and consultative parties who are bound to the consensus principle in both regional settings. Moreover, with the exception of the recent negotiation of three binding agreements, most policy making in the AC has remained non-binding even though Indigenous peoples along with state and non-state observers have often argued in favour of regionally binding regulations.

As at the regional level, so also at the national level, the inclusion of non-state actors in policy making that concerns the Polar Regions has been generally promoted and perceived as legitimate in the American polar-rim states. With regard to the Arctic in particular, the practical, popular and formal geopolitical discourses examined in the book encouraged the establishment of permanent participant status for Indigenous peoples' organisations as early as the 1990s. In the Antarctic, on the other hand, the inclusion of non-state actors in policy making has usually been described as "intended" by the southern rim states – if it is mentioned at all. Moreover, newspaper articles published in Chile in the 1990s tended to adopt a critical stance towards the campaigning work of ENGOs who were perceived as promoting an ideology developed "outside" Chile (in "rich, industrialised, western countries") that did not acknowledge the circumstances prevailing in Chile itself because conservation would limit Chile's economic growth and prosperity.

The ENGO Greenpeace and its campaigning in the Antarctic and Arctic, on the other hand, has been widely perceived in positive terms by newspapers in all American polar-rim states. Often, for example, Greenpeace itself has been described as a "global environmental watchdog" (like the Indigenous peoples in the North who have been represented as "the frontline environmental watchdogs and police"), and its members are seen as activists counteracting "los poderosos" [the powerful] and "las petroleras" [the oil companies] and protecting the "frágil ambiente" [fragile environment] particularly from oil and gas development in the Arctic. In the Antarctic too, Greenpeace has received recognition for having encouraged opposition to the Convention on the Regulation of Antarctic Mineral Resource Activities (CRAMRA), for the establishment of the Antarctic Whaling Sanctuary and for having contributed to the growing awareness of the instability of the Larsen Shelf through its research. In all American polar-rim states, moreover, Greenpeace is represented as an ENGO that raises awareness of climate change impacts through powerful campaigning such as that conducted, for instance, by the "Arctic 30", and when arguing that the world is turning into a boiler ("una caldera") that "humans have switched on". Although the political say of ENGOs such as Greenpeace and their formal inclusion in policy making on the Polar Regions are still not much addressed in the American polar-rim states, the positive assessments referred to before show that, in the public mind, non-state actors such as Greenpeace are thought of as legitimately aiming to protect the changing Polar Regions.

Overall, the political say of non-state actors and their inclusion in the political processes concerning the Polar Regions have been growing during the period of investigation. The predominant influence of states as the principal actors steering the national and regional politics of the Polar Regions, however, explains why, for instance, national interests (particularly those that concern sovereignty and international relations) dominate their politics. Moreover, in Argentina and Chile, in contrast to Canada and the U.S., the limited number of natural and social scientists and the far fewer resources that the countries have at their disposal have resulted in information and interpretations most often not being developed there so that the two countries tend not to challenge but to reproduce dominant understandings (particularly concerning the impact of environmental changes but also the race for resources, non-applicable in Antarctica but inspired by the equivalent interpretation of the situation prevailing in the Arctic). Similarly, the number of ENGOs focusing on the Antarctic in Argentina and Chile has always been very low, which was likewise explained as being entangled with a lack of the resources (expenditures for campaigns in Antarctica are very high) and a lack of the knowledge that would allow them to challenge interpretations mostly provided by "an elite" focusing on topics related to Antarctica. These topics are, moreover, (still) not perceived as priority concerns compared to other issue areas addressed in policy making in Argentina and in Chile.

To put it succinctly: In the politics of the Polar Regions a predominance of certain understandings is enforced by hierarchical policy making steered by state actors, by a lack of access to knowledge and resources that particularly affects Indigenous peoples' organisations in the Arctic, the Antarctic-rim states Argentina and Chile and ENGOs in these countries, and by a different prioritisation of concerns in both Polar Regions and in the four American polar-rim states that is also touched upon in the following.

6.2 Entangled geopolitical discourses in the politics of the Polar Regions?

In discourses on the Arctic and Antarctic, both similar and distinct concerns have been addressed and prioritised reflecting the disparities between the two regions and their politics. However, since 1989 across national and regional levels environmental concerns and international cooperation have been the issues predominantly discussed in the AC, the ATS and within the American polar-rim states themselves. The various entanglements among practical, formal and popular geopolitical discourses in and beyond the American polar-rim states have contributed to the dominant interpretations of these themes in the politics of the Polar Regions as the recurring geopolitical imaginaries and representations in particular have illustrated.

In these discourses loaded imaginaries and catchy statements have been frequently (re)produced to emphasise central arguments that conflict with others and at the same time enforce friend–foe modes of thought. These entanglements correspond, therefore, to the observation made by other scholars focusing on Critical Geopolitics who have pointed out that the "political and media landscape [. . .] increasingly privileges rapid, polarizing, sweeping and polemical constructions of geopolitical realities, particularly in cases of conflict" (Albert et al., 2014, p. 329). Most notably the representation of the Arctic and Antarctic environment, for instance, as a "national [!] treasure" and as a "unique and pristine environmental cathedral" constitute an interpolar entanglement in the discourses I have examined and an understanding that is often contrasted with the representation of the Arctic and Antarctica as "our planet's last great frontiers" containing untapped resource reserves where the management of activities in a harsh "unforgiving" environment is also seen as contributing to a prestigious reputation. Resource development in the Arctic, specifically, has been described by those whose aim is to protect the environment as "ecorape" and a "threat" to life, culture and tradition, whereas those in favour of the (sustainable) "development of the North" have spoken about these same activities as means of "empowering" the people living in remote regions by providing better infrastructure, substantial oil dividends (money needed to maintain and improve lifestyles and to adapt to impacts of climate change), as a way of reducing dependence on foreign oil and gas and more generally as contributing to "nation-building".

The dominant understanding of the Arctic and the Antarctic as "shared" regions, regions of transnational significance where the protection of the environment is more often interpreted as being for the "sake of humanity at large" and determining the "health of the planet" and preserving the "planetary home" for "succeeding generations" stands as another interpolar entanglement in practical, popular and formal geopolitics in all American polar-rim states as well as in regional politics. At the national level, this understanding has encouraged the recurring perception of an escalating conflict as shown by, for instance, the controversy surrounding the opening of the Arctic National Wildlife Refuge in Alaska. Not only the disputed "facts" but also the interpretation of this controversy as a prelude to "battles" that ought to be defused through "killing the drilling" were later reproduced in the Arctic to underpin opposite positions particularly after the 2007 planting of the Russian flag at the North Pole. Popular geopolitical discourses in all American polar-rim states referred to the latter incident as a clear sign of a "resource race", as "la carrera por riqueza del Ártico" [the race for the Arctic's riches] and "una carrera geopolitical" [a geopolitical race]. "Russia's appetite for Arctic riches" was further interpreted as bringing about the revival of a "military theatre" in the Arctic and fuelled similar interpretations of the situation in the Antarctic. At the same time in newspaper reporting in Argentina and Chile, for example, representations such as "la conquista de la Antártida", "puja por recursos energéticos", and "las luchas por los recursos naturales" [the conquest of Antarctica; the battle for energy resources; the struggles for natural resources] circulated again despite treaty provisions that regulate and prohibit mineral resource development at least until 2048. Of the formal and practical geopolitical discourses in the American polar-rim states, some upheld and others challenged these representations and related interpretations, particularly by describing resource forecasts as "overestimated" or representations of emerging conflicts as "alarming" or "misleading" and likely to undermine cooperative efforts and diplomatic dialogue more generally.

Polarising representations of environmental changes in the Polar Regions related particularly to their being understood as "local/national" or as "global" concerns. Representations of the Arctic and Antarctic as the "planet's air conditioners", for example, or "canaries in the coal mine" or "barometers for climate change" have often been introduced to encourage negotiations on regulations to protect the environment. At the same time, such representations have also been perceived as enforcing an understanding of the Polar Regions as "global commons", neglecting the fact that the Arctic is not uninhabited and ignoring the governance frameworks that apply to both Polar Regions. In this regard, geographic proximity and sovereignties are the main factors emphasised by those who demand local, national and regional solutions and construe global measures as leading to colonisation and domination of the "primeras víctimas" [first victims] by "outside countries".

Also because of growing awareness at national level of the global impacts of environmental problems, all American polar-rim states emphasise the protection of their sovereignty in practical geopolitics concerning the Polar Regions while at the same time promoting multilateral cooperation (even among countries with controversial sovereignty claims, as with Argentina and Chile in Antarctica, and Canada and the U.S. in the Arctic). Particularly the representation of an "Arctic" and of an "Antarctic" identity (and the related understanding that the Polar Regions "matter" for the countries under analysis, underlined by imaginaries such as "the North as our home", the "Indigenous homeland", and "gateways to Antarctica") has often been promoted in practical geopolitics to underpin sovereignty claims. The construction of such "polar identities", however, was disputed in popular and formal geopolitical discourses in the American polar-rim states, for example, when newspaper articles and scholars described the Polar Regions as not being of significance to the majority of the citizens of these states.

In regional settings and the American polar-rim states alike, references to the geographic proximity of the states to the poles are most often used by those who emphasise their "Arctic" or "Antarctic" identity. Both factors are stressed in all American polar-rim states and in practical, popular and formal geopolitics by actors who wish to strengthen their political say by claiming a special responsibility and leadership role in shaping developments in the Arctic and the Antarctic. This pattern of interpretation is also reproduced by Indigenous peoples who perceive themselves as "rightsholders". In practical geopolitical discourses at regional levels, both factors are further picked up to enforce a "spirit of cooperation" among the different actors engaged in the AC and the ATS, which is also encouraged by family analogies (in allusions to such things as the "fraternal spirit" and "hermanidad" [brother/sisterhood]) and references to "the very best tradition" that recurred in both the AC and the ATS. Indeed, in the cases of Argentina and Chile, this understanding of a shared geographic proximity encouraged cooperation between Latin American and other Antarctic-rim states ("América se une en la Antártica" [America unites in the Antarctic]) who are viewed as having a more legitimate say in Antarctica also because of their historical connections.

Accordingly, environmental changes were not the primary drivers of this growing cooperation over Antarctica but the sense that cooperation is necessary to defend sovereignty rights that are based on geographic proximity. Geographical distance, by contrast, is seen as a hindrance to finding common solutions in the Arctic as in the case of the U.S., for instance, where different perspectives (and priorities) concerning the AC chairmanship agenda and regional cooperation are to be found at the federal level and in the State of Alaska. Geographic relationships with the Arctic and the Antarctic are thus stressed as vital factors underpinning political influence – particularly at times when growing significance is being attached to the changing Polar Regions by "outside actors". Moreover, in the discourses under analysis, a very significant role is ascribed to scientific findings that are seen as the means to provide evidence for "sovereignty" (for example through seabed mapping). At the same time, scholarly analyses also challenge sovereignty claims when they highlight the effects that environmental changes in the Polar Regions will have on "outside" actors too. Some, then, regard scientific cooperation as a means of defusing the "sovereignty time bomb", while others expect science to counter the "potential climate change time bomb".

Overall, despite the growing awareness of environmental concerns and climate change impacts in the discourses under analysis, most of the geopolitical reasoning has been sustained throughout the period of investigation and illustrates that policy making in the Polar Regions is strongly entangled with global, national and local politics as the reasoning drawn upon is represented, inspired and reinforced across these levels. Consequently, anyone who aims to change the direction of politics in the Polar Regions must consider political processes on these different levels instead of ascribing a general position to one actor or actor group. Reflecting on the positions with respect to different issue areas and prioritising them differently might thus solve conflicting positions between states and non-state actors. Moreover, in the discourses under analysis transparency was represented as a means of maintaining and strengthening the "spirit of cooperation" often emphasised in the politics of both Polar Regions. While transparency in (strategic) policy making more generally is likely to be a political challenge that is quite difficult to address, in scientific cooperation it was described as also requiring the "translation" of scientific results from a rather technocratic expert-language (so-called "scientese") into plain terms in order to inform wider publics who would then be better enabled to contribute to policy making.

6.3 From changing Polar Regions towards a new inter-American political space?

The politics of the Polar Regions are not only shaped by polar entanglements but also by inter-American ones. This is illustrated (1) by the growing cooperation among actors from the American Arctic- and Antarctic-rim states not only across both regional settings but particularly in neighbouring countries and (2) by the reproduction of similar patterns of interpretation in the discourses under analysis.

6.3.1 Cooperation

Regarding cooperation, specifically with respect to Antarctica, Argentina and Chile have further intensified multilateral cooperation in Latin America to meet shared logistical and scientific needs, on the one hand, and to develop a shared understanding in policy making, on the other. Canada and the U.S. too have been identified as important partners with whom Argentina and Chile have collaborated in scientific programmes and in maintaining scientific bases in Antarctica. Still focusing on Antarctica, inter-American collaboration among non-state actors is often shaped by a dependency of ENGOs from Argentina and Chile on organisations that are mostly headquartered in the U.S. This dependency is caused not only by the difficulty that Latin American organisations have in accessing funding and resources, but also because (even though it focuses exclusively on the Antarctic) the Antarctic and Southern Ocean Coalition, an ENGO alliance, has its headquarters in the U.S. and retains a formal status that enables it to participate in the ATS. By contrast, actors from Argentina and Chile are mostly absent from policy making in the Arctic as their infrequent participation in Arctic Council meetings and missing bilateral collaboration with the U.S. and Canada in the Arctic illustrate.

Unlike other regional organisations such as the European Union, the Organization of American States (OAS) has paid little attention to the changing Polar Regions so far. This is particularly remarkable for at least three reasons: (1) the OAS is a regional entity to which influential Arctic states belong and one-third of all OAS member countries have adopted the Antarctic Treaty and form part of the ATS; (2) the OAS has been approached by non-state actors with specific interests in the Polar Regions; and (3) discourses conducted at the OAS on environmental concerns, climate change and territorial disputes relate to those conducted under the auspices of the AC and ATS.

It remains to be seen whether or not the American polar-rim states will be encouraged by, or will encourage, other OAS members to include the Polar Regions in future negotiations on political actions that concern the Arctic and the Antarctic. It is already clear, however, that joint approaches are more likely to be discussed, particularly with regard to Antarctica, as the Antarctic-rim states Argentina and Chile have already brought about inter-American cooperation in this regard. With respect to the Arctic, on the other hand, it is more likely that Canada and the U.S. will try to keep discussion of Arctic concerns restricted to Arctic-specific settings. Canada and the U.S. clearly favour the AC as the premier forum for policy negotiations. Likewise, although the exchange of scientific knowledge that has been promoted encourages cooperation with "outside actors" more generally, they envision this kind of interchange being strengthened particularly with other Arctic-rim states. For other American states to be included in Arctic politics, much will depend on their prioritisation of environmental concerns and the interrelationships that they stress with the Arctic, which could provide them with a stake in the region that is increasingly perceived to be legitimate due to the growing awareness of global warming and climate change.

6.3.2 Similar patterns of interpretation

In the discourses under analysis, similar patterns of interpretation, views and imaginaries have been introduced at regional and domestic levels in the American polar-rim states that have contributed to dominant understandings of the changing Arctic and Antarctic, for instance, as places affected by growing internationalisation and a "resource race". These views have also contributed to the promotion of a more "global perspective" in the politics of the Arctic and the Antarctic and their being understood as regions of "shared responsibility". Although the different (practical, popular, formal geopolitical) discourses in which these views were introduced are entangled, further research is needed to investigate whether the changing Polar Regions are contributing to the emergence of a new inter-American political space, as it is beyond the scope of this book to assess the *intensity* of entanglements. The politics of the Polar Regions and entanglements among actors and discourses are, however, also entangled with politics, actors and discourses from other more distant regions. Those are only lightly touched upon in this book by referring to the United Nations Convention on the Law of the Sea and to the Organization of American States.

6.4 Further prospects and research

Climate change is widely considered to be not only altering the environment of both Polar Regions, but their political agendas and related policy making as well. In the ATS, the Consultative Parties have already agreed to deal only with the mitigation and not with the causes of climate change. Similarly, for all the attention that climate change and its impacts on the Arctic and beyond have recently received in the AC, it is unlikely that "responsibilities" or measures to limit the causes of climate change will be negotiated under the Council's auspices in the future. The discourses examined in this book have evidenced the ever controversial nature of positions in this regard (most notably illustrated by the criticism expressed of the long "reluctance" of Canada, Russia and the U.S. to adopt appropriate international frameworks, which led to their being represented as causing the South to "turn into hell" or being responsible for "el mundo convertido en un infierno sin ozono" [the world being turned into an inferno without ozone]. Even if the causes of climate change were to receive more attention in AC policy making, and the Council were perhaps even encouraged to negotiate agreements someday in the future, these would likely be shallow. It is even more likely that conflictive positions would impact cooperation in other issue areas. Against this backdrop, Arctic Council member states will probably continue to avoid any escalations in conflict areas, which – as Humrich and Wolf (2012, p. 19) point out – could cause the collapse of "not legally institutionalized" cooperative efforts. Controversial issues that are not considered to be of significance exclusively for the Arctic region are thus expected to continue to be dealt with elsewhere. Controversies among AC member states relating to processes in other regions have not yet impacted policy agendas in the Council (as I show elsewhere with respect to the Ukraine crisis, see Wehrmann, 2017). However, like the "Antarctic factor in international relations" identified by Beck (1990), the AC has clearly been used as a platform for "warning": Canada, for instance, officially censured Russia's intervention in Ukraine there. Against this backdrop, the consensus principle is also likely to maintain in both governance settings as it serves as a "double bottom" to secure cooperative efforts even in cases of controversial positions that concern "outside" conflicts.

Nevertheless, climate change and the related environmental changes in the Polar Regions will remain a prominent topic in both. The U.S. chairmanship of the Arctic Council that prioritised the "global Arctic" ended in May 2017. Finland succeeded to the chairmanship and has since then obtained a special agenda-shaping power (cf. Wehrmann, 2016) that has enabled it, amongst other things, to focus on the implementation of the COP-21 Agreement and the UN Sustainable Development Goals (SDGs) "as part of Arctic cooperation" (Ministry of Foreign Affairs of Finland, 2017). The mitigation of climate change remains a topic of particular significance in the Arctic Council. However, as the prioritisation of the SDGs also illustrates, it is expected to be related to economic questions. In this regard, the Finnish Ministry of Foreign Affairs has already emphasised that in "issues related to the economy, close collaboration with the Arctic Economic Council will be sought", although the Council is still a controversial institution among stakeholders in the Arctic who are either excluded from participating in it or demand more transparency from it.

While the prioritisation of climate change and sustainable development underpins Knecht and Keil's observation of the "coexistence of Arctic regional and Arctic global concerns" and the strengthening of the "global Arctic paradigm" (2017, p. 308f.), the entanglement and Critical Geopolitics perspectives applied in this book make it clear that to avoid political disputes it will be important not to lose sight of (seemingly) more local concerns and knowledge (that are, in fact, also applicable in other places), which are crucially needed in any endeavour to truly transform policy making in the Polar Regions from a "top-down" to a "bottom-up" steering strategy that also includes non-state actors. As with the AC, the institutional structure of the ATS places decision-making rights with respect to the Antarctic exclusively in the hands of states and it is in any case a less open framework for the inclusion of non-state actors. However, for far longer than in policy making in the Arctic Council (where the Agreement on Enhancing Scientific Cooperation was only recently negotiated), the prioritisation of scientific cooperation in Antarctica has provided opportunities for non-state actors to introduce and share their knowledge (once they are accepted as offering valuable insights and resources). Since scientific knowledge is considered to guide policy making in the politics of the Polar Regions, particularly with regard to climate change, non-state actors may come to contribute significantly to the ordering of priorities (particularly in regional policy making). ENGOs have already realised that this opportunity exists and are increasingly trying to seize it by stressing their scientific expertise and transforming and enhancing their institutional profiles. In this regard, however, the cases of Argentina and Chile illustrate that particularly smaller ENGOs (and other non-state actors) with more difficult access to external funding are kept out, especially if they are based in countries that do not belong to the group of leading industrialised countries. During times in which "shared responsibility" and the "global dimension of climate change" are being stressed, this perpetuation of inequality that is entangled with the means of participation in policy making increases the likelihood of conflicts.

With this in mind, further research is needed that considers findings, for instance, from the fields of political ecology and democratisation studies that contribute to identifying political measures that encourage environmental justice and the inclusion of actors in the politics of the Polar Regions. The limited time allowed for negotiations in regional settings has often been mentioned as a hindrance to the inclusion of more actors. This claim was reinforced by the urgency required from policy making given the rapidity of changes in the Polar Regions. Although, generally, more diverse thinking and "ideas from everywhere" have been widely promoted in the discourses under analysis, against the backdrop of a lack of time for negotiations coupled with an urgent need for policy making with respect to climate change, the inclusion of a multitude of perspectives has been represented as problematic. Further research is thus also needed that will analyse how across the local, national, regional and global levels political instruments can facilitate inclusive, just and efficient policy making with respect to the changing Polar Regions and, in this regard, draw on other findings from global governance approaches as well.

7 Annex

Definitions of the Arctic and the Antarctic

The U.S. defines the Arctic in statute 15 USC § 4111 as

> all United States and foreign territory north of the Arctic Circle and all United
> States territory north and west of the boundary formed by the Porcupine,
> Yukon, and Kuskokwim Rivers; all contiguous seas, including the Arctic
> Ocean and the Beaufort, Bering, and Chukchi Seas; and the Aleutian Chain.
> (Arctic Research and Policy Act of 1984,
> cf. e.g. U.S. Government Publishing Office, 2011)

Canada, on the other hand, "has traditionally considered the lower 60° line of [north-
ern] latitude rather than the polar circle, to demark the Arctic" (Hough, 2013, p. 5;
Transport Canada, 2015). In view of these initial definitions, it is also not surprising
that the working groups under the auspices of the Arctic Council, which is the main
intergovernmental forum for addressing issues related to the Arctic region, apply different definitions to the Arctic region "that reflect each of their interests" (GRID-Arendal,
2013). For the purposes of the Arctic Monitoring Assessment Program (AMAP), for
instance, the "Arctic region" was defined in accordance with "a compromise among
various definitions" (AMAP, 1998, p. 10): AMAP relates its work to the area generally
located north of the Arctic Circle (66°32'N), and north of 62°N in Asia and 60°N in
North America, which covers the marine areas north of the Aleutian chain, Hudson
Bay and parts of the North Atlantic Ocean including the Labrador Sea, but excludes
the Baltic Sea (AMAP, 1998, p. 10). The Conservation of Arctic Flora and Fauna
(CAFF), on the other hand, follows the Circumpolar Arctic Vegetation Map's definition of the Arctic (CAFF, 2013), which builds on scientific criteria for Arctic habitats.

Similarly, the Antarctic Treaty refers to all territory below the 60° line of southern latitude while the Commission for the Conservation of Antarctic Marine Living
Resources (CCAMLR), which forms an integral part of the Antarctic Treaty System, encompasses an area reaching partly to the 48° line of southern latitude. The
Scientific Committee on Antarctic Research (SCAR), on the other hand, defines
Antarctica for its purposes as being bounded by the convergence zone but also
including the sub-Antarctic islands. As these examples indicate, on the international level, all signatories of the Antarctic Treaty, of the CCAMLR-Convention
as well as the members of the different Arctic Council Working Groups, work on

the basis of consolidated but distinct geographical definitions of the Polar Regions in the respective committees.

According to the Chilean Decreto Supremo N° 1747, for example, the Chilean Antarctic territory includes

> todas las tierras, islas, islotes, arrecifes, glaciales y demás conocidos o por conocerse, en el mar territorial respectivo existente dentro de los límites del casquete constituido por los meridianos 53° y 90° de longitude oeste de Greenwich.
>
> (All land, islands, islets, reefs, glaciers and other territory known or to be known located in the respective ocean between the limitations assigned to the longitudinal meridians 53° and 90° East of Greenwich.)

The Argentine government, however, defines "La Antártida Argentina o el Sector Antártico Argentino" as "un área comprendida entre los meridianos 74° Oeste y 25° Oeste y el paralelo 60° Sur y el Polo Sur" (Armada Argentina, 2006 – [An area between the meridians 74° and 25° East and the parallel 60° South and the South Pole]). These definitions provided by Chile and Argentina (and the areas of territory they define) partly overlap – as do the definitions of the Canadian and the U.S.-American Arctic territories (e.g. with regard to the Beaufort Sea).

Box 7.1 National policies and strategy papers

Policies and Strategy Papers	national level
Canada	• Pan-Territorial Adaptation Strategy (Northwest Territories, Yukon, Nunavut, 2011) • "Statement on Canada's Arctic Foreign Policy" (Government of Canada, 2010) • "Canada's Northern Strategy – Our North, Our Heritage, Our Future" (Government of Canada, 2009) • "Rising to the Arctic Challenge: Report on the Canadian Coast Guard" (Canadian Standing Senate Committee on Fisheries and Oceans, 2009) • "A Northern Vision: A Stronger North and a Better Canada" (Northwest Territories, Yukon, Nunavut, 2007)
U.S.	• Arctic Policy for Alaska (Government of Alaska, 2015a and 2015b) • "Enhancing Coordination of National Efforts in the Arctic" (The White House, 2015) • Implementation Plan for "The National Strategy for the Arctic Region" (The White House, 2014) • "National Strategy for the Arctic Region" and "United States Coast Guard Arctic Strategy" (The White House, 2013b) (United States Coast Guard, 2013) • "Changing Conditions in the Arctic – Strategic Action Plan" (National Ocean Council, 2011) • "National Security Presidential Directive and Homeland Security Presidential Directive" (The White House, 2009) • United States Policy on the Arctic and Antarctic Regions (The White House, 1994)

Policies and Strategy Papers	*national level*
Chile	• Plan Estratégico Antártico (Consejo de Política Antártica, 2014) (Ministero de Relaciones Exteriores de Chile, 2015) • Decreto N°429: Política Antártica Nacional (Instituto Antártico Chileno, 2000)
Argentina	• Plan Anual Antártico (Dirección Nacional del Antártico, 2006) (Dirección Nacional del Antártico, 2010) (Dirección Nacional del Antártico, 2014) (Dirección Nacional del Antártico, 2015) • Proyecto de Ley (Senado de la Nación, 2014) • Decisión Administrativa 509/2004 (Dirección Nacional del Antártico, 2004) • Antártida Argentina: Política Provincial (Argentina Ambiental, 1996) • Decreto N° 1041/95 (Ministerio de Defensa, 1995) • Política Nacional Antártica (El president de la Nación Argentina, 1990)

Box 7.2 Regional policies

	Arctic Council	*Antarctic Treaty System*
State and Non-State Actors	• Declarations of the Ministerial Meetings of the AC (1996–2015) • SAO reports (1996–2014) • Lists of Participants, Agendas, Minutes, Statements, Press Releases, Meeting Summaries	• Final reports of the Antarctic Treaty Consultative Meetings (1989–2014) • Final Meeting Reports of the Committee for Environmental Protection (1989–2014) • Lists of Participants, Agendas, Minutes, Final Lists of Documents, Guidelines and Lists of Recommendations, Opening Addresses and Reports, Measures, Resolutions, Declarations

Box 7.3 Overview of crucial events – Part I

Date	*'Crucial' Event*
28 January 1989	• Sinking of the Bahía Paraíso
19 November 1990 (Chile)	• Special ATCM on protocol on Environmental Protection to the Antarctic Treaty in Chile
5 November 1990 (Argentina)	• "Política Nacional Antártica" (Decreto Presidential No. 2316, Argentina)
4 October 1991	• Adoption Madrid Protocol

Date	'Crucial' Event
28 August 1992	• "Declaración Connunta" between Chile and Argentina
11 January 1993 (Chile)	• Formation of the Barents Euro-Arctic Council
7 May 1993 (Argentina)	• Statement on the extension of the exclusive fishing zone of the South Georgia and the South Sandwich Islands by the United Kingdom (Argentina)
21 December 1994	• The International Whaling Convention approves "Southern Ocean Whale Sanctuary" as advocated by Greenpeace
15 January 1995 and 5 February 1995	• Loss of Larsen A Ice Shelf; implementation of Chilean base Prof. Julia Escudero in Antarctica
19 September 1996	• Ottawa Declaration
11 December 1997	• Adoption of the Kyoto Protocol
14 January 1998 and 23 July 1998	• Madrid Protocol enters into force; RAPAL meeting in Chile
10 April 1999	• Ice-loss in Antarctica, global warming
28 March 2000	• Adoption of the "Política Antártica Chilena Nacional" (Decreto N° 429)

Box 7.4 Overview of crucial events – Part II

Date	'Crucial' Event
18 April 2001	• Meeting of the Chilean "Consejo de Política Antártica (CPA)" in Puerto Williams
5 March 2002	• Loss of Larsen B Ice Shelf
20 July 2003	• Chile and Argentina open the "Refugio Abrazo de Maipú" in Antarctica
15 November 2004	• Australian submission to CLCS
16 February 2005	• The Kyoto Protocol enters into force
19 April 2006	• New Zealand submits territorial claim to CLCS
2 August 2007 and 23 November 2007	• Russian flag planting and Sinking of the MS Explorer
28 May 2008	• Ilulissat Declaration
21 April, 5 May, 6 May, 7 May, 11 May 2009	• CLCS submissions by Argentina, South Africa, France and South Africa, Chile (preliminary note), GB (Falkland and South Sandwich)
27 February 2010	• Giant iceberg collapses in Antarctica
20 June – 1 July, 2011	• ATCM in Buenos Aires
16 March 2012	• Argentina and Chile sign the agreement forming a "Comité Ad-hoc Sistema del Tratado Antártico"
27 September 2013	• Summary of IPCC report released
30 January 2014	• Final version of IPCC report released

Table 7.1 Overview of interviews conducted

Representatives of . . .	Canada and the U.S.	Chile and Argentina	Contacted	Interviewed	Written Response
Permanent Participants	X		6	2	2
ENGOs	X		28	9	1
		X	7	1	0
Industry (oil, gas, shipping, tourism)	X		6	0	1
		X	6	1	0
Governmental agencies	X		12	4	1
		X	17	2	0
International fora	X		1	0	1
		X	2	1	0
Scientists		X	5	2	1

Methods

In social science a combination of quantitative and qualitative analyses is increasingly applied especially in studies dealing with large amounts of data and in those aiming at answering different types of research questions (Mayring, 2007; Tinati et al., 2014). As Mayring (2014, p. 8) points out, however, this "has not led to a new methodology", instead, "different steps of analysis [are put together] with their different logics, mainly following a pragmatic theory of science (the methodology is adequate if it leads to the solution of the research question)". Following Mayring, the qualitative and quantitative methods applied in this study were selected in light of the different types of material accessed and collected.

As previously set out, the aims of this research and the selection of empirical material were based on a mixed relational and explorative approach. Accordingly, for the analysis of all selected policies, statements and strategy reports, think tank and academic works, for the newspaper articles and for the interviews different stages of qualitative analysis with their distinct logics applied. More explicitly, the analyses presented in Chapters 3 and 4 always relate to the broader context before examining the different geopolitical perspectives (e.g. the items placed on the agendas of the Arctic Council and in the Antarctic Treaty System or the main thematic contexts predominantly addressed in the newspapers and academic works under analysis). This is done to better classify the relevance of the different geopolitical perspectives introduced within a particular context (relational approach). Following the method of Qualitative Content Analysis all information retrieved from the material has been structured, analysed and evaluated based on a rule-bound coding procedure (different codes were used to organise all the information in categories that relate to the focus of the study, cf. Mayring, 2014, p. 39). More specifically, categories were first constructed and their specific variables and dimensions were outlined to predefine coding rules in order to secure consistency during the coding process. The categories were developed inductively after a first screening of the material collected during the field research in North America (explorative approach). In a second step, the codes were reviewed, specified and newly applied to the material collected in South America. Then the complete corpus was coded (according to the rules defined for each code). Later the findings within categories and between categories were compared to each other (relational approach), cases were summarised and typical arguments were typologised (cf. Kuckartz, 2014, p. 118). In a next step, all relevant information was obtained from the empirical material after it had been narrowed down by working through the text corpus line by line. New subordinate categories were formulated to which more specific information was assigned. The whole code system was revised, once no more new categories were being formulated. In order to assure intracoder reliability, definitions and anchor samples were provided for each code. These guided the coding process and were reviewed before the reduced material was examined one more time (cf. Mayring, 2014, p. 79).

To provide a solid orientation within the vast corpus of newspaper material a slightly different coding procedure was applied. In a first step, all newspaper articles addressing similar topics were organised into main categories. To articles available in electronic format, a full-text analysis was applied; all newspaper articles collected in print format were manually categorised into groups with similar topics. In a second step, the categories and classifications of all articles were reviewed manually to avoid the ascription of unsuitable codes. Newspaper articles from both regions were coded simultaneously to eliminate any possible bias that could have arisen as a result of the author reading the articles published in North American newspapers first. In this coding step main categories (representing the major thematic contexts addressed in the respective articles) were assembled and sub-codes were also assigned to the articles to identify linkages to other topics and include information on the contexts in which certain topics were discussed. Third, to identify key topics in each year of the period of investigation a Quantitative Frequency Analysis was applied, whereby all newspaper articles collected were aggregated according to the code assigned to them and the year in which they were published in order to depict the quantitative significance of topics reported on. This quantitative comparison served as a starting point from which the in-depth analysis provided in Chapter 4 evolved.

The interviews followed the focused, guideline-based structure developed by Merton and Kendall (1946). This interview format is based on preformulated assumptions that are verified or falsified during the interview by the subjective experience of the person questioned and at the same time provide the content structure. This format also allows unexpected responses and new points of view or aspects to be included in further analysis as the questionnaires are only partly standardised, and also contain individual questions concerning the personal and institutional background of the interviewees. Questionnaires were generally adapted to the latter. Additional questions were developed spontaneously whenever the interviewees addressed other topics relevant to the research focus. In general, most interviews were conducted at offices or in institution's conference rooms and questions were answered without third-person involvement, with the exception of one focused group interview.

Table 7.2 Topics prominently addressed in Ministerial Meetings (1998–2015)*

Main Topics Outlined in the Arctic Council Declarations	1998	2000	2002	2004	2006	2009	2011	2013	2015
Cooperation with other "relevant bodies"						■			
Structure of the AC	■	■	■		■	■	■	■	■
Research		■							
Energy						■			
Human conditions			■	■				■	
Climate change			■						
Environmental threat: Shipping	■				■	■			
Environmental threat: Offshore oil and gas							■		
Environmental threat: Pollution					■				
Environmental Protection	■	■	■	■				■	■
Sustainable development									■

*The categories are named according to the topics addressed in Arctic Council Declarations, and were either repeatedly addressed or prominently placed as a headline in a declaration. For more information on the different categories, see also the codebook. The name of the category "cooperation with other 'relevant bodies'" is an expression taken from the Barrow Declaration (2000).

Table 7.3 Topics prominently addressed in SAO Meetings (1999–2014)*

Topics at SAO Meetings	1999	2000	2001	2002	2003	2004	2005	2006	2007	2008	2009	2010	2011	2012	2013	2014
Human development										■			■		■	
Energy									■						■	
Sustainable development				■						■				■		
Ocean management				■											■	
Gender equality			■													
Infrastructure			■				■							■		
Pollution		■					■	■		■						
Public awareness, education and outreach						■	■									
Arctic Military Environmental Cooperation					■											
Arctic Council Action Plan		■		■												
Climate change							■			■			■	■	■	■
Arctic research		■					■			■			■	■	■	■
Financing		■					■						■	■	■	■
Structure of the AC			■			■	■						■	■	■	■
Presentations by PPs, observers and experts		■		■		■	■			■			■	■	■	■
Coordination with other international bodies		■					■			■			■	■		■
Coordination with working groups		■					■			■			■	■		■

*This table provides an overview of all topics addressed in the two SAO meetings regularly taking place per year (with the exception of the years 2000 and 2006, in which only one SAO meeting was arranged).

Table 7.4 Topics prominently addressed in ATCMs (1989–2014)

Topics at ATCMs	1989	1991	1992	1993	1994	1995	1996	1997	1998	1999	2001	2002	2003	2004	2005	2006	2007	2008	2009	2010	2011	2012	2013	2014
Operation of the Antarctic Treaty System	■	■	■	■	■	■	■	■	■	■														
Measures for the protection of the Antarctic environment and ecosystems	■																							
Antarctica and global (climate) change	■																			■	■			
Antarctic Protected Area System	■																							
International (Antarctic) scientific (and logistic) cooperation	■																							
(Effects of) tourism and non-governmental expeditions	■	■	■	■	■	■	■	■	■	■	■	■	■	■	■	■	■	■	■	■	■	■	■	■
Antarctic meteorology and telecommunications	■	■	■																					
Marine hydrometeorological services	■																							
Air safety in Antarctica	■	■																						
Jurisdiction in Antarctica							■		■															
Data and information management											■	■	■					■	■	■	■	■	■	■
Antarctic infrastructures																								
Antarctic science – initiatives and cooperation												■	■	■	■									
Cultural and aesthetic values of the Antarctic																								
Education and training												■		■	■	■								
Safety and operation in Antarctica													■											
International Polar Year															■	■	■	■	■					
Relevance of developments in the Arctic and Antarctic					■	■	■	■	■	■	■	■	■	■	■	■	■	■	■	■	■	■	■	■
Emergency response and contingency planning									■															
Question of liability as referred to in the Environmental Protocol											■	■	■	■	■	■	■	■	■	■	■	■	■	■
Cooperation in the Antarctic Treaty System	■	■																		■	■			
Biological prospecting in Antarctica										■	■	■	■	■	■	■	■	■	■	■	■	■	■	■

Table 7.5 Topics prominently addressed in CEPs (1998–2014)

Topics at CEPs	1998	1999	2000	2001	2002	2003	2004	2005	2006	2007	2008	2009	2010	2011	2012	2013	2014
Operation of the CEP	X	X	X	X	X	X	X	X	X	X	X	X	X	X	X	X	X
Compliance with the Madrid Protocol	X	X	X	X	X	X	X	X	X		X	X	X				
Environmental monitoring	X	X	X	X	X	X	X	X	X	X	X	X	X	X	X	X	X
State of the Antarctic Environment Report	X	X	X	X	X	X	X										
Data and exchange of information	X	X	X	X	X	X	X	X									
Emergency response and contingency planning			X	X	X	X	X	X	X	X							
Cooperation with other organisations				X	X	X	X	X	X	X	X	X	X	X	X	X	X
Biological prospecting							X	X	X	X							
The future of the CEP									X	X		X	X	X	X	X	X
International Polar Year									X	X		X					X
Area protection and management							X	X	X	X	X	X	X	X	X	X	X
Conservation of Antarctic fauna and flora								X	X	X	X						
Inspection reports								X	X	X	X						
Waste management									X	X	X						X
Prevention of marine pollution										X	X						
Climate change implications for the environment												X	X	X	X	X	X
Repair and remediation of environmental damage																X	X

Box 7.5 List of codes and sub-codes assigned to newspaper articles under analysis

codes *(= main thematic context addressed)*	*sub-codes* *(= topics to which the article also relates)*
territorial/maritime dispute (any form of dispute over territory, being land, water or ice)	Beagle Channel, Islas Malvinas/ Falkland Islands, Southern Patagonian Ice Field, territorial claim, bilateral cooperation, CLCS
resources (any article that relates to resource deposits, – exploration or – extraction)	illegal, unreported and unregulated fishing (IUU), oil, gas, mining, fishing, whaling, non-state actors
environment (any article referring to any form of environmental change, addressing environmental protection or contamination)	ozone layer, global warming, climate change, Kyoto Protocol, pollution, protection, policy, marine protected area (MPA), ocean sanctuary, whaling sanctuary, flora & fauna, Gondwana, iceberg, non-state actor
shipping (any article that refers to the use of new shipping routes, shipping in general in Arctic/Antarctic waters, shipping infrastructure, SAR measures, shipping regulations, shipping with the purpose to transport tourists or commodities)	accident, search and rescue (SAR), tourism, infrastructure (port), cooperation, environment, governance, non-state actors
science (any article that focuses on science, being scientific studies and results, conditions under which science is conducted and scientific infrastructure)	scientific expedition, establishment of Antarctic bases, krill, penguin, International Polar Year (IPY), militarisation, cooperation
governance (any article that refers to explicit agreements or governance structures)	regulation, agreement, cooperation, tourism, resources, shipping, fishing, environment, peace, marine protection, national, international, Arctic Council (AC), Antarctic Treaty System (ATS), Antarctic Treaty Secretariat
cooperation (any article relating to bilateral or multilateral cooperation)	science, search and rescue (SAR), conflict, internal (investments), non-state actor, Mercado Común del Sur (Mercosur), Organization of American States (OAS), Cono Sur

codes (= main thematic context addressed)	sub-codes (= topics to which the article also relates)
last frontier (any article emphasising the Polar Regions' pristineness, focusing on the narrative of conquest, and on athletic expeditions)	pristineness, expedition, sovereignty, infrastructure, remoteness
identity (any article relating to the special relationship of an actor with the polar environment)	national campaigns, education, maps, devolution, flag planting, inter-American cooperation, "Puerta de la Entrada"/Gateway City (Punta Arenas, Ushuaia), citizenship, día de la Antártida, anniversary of base, culture, sovereignty, religion, national Antarctic history

Box 7.6 Codebook – Part I – Topics at Arctic Council Ministerial Meetings

Code	Coding examples
sustainable development	thematic headline addressed in the declarations, independent topic, related to the work of the Sustainable Development Working Group, circumpolar cooperation
environmental protection	biodiversity conservation, sustainable use of resources, emergency prevention preparedness and response, protection of the Arctic marine environment, conservation of flora and fauna offshore oil and gas
environmental threat	shipping (thematic headline addressed in the declarations, independent topic, related to the work of working groups), pollution (global emissions of persistent organic pollutants), action on contaminants in the Arctic (thematic headline addressed in the declarations, independent topic, related to the work of working groups)
climate change	thematic headline addressed in the declarations, independent topic, related to the work of working groups
human conditions	thematic headline addressed in the declarations, independent topic, related to the work of working groups, human health and human development, improving economic and social conditions
Energy	thematic headline addressed in the declarations, independent topic, related to the work of working groups
research	research projects conducted besides the working groups and task forces, e.g. Arctic monitoring and assessment, science and monitoring in more general terms
structure of the AC	AEPS, financing, inclusion of permanent participants, establishment of new working groups, capacity building, efficiency, prioritisation, administration and organisation of the AC

Box 7.7 Codebook – Part II – Topics at Arctic Council SAO meetings

Code	Coding examples
coordination of working group activities	thematic headline addressed in the meeting protocol, independent topic, e.g. working group reports and operating guidelines, work plans
coordination with other international bodies	thematic headline addressed in the meeting protocol, independent topic, World Summit on Sustainable Development (e.g. discussion of the report of the 10th anniversary of the Arctic Environmental Cooperation preparation for the WSSD), report on the four regional councils of the Northern Region (AC, BEAC, CBSS, NCM), Barents Euro-Arctic Summit, Arctic Council involvement in the international Polar Year, preparation for the Petersburg International Economic Forum, contribution of the AC to the World Water Forum, the AC and the UN Climate Change Conference, AC cooperation with the Northern Forum, the Nordic Council of Ministers and the EU, presentation by IMO
presentations by permanent participants, observers and experts	the Northern Forum, University of the Arctic, International Red Cross, presentation of Arctic policies and activities of the EU, Arctic policies of the observer countries, the EU's Northern Dimension Action Plan, Fifth Conference of Parliamentarians of the Arctic Region, Arctic Athabaskan Council
Arctic Council structure	e.g. report on possible ways to improve the structure of work in the Arctic Council, organising the work of the Arctic Council and its subsidiary bodies based on findings of the ACIA, new staff and responsibilities of the Arctic Council Secretariat, Indigenous Peoples Secretariat, effectiveness and efficiency, chairmanship-led initiatives
Arctic research	thematic headline addressed in the meeting protocol, Arctic science summit week, environmental monitoring, report on Arctic Human Development
climate change	thematic headline addressed in the meeting protocol, independent topic of concern, Arctic Climate Impact Assessment, biodiversity, environment
Arctic Council Action Plan	thematic headline addressed in the meeting protocol, independent topic of concern
public awareness, education and outreach	thematic headline addressed in the meeting protocol, higher education day, meeting of Ministers of Education and Science of the Arctic states, presentation of the Northern Encyclopaedia, cooperation in education, cultural dimension of cooperation in the AC
financing	thematic headline addressed in the meeting protocol, international financing of (Arctic Council) projects, Nordic Environment Finance Corporation (NEFCO), financing of permanent participants' participation, Arctic Council Project Support Instrument (PSI), financial situation of secretariats of AC working groups

**Box 7.8 Codebook – Part III – Topics at Antarctic Treaty
Consultative Meetings**

Code	Coding examples
operation of the Antarctic Treaty system	reports, for instance, by subsidiary bodies CCAMLR, CCAS, SCAR, depositary governments of the treaties reporting on current status of adoption and implementation and organisational/management aspects (e.g. related to the Secretariat), Protocol on and Committee for Environmental Protection to the Antarctic Treaty 1992–2014, development of a multi-year strategic work plan (2010–2014)
measures for the protection of the Antarctic environment and ecosystems	measures for the protection of the Antarctic environment and ecosystems, environmental monitoring, implementation of environmental impact assessment procedures (1992, 1994, 1995)
Antarctica and global (climate) change	Antarctic and global change, significance of the ozone layer, implications of climate change for the management of the Antarctic Treaty area
Antarctic Protected Area System	discussions relating to the designation of Antarctic Special Protected Areas (ASPAs) or Antarctic Special Marine Areas (ASMAs)
international (Antarctic) scientific (and logistic) cooperation	forms and promotion of scientific cooperation, environmental standards and international scientific cooperation
(effects of) tourism and non-governmental expeditions	tourism and non-governmental expeditions, human impact (1989, 1991), uses of Antarctic ice (1989)
Antarctic meteorology and telecommunications	improvement of the availability of Antarctic meteorological data via improved communication systems and practices
marine hydrometeorological services	improvement of the availability and use of Antarctic hydrometeorological data
air safety in Antarctica	safety measures to improve flight activities in Antarctica
jurisdiction in Antarctica	questions related to the exercise of jurisdiction in Antarctica
data and information management	data management, exchange of information
Antarctic infrastructures	infrastructure investments, infrastructure for vessels and tourism, permanent infrastructure
Antarctic science – initiatives and cooperation	scientific achievements
cultural and aesthetic values of the Antarctic	highlighting aesthetic values of the Antarctic in art and writings to outline the special value of Antarctica to the public for promoting a better understanding and appreciation of the values of Antarctica

Code	Coding examples
education and training	educational opportunities (e.g. the Antarctic Environmental Management Course "Classroom Antarctica") to support public knowledge on Antarctica
safety and operations in Antarctica	shipping guidelines (e.g. International Code of Safety for Ships, Polar Code), safety of navigation and for the Antarctic community
International Polar Year	the role of the Polar Regions in driving and responding to Global Climate Change, priority polar science issues of global relevance
relevance of developments in the Arctic and the Antarctic	reports on the Arctic Council and on the Arctic Environmental Protection Strategy, contamination in both regions, bipolar science, strengthening the effectiveness of the Antarctic Treaty System
Emergency response and contingency planning	scientific and logistical support activities, fuel spills, responsibilities of parties
question of liability as referred to in the Environmental Protocol	as referred to in Article 16 of the Protocol
cooperation in the Antarctic Treaty System	inspections under the Antarctic Treaty and the Environmental Protocol, compliance with the Protocol on Environmental Protection, cooperation among parties with respect to Article 6 of the Protocol, commemoration of the 50th Anniversary of the Antarctic Treaty, cooperation in the hydrographic charting of Antarctic waters
biological prospecting in Antarctica	discussions on the need to regulate biological prospecting and to assess its environmental impacts

Box 7.9 Codebook – Part IV – Topics at meetings of the Committee for Environmental Protection

Code	Coding examples
operation of the CEP	discussion of working papers dealing with the prioritisation of work for the CEP, organisation of workshops, establishment of open-ended contact groups, Environmental Impact Assessment (Annex I), conservation of Antarctic fauna and flora (Annex II), waste disposal and waste management (Annex III) prevention of marine pollution (Annex IV), area protection and management (Annex V), data and information exchange, environmental monitoring, state of the Antarctic Environment Report, CEP-structure

Code	Coding examples
compliance with the Protocol on Environmental Protection	discussion of reports provided by the parties on Annex I-V, by the U.S. as the depository government for the Antarctic Treaty and its Protocol, information papers by other parties (e.g. ASOC) monitoring the legal implementation of the Protocol
environmental monitoring /impact assessment	discussion of working papers on the monitoring of environmental impacts of scientific activities and operations in Antarctica
state of the Antarctic Environment Report	Report by SCAR
emergency response and contingency planning	assessment of environmental emergencies arising from tourism activities in Antarctica, e.g. by IAATO
data and exchange of information	annual exchange of information via the world wide web, development of website, format of reports
cooperation with other organisations	in accordance with Article 11 of the Protocol
biological prospecting	discussion of information papers on biological data collected in Antarctica
strategic discussions on the future of the CEP	work plans
International Polar Year	information exchange with IPY Programme Office and general discussion on the IPY
area protection and management	historic sites and monuments, human footprint and wilderness value, site guidelines, marine spatial protection and management
conservation of Antarctic fauna and flora	quarantine and non-native species, specially protected species, marine acoustics
inspection reports	reports on the inspection of protected areas and sites in Antarctica
waste management	Management Policy for disposal of waste by vessels
prevention of marine pollution	guidelines for ships operating in ice-covered areas, of baseline pollution
climate change implications for the environment	strategic approach
repair and remediation of environmental damage	assessment and monitoring of damage, definition of what constitutes environmental damage in Antarctica

Bibliography

AAC. (2007). *Improving the Efficiency and Effectiveness of the Arctic Council: A Discussion Paper*. [Online] Available from: https://oaarchive.arctic-council.org/bitstream/handle/11374/694/ACSAONO01_10_1_AAC_AC_Future.pdf?sequence=1&isAllowed=y. [Accessed: 1st May 2016].

Abdel-Motaal, D. (2016). *Antarctica: The Battle for the Seventh Continent*. Santa Barbara, Denver, Oxford: Praeger.

Agnew, J. (1994). The Territorial Trap: The Geographical Assumptions of International Relations Theory. *Review of International Political Economy*. 1(1). pp. 53–80.

Aguas, M. (2008). Coalición por la Antártida. *Clarín*. [Print] 28th June. Available from: Biblioteca Nacional Mariano Moreno.

Agüero Garcés, F. (1996). Prioridad del Tema Medioambiental. *El Mercurio*. [Print] 18th September. Available from: Biblioteca Nacional de Chile.

AHDR. (2004). *Arctic Human Development Report*. [Online] Available from: https://oaarchive.arcticcouncil.org/handle/11374/51. [Accessed: 1st June 2016].

Åhrén, M. (2009) *Statement by Mattias Åhrén, President of the Sámi Council*. [speech] Arctic Council 6th Ministerial Meeting in Tromsø, 29 April 2009. [Online] Available from: https://oaarchive.arctic-council.org/handle/11374/1562. [Accessed: 1st April 2016].

Ahmad, M. Y., Jaya, S. (2005). Many Maps of Canada Ignore Canadian Arctic. *Toronto Star*. [Print] 23rd August. Available from: ProQuest Database. [Accessed: 30th March 2014].

Albert, M., Bluhm, G., Helmig, J., Leutzsch, A., Walter, J. (2009). Introduction: The Communicative Construction of Transnational Political Spaces. In Albert, M., et al. (eds.) *Transnational Political Spaces: Agents-Structures-Encounters*. Frankfurt, New York: Campus Verlag. pp. 7–31.

Albert, M., Reuber, P., Wolkersdorfer, G. (2014). Critical Geopolitics. In Schieder, S. & Spindler, M. (eds., translated by A. Skinner) *Theories of International Relations*. London, New York: Routledge. pp. 321–336.

Albert, M., Wehrmann, D. (2015). Polarpolitik. Ein Bericht zur politikwissenschaftlichen Arktis- und Antarktisliteratur. In *Neue Politische Literatur. Berichte aus Geschichts- und Politikwissenschaft*. 60(1). Frankfurt am Main: Peter Lang Verlag. pp. 63–89.

Aleut International Association. (2013). Arctic Council: Bold Steps Needed to Support Indigenous Participation at the Arctic Council. *Northern Public Affairs*. September. pp. 55–58.

Aleut International Association. (2015). *Arctic Council Ministerial Meeting-AIA*. [Speech] Arctic Council 9th Ministerial Meeting in Iqaluit, Canada, 24th–25th April. [Online]

Available from: https://oaarchive.arctic-council.org/ handle/11374/901. [Accessed: 1st April 2016].

Alexander, C. (2009). People of the Arctic: Use Them or Lose Them. *Globe and Mail.* [Print] 31st March. Available from: ProQuest Database. [Accessed: 30th March 2014].

Alexander, S. (1995). Security Doesn't Justify Arctic Drilling. *New York Times.* [Print] 11th September. Available from: ProQuest Database. [Accessed: 30th March 2014].

Ali, S. H., Pincus, R. (2015). Introduction. A Cold Prelude to a Warming World. In: Ali, S. H., & Pincus, R. (eds.) *Diplomacy on Ice: Energy and the Environment in the Arctic and Antarctic.* New Haven, London: Yale University Press. pp. 1–10.

Ali, S. H., Pincus, R. (2015). *Diplomacy on Ice: Energy and the Environment in the Arctic and Antarctic.* New Haven, London: Yale University Press. pp. 151–160.

Alia, V. (2007). *Names & Nunavut: Culture and Identity in the Inuit Homeland.* New York: Berghahn Books.

Allen, M. (2000). Bush Supports Oil Exploration in Arctic Refuge. *The Washington Post.* [Print] 30th September. Available from: ProQuest Database. [Accessed: 30th March 2014].

Alvarez, L. (2001). Industry Has Powerful Allies on Drilling Bill. *New York Times.* [Print] 18th March. Available from: ProQuest Database. [Accessed: 30th March 2014].

AMAP. (1998). *Physical/Geographical Characteristics of the Arctic.* [Online] Available from: http://amap.no/documents/download/88. [Accessed: 9th December 2015].

Anderson, B. (2006[1983]). *Imagined Communities: Reflections on the Origin and Spread of Nationalism.* London: Verso. [Online] Available from: http://energy.usgs.gov/Coal/ AssessmentsandData/CoalAssessments.aspx. [Accessed: 6th February 2018].

Antarctic Treaty Secretariat. (2011). *Final Report of the Thirty-Fourth Consultative Meeting.* Buenos Aires, 20 June–1 July 2011. Buenos Aires: Secretariat of the Antarctic Treaty.

Antarctic Treaty Secretariat. (2016). *Antarctic Treaty.* (1959) [Online] Available from: http:// ats.aq/documents/keydocs/vol_1/vol1_2_AT_Antarctic_Treaty_e.pdf. [Accessed: 11th January 2016].

Appadurai, A. (2003[1996]). *Modernity at Large: Cultural Dimensions of Globalization.* Minneapolis: University of Minnesota Press.

Arctic Council. (2013a). *Arctic Council Rules of Procedure.* [Online] Available from: https://oaarchive.arcticcouncil.org/bitstream/handle/11374/940/2015-09-01_Rules_of_ Procedure_website_version.pdf?sequence=1&isAllowed=y. [Accessed: 11th January 2016].

Arctic Council. (2013b). *Observer Manual.* [Online] Available from: https://oaarchive. arcticcouncil.org/bitstream/handle/11374/939/EDOCS-3020-v1APDF?sequence=5. [Accessed: 11th January 2016].

Arctic Council. (2015a). *Observer Manual for Subsidiary Bodies as Adopted by the Arctic Council at the 8th Arctic Council Ministerial Meeting, Kiruna, Sweden, 15 May 2013, and Addendum Approved by the Senior Arctic Officials at the Meeting of the Senior Arctic Officials.* Tromsø: Arctic Council Secretariat. [Online] Available from: https:// oaarchive.arcticcouncil.org/handle/11374/939. [Accessed: 3rd March 2016].

Arctic Council. (2015b). *Arctic Human Health Expert Group.* [Online] Available from: www.sdwg.org/expert-groups/arctic-human-health-expert-gro. [Accessed: 11th January 2016].

Arctic Council. (2015c). *Iqaluit Intervention by the Danish Foreign Minister.* Arctic Council's 9th Ministerial Meeting, Iqaluit, Canada, 24th–25th April. [Online] Available from: https://oaarchive.arctic-council.org/handle/11374/376. [Accessed: 18th January 2016].

Arctic Science Portal. (2016). *Organizations: US Government.* [Online] Available from: https://arctic.gov/portal/us_gov.html. [Accessed: 11th November 2016]. Dirección

Nacional del Antártico. (2004). *Decisión Administrativa 509/2004.* [Online] Available from: http://www.dna.gob.ar/userfiles/12_decisionadministrativa509_2004.pdf. [Accessed: 24th November 2016].

Argentine Republic. (2015). *Statement by the Minister of Foreign Affairs and Worship of the Argentine Republic before the United Nations Special Committee on Decolonialisation.* [Online] Available from: http://mrecic.gov.ar/discurso-del-canciller-timerma. [Accessed: 6th February 2018].

Armada Argentina. (2006). *Sector Antártico Argentino.* [Online] Available from: http://ara. mil.ar/pag.asp?idItem=168. [Accessed: 6th February 2018].

Armas Barea, C. A., Beltramino, J. C. (1992). *Antártida al iniciarse la década de 1990.* Buenos Aires: CARI.

Arnaudo, R. V. (2013). United States Policy in the Arctic. In Berkman, P. A. & Vylegzhanin, A. N. (eds.) *Environmental Security in the Arctic Ocean.* Dordrecht: Springer. pp. 81–92.

ATCM. (1991). *Final Report of the Sixteenth Antarctic Treaty Consultative Meeting.* [Online] Available from: http://ats.aq/devAS/info_finalrep.aspx?lang=e&menu=5. [Accessed: 15th June 2016].

ATCM. (1994). *Final Report of the Eighteenth Antarctic Treaty Consultative Meeting.* [Online] Available from: http://ats.aq/devAS/info_finalrep.aspx?lang=e&menu=5. [Accessed: 15th June 2016].

ATCM. (1995). *Final Report of the Nineteenth Antarctic Treaty Consultative Meeting.* [Online] Available from: http://ats.aq/devAS/info_finalrep.aspx?lang=e&menu=5. [Accessed: 15th June 2016].

ATCM. (1996). *Final Report of the Twentieth Antarctic Treaty Consultative Meeting.* [Online] Available from: http://ats.aq/devAS/info_finalrep.aspx?lang=e&menu=5. [Accessed: 15th June 2016].

ATCM. (1997). *Final Report of the Twenty-First Antarctic Treaty Consultative Meeting.* [Online] Available from: http://ats.aq/devAS/info_finalrep.aspx?lang=e&menu=5. [Accessed: 15th June 2016].

ATCM. (1998). *Final Report of the Twenty-Second Antarctic Treaty Consultative Meeting.* [Online] Available from: http://ats.aq/devAS/info_finalrep.aspx?lang=e&menu=5. [Accessed: 15th June 2016].

ATCM. (1999). *Final Report of the Twenty-Third Antarctic Treaty Consultative Meeting.* [Online] Available from: http://ats.aq/devAS/info_finalrep.aspx?lang=e&menu=5. [Accessed: 15th June 2016].

ATCM. (2001). *Final Report of the Twenty-Fourth Antarctic Treaty Consultative Meeting.* [Online] Available from: http://ats.aq/devAS/info_finalrep.aspx?lang=e&menu=5. [Accessed: 15th June 2016].

ATCM. (2009). *Final Report of the Thirty-Second Antarctic Treaty Consultative Meeting.* [Online] Available from: http://ats.aq/devAS/info_finalrep.aspx?lang=e&menu=5. [Accessed: 15th June 2016].

ATCM. (2011). *Final Report of the Thirty-Fourth Antarctic Treaty Consultative Meeting.* [Online] Available from: http://ats.aq/devAS/info_finalrep.aspx?lang=e&menu=5. [Accessed: 15th June 2016].

ATCM Rules of Procedure. (2015). *Revised Rules of Procedure.* [Online] Available from: http://ats.aq/documents/keydocs/vol_2/Rules_atcm_e. pdf. [Accessed: 30th April 2016].

ATS. (2015). *Parties.* [Online] Available from: http://ats.aq/devAS/ats_parties.aspx?lang=e. [Accessed: 6th February 2018].

ATS. (2016). *The Committee for Environmental Protection.* [Online] Available from: http://
    ats.aq/e/cep.htm. [Accessed: 28th April 2016].
Aubry, J. (1996). Inuit Receive $ 10 Million for Arctic Relocation. *Toronto Star.* [Print] 8th
    March. Available from: ProQuest Database. [Accessed: 30th March 2014].
Axworthy, T. S. (1992). Rallying Around the North Pole: The Arctic: Canada Lacks a
    Northern Foreign Policy: The Arctic Usually Sits Forgotten: But Now a New Initiative
    Is under Way. *Globe and Mail.* [Print] 13th November. Available from: ProQuest Data-
    base. [Accessed: 30th March 2014].
Babbitt, B. (2004). Another Attack on the Arctic. *New York Times.* [Print] 8th July. Avail-
    able from: ProQuest Database. [Accessed: 30th March 2014].
Baker, J. A. (2011). Cut the Red Tape, Already: Open Up Drilling in Alaska. *USA Today.*
    [Print] 1st June. Available from: ProQuest Database. [Accessed: 30th March 2014].
Banerjee, N. (2000). Can Black Gold Ever Flow Green? *New York Times.* [Print] 12th
    November. Available from: ProQuest Database. [Accessed: 30th March 2014].
Banerjee, N. (2004). Would More Drilling in America Make a Difference? *New York Times.*
    [Print] 20th June. Available from: ProQuest Database. [Accessed: 30th March 2014].
Bär, N. (2003). Viceversa: Un paso hacia lo desconocido. *La Nacion.* [Print] 3rd December.
    Available from: Biblioteca Nacional Mariano Moreno.
Barra, G. (2007). Tres expertos, entre el apoyo y las dudas. No creen que avance el rec-
    lamo argentino. *La Nación.* [Print] 13th December. Available from: Biblioteca Nacional
    Mariano Moreno.
Bartsch, G. M. (2015). *Klimawandel und Sicherheit in der Arktis. Hintergründe, Perspe-
    ktiven, Strategien.* Wiesbaden: Springer VS.
Bastmeijer, K. (2015). Shared Responsibility for Cumulative Environmental Impacts in
    Antarctica. In Nollkaemper, A. (ed.) *The Practice of Shared Responsibility, Vol. 2: Law
    of the Sea and Environmental Law.* Cambridge: Cambridge University Press.
Batterbee, K., Fossum, J. E. (2014). *The Arctic Contested.* Brussels: P.I.E. Peter Lang S.A.
Bauck, S., Maier, T. (2015). Entangled History. *InterAmerican Wiki: Terms-Concepts-
    Critical Perspectives.* [Online] Available from: http://wiki.elearning.unibielefeld.de/
    wikifarm/fields/ges_cias/field.php/Main/Unterkapitel5. [Accessed: 17th December
    2015].
Beach, L. (2005). Who Decides about the Arctic Refuge? *The Washington Post.* [Print] 29th
    September. Available from: ProQuest Database. [Accessed: 30th March 2014].
Beach, L. (2010). Don't Drill, Baby, Drill in the Arctic. *The Washington Post.* [Print] 5th
    June. Available from: ProQuest Database. [Accessed: 30th March 2014].
Becerra, A. (1989). ¿Para qué sive la Antárdia? *Clarín.* [Print] 4th February. Available
    from: Biblioteca Nacional Mariano Moreno.
Beck, P. J. (1990). International Relations in Antarctica: Argentina, Chile and the Great
    Powers. In Morris, M. (ed.) *Great Power Relations in Argentina, Chile and Antarctica.*
    Basingstoke, New York: Palgrave Macmillan. pp. 101–130.
Beck, P. J. (1995). Through Arctic Eyes: Canada and Antarctica: 1945–62. *Arctic.* 48(2).
    pp. 136–146.
Beck, P. J. (2003). Twenty Years on: The UN and the 'Question of Antarctica,' 1983–2003.
    Polar Record. 40(214). pp. 205–212.
Beck, P. J. (2004). Twenty Years on: The UN and the 'Question of Antarctica,' 1983–2003.
    Polar Record. 40(3). pp. 205–212.
Beck, P. J. (2014). *The International Politics of Antarctica.* New Ed., First Published 1986.
    London, New York: Routledge.

Benedetto, W. (1990). Put the Brakes on Our Reckless Use of Oil. *USA Today*. [Print] 23rd October. Available from: ProQuest Database. [Accessed: 30th March 2014].

Bennett, M. (2014). Global Media Interpretations of China's Rescue of Stranded Passengers Off Antarctica Vary. *Cryopolitics*. [Online] Available from https://cryopolitics. com/tag/pola orientalism/. [Accessed: 30th April 2016].

Berger, A. C. (2015). *Agenda-Shaping in the Arctic Council: Projecting National Agendas in a Consensus-Based Regime*. Master's thesis, University of Oslo. [Online] Available from: www.duo.uio.no/handle/10852/45355. [Accessed: 6th February 2018].

Bergh, K. (2012). The Arctic Policies of Canada and the United States: Domestic Motives and International Context. In *SIPRI Insights on Peace and Security*. Vol. 1. Sweden: Stockholm International Peace Research Institute.

Berkman, P. A. (2013). Preventing an Arctic Cold War. *New York Times*. [Print] 13th March. Available from: ProQuest Database. [Accessed: 30th March 2014].

Bermúdez Liévano, A. (2012). Una carrera geopolítica. El Ártico se derrite y desata otra Guerra Fría entre varios países. *La Nación*. [Print] 30th September. Available from: Biblioteca Nacional de Chile.

Bert, M. (2012). The Arctic Is Now: Economic and National Security in the Last Frontier. *American Foreign Policy Interests*. 34(1). pp. 5–19.

Bilefsky, D. (2013). Talks on Antarctic Marine Reserve Fail to Reach Agreement. *New York Times*. [Print] 2nd November. Available from: ProQuest Database. [Accessed: 30th March 2014].

Bittner, J. (2016). East vs. West in the Arctic Circle. *New York Times*. [Online] 28th April. Available from: http://nytimes.com/2016/04/28/opinion/east-vs-west-in-the-arctic-circle.html ?_r=0. [Accessed: 6th February 2018].

Blanco Wells, G., Fuenzalida, M. I. (2013). La construcción de agendas científicas sobre cambio climático y su influence en la territorialización de políticas públicas: reflexiones a partir del caso chileno. In Postigo, J. C., et al. (eds.) *Cambio climático, movimientos sociales y políticas públicas*. Santiago: Instituto de Ciencias Alejandro Lipschutz. pp. 75–102.

Bloom, E. (2016). *The History, Vision behind and Impact of the Protocol on Environmental Protection to the Antarctic Treaty*. [Online] Available from: https://state.gov/e/oes/rls/ remarks/2016/258286.htm. [Accessed: 21st November 2016].

Bocchicchio, S. (2006). En el Malba. Depaten los efectos del cambio climático. *La Nación*. [Print] 13rd September. Available from: Biblioteca Nacional Mariano Moreno.

Bombin Sanhueza, J. S. (2009). La Política Antártica Chilena. *Revismar*. 5. pp. 446–454.

Borgerson, S. G. (2008). Arctic Meltdown: The Economic and Security Implications of Global Warming. *Foreign Affairs*. 87(2). pp. 63–77.

Borgerson, S. G., Antrim, C. (2009). An Arctic Circle of Friends. *New York Times*. [Print] 28th March. Available from: ProQuest Database. [Accessed: 30th March 2014].

Bostrom, M. (2010). A Climate Plan for Climate-Change Deniers. *The Washington Post*. [Print] 14th November. Available from: ProQuest Database. [Accessed: 30th March 2014].

Bottici, C. (2014). *Imaginal Politics: Images beyond Imagination and the Imaginary*. New York: Columbia University Press.

Boyer, V. (2008). Drilling Offshore Won't Necessarily Help U.S. *USA Today*. [Print] 18th July. Available from: ProQuest Database. [Accessed: 30th March 2014].

Brady, A.-M. (2012). *The Emerging Politics of Antarctica*. London, New York: Routledge.

Bratspies, R. (2015). Chapter 11: Using Human Rights to Improve Arctic Governance. In: Ali, S. H. & Pincus, R. (eds.) *Diplomacy on Ice: Energy and the Environment in the Arctic and Antarctic*. New Haven, London: Yale University Press. pp. 171–185.

Brede, F., Schultze, R. O. (2008). Das politische System Kanadas. In Stüwe, K. & Rinke, S. (eds.) *Die politischen Systeme in Nord und Lateinamerika*. Wiesbaden: Verlag für Sozialwisenschaften. pp. 315–340.

Brigham, L. W. (2010). Think Again: The Arctic. *Foreign Policy*. [Online] Available from: https://foreignpolicy.com/2010/08/06/think-again-the-arctic/. [Accessed: 1st April 2016].

British Antarctic Survey. (2015). *Mining*. [Online] Available from: www.bas.ac.uk/about/antarctica/environmental-protection/mining/. [Accessed: 6th February 2018].

Broadhead, L.-A. (2010). Canadian Sovereignty Versus Northern Security: The Case for Updating our Mental Map of the Arctic. *International Journal*. 65(4). pp. 913–930.

Broder, J. M. (2012a). Ice Forces Shell to Halt Work on Arctic Well. *New York Times*. [Print] 11th September. Available from: ProQuest Database. [Accessed: 30th March 2014].

Broder, J. M. (2012b). U.S. Approves an Initial Step in Oil Drilling Near Alaska. *New York Times*. [Print] 31st August. Available from: ProQuest Database. [Accessed: 30th March 2014].

Broder, J. M. (2012c). Alaska: Groups Sue over Drilling in Arctic Ocean. *New York Times*. [Print] 11th July. Available from: ProQuest Database. [Accessed: 30th March 2014].

Broder, J. M. (2013). To Resume Drilling in Arctic, Shell Is Told to Address Safety and Other Problems. *New York Times*. [Print] 15th March. Available from: ProQuest Database. [Accessed: 30th March 2014].

Broder, J. M., Krauss, C. (2012). Offshore Oil Drilling's New and Frozen Frontier. *New York Times*. [Print] 24th May. Available from: ProQuest Database. [Accessed: 30th March 2014].

Brooke, J. (2000). A Big Push Is on for Natural Gas under the Arctic. *New York Times*. [Print] 28th September. [Accessed: 31st March 2014].

Browne, M. W. (1997). Ice Shifts May Be Tied to Warming. *New York Times*. [Print] 18th November. Available from: ProQuest Database. [Accessed: 31st March 2014].

Bruun, J. M., Medby, I. A. (2014). Theorising the Thaw: Geopolitics in a Changing Arctic. *Geography Compass*. 8(12). pp. 915–929.

Butts, G. (2010). An Arctic Spill Would Be Even Worse. *Globe and Mail*. [Print] 11th June. Available from: ProQuest Database. [Accessed: 30th March 2014].

Buzza Smith, P. (2010). More Harper Hypocrisy on Arctic. *Toronto Star*. [Print] 1st September. Available from: ProQuest Database. [Accessed: 30th March 2014].

Byers, M. (2007a). Sovereignty Will Solve the Northwest Passage Dispute. *Globe and Mail*. [Print] 11th August. Available from: ProQuest Database. [Accessed: 30th March 2014].

Byers, M. (2007b). Climate Change: How Much Does Canada Care? *Globe and Mail*. [Print] 29th September. Available from: ProQuest Database. [Accessed: 30th March 2014].

Byers, M. (2009). *Who Owns the Arctic: Understanding Sovereignty Disputes in the North*. Vancouver: Douglas & McIntyre.

Byers, M. (2009a). Unleashing the Human Potential in Canada's North. *Globe and Mail*. [Print] 20th August. Available from: ProQuest Database. [Accessed: 30th March 2014].

Byers, M. (2009b). The Northwest Passage is Already Canadian. *Globe and Mail*. [Print] 27th October. Available from: ProQuest Database. [Accessed: 30th March 2014].

Byers, M. (2010). It's Time to Resolve Our Arctic Differences. *Globe and Mail*. [Print] 30th April. Available from: ProQuest Database. [Accessed: 30th March 2014].

Byers, M. (2011). Russia Pulling ahead in the Arctic. *Toronto Star*. [Print] 29th December. Available from: ProQuest Database. [Accessed: 30th March 2014].

CAFF. (2013). *Arctic Biodiversity Assessment. Status and trends in Arctic biodiversity.* [Online] Available from: http://caff.is/assessment-series/10-arctic-biodiversity-assessment/233-arctic-biodiversity-assessment-2013. [Accessed: 9th December 2015].

Calamai, P. (2003a). Massive Ice Shelf Breaks Up as Arctic Temperatures Climb. *Toronto Star*. [Print] 23rd September. Available from: ProQuest Database. [Accessed: 30th March 2014].

Calamai, P. (2003b). Coldest Continent Heating Up: Scientific Interest in Antarctica Is on the Rise and Effects of Global Warming Taking a Toll. *Toronto Star*. [Print] 11th October. Available from: ProQuest Database. [Accessed: 30th March 2014].

Campbell, M. (2008). Huge Sections of Northern Ice Shelf Lost in August, Researchers Report. *Globe and Mail*. [Print] 3rd September. Available from: ProQuest Database. [Accessed: 30th March 2014].

Canadian Standing Senate Committee on Fisheries and Oceans. (2009). *Rising to the Arctic Challenge: Report on the Canadian Coast Guard*. [Online] Available from: http://parl.gc.ca/Content/SEN/Committee/402/fish/rep/rep02may09-e.pdf. [Accessed: 23rd November 2016].

Carr, D. A. (1999). Protect Wilderness from Drilling. *New York Times*. [Print] 28th March. Available from: ProQuest Database. [Accessed: 30th March 2014].

Castoriadis, C. (1998). *The Imaginary Institution of Society*. Cambridge, MA: MIT Press.

Castree, N. (2003). The Geopolitics of Nature. In Agnew, J., Mitchell, K. & Ó Tuathail, G. (eds.) *A Companion to Political Geography*. Maiden, Oxford, Melbourne, Berlin: Blackwell Publishers Ltd. pp. 423–439.

Cavaney, R. (2002). Arctic Oil Security. *New York Times*. [Print] 20th March. Available from: ProQuest Database. [Accessed: 30th March 2014].

Cavaney, R. (2006). Oil in Arctic Refuge. *New York Times*. [Print] 20th March. Available from: ProQuest Database. [Accessed: 30th March 2014].

Cecchi, H. (2002). Se desprendio una masa de hielo de 3250 km2. Una ruptura escalofriante. *Página 12*. [Print] 20th March. Available from: Biblioteca Nacional Mariano Moreno.

CEP. (2011). *Revised Rules of Procedure for the Committee for Environmental Protection*. [Online] Available from: http://ats.aq/documents/keydocs/vol_2/Rules_cep_e.pdf. [Accessed: 30th April 2016].

Chang, K. (2002). Patrones climáticos contradictorios. La Antártida confunde a los científicos. *La Nación*. [Print] 3rd April. Available from: Biblioteca Nacional Mariano Moreno.

Charest, J., Doer, G. (2005). Seize the Climate-Friendly Day: Reversing Climate Change Isn't Just about Sacrifices. *Globe and Mail*. [Print] 7th December. Available from: ProQuest Database. [Accessed: 30th March 2014].

Charron, A. (2012). Canada and the Arctic Council. *International Journal*. 67(3). pp. 765–783.

Charron, A. (2014). *Has the Arctic Council Become Too Big? International Relations and Security Network (ISN)*. [Online] Available from http://isn.ethz.ch/ Digital-Library/Articles/Detail/?id=182827. [Accessed: 18th January 2016].

Charter of the Organization of American States. (1948). *Organization of American States*. [Online] Available from: http://oas.org/en/sla/dil/docs/inter_american_treaties_A-41_charter_OAS.pdf. [Accessed: 30th July 2016].

Chaturvedi, S. (2011). *Antarctica: 'A Global Knowledge Commons'.* Position paper for the 6th Open Assembly of the Northern Research Forum, Hveragerdi, Iceland, 4th–6th September. [Online] Available from: http://rha.is/static/files/NRF/OpenAssemblies/Hveragerdi2011/position_papers/chaturvedi-antarcticaa_global_knowledge_commons-nrf_2011.pdf. [Accessed: 18th April 2016].

Child, J. (1988). *Antarctica and South American Geopolitics: Frozen Lebensraum.* Westport, CT, London: Praeger.

Chivian, E. (2007). Climate Change: No Time to Debate. *New York Times.* [Print] 2nd January. Available from: ProQuest Database. [Accessed: 30th March 2014].

Christensen, M., Nilsson, A. E., Wormbs, N. (2013). Globalization, Climate Change and the Media: An Introduction. In Christensen, M., Nilsson, A. E. & Wormbs, N. (eds.) *Media and the Politics of Arctic Climate Change: When the Ice Breaks.* London: Palgrave Macmillan. pp. 1–25.

Claes, D. H., Moe, A. (2014). Arctic Petroleum Resources in a Regional and Global Perspective. In Tamnes, R. & Offerdal, K. (eds.) *Geopolitics and Security in the Arctic.* London: Routledge. pp. 97–120.

Clarín. (1998). *Ecologia: Campaña contra la contaminación del mar y los rios Argentinos. Un barco de Greenpeace vino en viaje de protesta.* [Print] 1st December. Available from: Biblioteca Nacional Mariano. Moreno.

Clarín. (1999). *Los Cambios Climaticos: Advertencia por el Calentamiento de la tierra. Clinton revelará fotos secretas de la Antártdia para alertar por el clima.* [Print] 16th September. Available from: Biblioteca Nacional Mariano Moreno.

Clarín. (2001). *Quieren impedir que barco japoneses cacen cetaceso en el mar antartico. Greenpeace interceptó a balleneros.* [Print] 15th December. Available from: Biblioteca Nacional Mariano Moreno.

Clarín. (2003). *El frente externo. Discurso completo de Kirchner en la ONU.* [Print] 26th September. Available from: Biblioteca Nacional Mariano Moreno.

Clarín. (2004). *Los cien años en la Antártida.* [Print] 26th February. Available from: Biblioteca Nacional Mariano Moreno.

Clarín. (2007). *Influencias del cambio climático en la Patagonia.* [Print] 27th April. Available from: Biblioteca Nacional Mariano Moreno.

Clarín. (2008a). *Editorial: Las luchas por los recursos naturales.* [Print] 27th June. Available from: Biblioteca Nacional Mariano Moreno.

Clarín. (2008b). *Editorial: Las pretensiones sobre la Antártida.* [Print] 17th May. Available from: Biblioteca Nacional Mariano Moreno.

Clarín. (2009). *Antártida: diputados de Argentina y Chile rechazan el reclamo inglés.* [Print] 7th March. Available from: Biblioteca Nacional Mariano Moreno.

Clarín. (2010). *Con la Antártida en su real proporción, hay un nuevo mapa oficial de la Argentina.* [Print] 24th November. Available from: Biblioteca Nacional Mariano Moreno.

Clarín. (2012a). *Cambios que preocupan a los investigadores.* [Print] 26th July. Available from: Biblioteca Nacional Mariano Moreno.

Clarín. (2012b). *Un iceberg que mide más de la mitad de la Capital, a la deriva: En el Norte de Groenlandia.* [Print] 20th July. Available from: Biblioteca Nacional Mariano Moreno.

Clarín. (2012c). *Aumentar la presencia argentina en la Antártida.* [Print] 29th March. Available from: Biblioteca Nacional Mariano Moreno.

Clarín. (2012d). *Cristina llegó a Chile y agradeció el apoyo por las Malvinas.* [Print] 16th March. Available from: Biblioteca Nacional Mariano Moreno.

Clarín. (2013). *Malvinas guardará por siempre nuestra memoria.* [Print] 2nd April. Available from: Biblioteca Nacional Mariano Moreno.

Clausen, M., Clausen, D. (2013). Conceptualizing Climate Security for a Warming World: Complexity and the Environment-Conflict Linkage. In Zellen, B. S. (ed.) *The Fast-Changing Arctic: Rethinking Arctic Security for a Warmer World.* Calgary: Calgary University Press. pp. 57–82.

Clavin, P. (2005). Defining Transnationalism. *Contemporary European History.* 14(4). pp. 421–439.

Coates, K. S., et al. (2008). *Arctic Front: Defending Canada in the Far North.* Toronto: T. Allen Publishers.

Coates, P. (1989). Project Chariot. Alaskan Roots of Environmentalism. *Alaska History.* 4. pp. 1–31.

Cochran, P. (2009). *Introductory Remarks by Patricia Cochran, Chair, Inuit Circumpolar Council.* [Speech] Arctic Council 6th Ministerial Meeting in Tromsø, 29 April. [Online] Available from: https://oaarchive.arctic-council.org/handle/11374/1562. [Accessed: 1st April 2016].

Cohen, H. (2011). *Public Participation in Antarctica: The Role of Non-Governmental and Intergovernmental Organizations.* International Union for Conservation nature. Smithsonian Libraries. [Online] Available from: https://repository.si.edu/handle/10088/16184. [Accessed: 28th April 2016].

Colacrai, M. (2012). La Política Antártica Argentina y su compromiso con el Tratado Antártico. *Anuario de la Asociación Argentina de Derecho Internacional.* Septiembre. pp. 267–276.

Colacrai de Trevisan, M. (1994). La política exterior Argentina y la cuestión Antártica: Un ejemplo de negociación permanente dentro del regimen Antártico. In Centro de Estudios en Relaciones Internacionales de Rosario (ed.) *La Política exterior del gobierno de Menem.* Rosario: Ediciones CERIR. pp. 337–355.

Collins, M. (2012). Investing in the North. *Toronto Star.* [Print] 2nd January. Available from: ProQuest Database. [Accessed: 30th March 2014].

Comier, Z. (2007). Look out below/Antarctica: The New Hot Real Estate: There's Oil and Gas in the Antarctic, Too, Which Global Warming May Open Up. *Toronto Star.* [Print] 18th November. Available from: ProQuest Database. [Accessed: 30th March 2014].

COMNAP. (2016). *About the Chilean National Antarctic Program.* [Online] Available from: https://comnap.aq/Members/INACH/SitePages/Home.aspx. [Accessed: 22nd November 2016].

Conley, H., Kraut, J. (2010). *U.S. Strategic Interests in the Arctic: An Assessment of Current Challenges and New Opportunities for Cooperation.* Washington, DC: Center for Strategic and International Studies.

Conley, H., Toland, T., David, M., Jegorova, N. (2013). *The New Foreign Policy Frontier: U.S. Interests and Actors in the Arctic.* Washington, DC: Center for Strategic and International Studies.

Conrad, S., Randeria, S. (2002). Geteilte Geschichten – Europa in einer postkolonialen Welt. In Conrad, S., Randeria, S. & Sutterlüty, B. (eds.) *Jenseits des Eurozentrismus. Postkoloniale Perspektiven in den Geschichts- und Kulturwissenschaften.* Frankfurt am Main, New York: Campus. pp. 9–49.

Consejo de Política Antártica. (2014). *Plan Estratégico Antártico 2015–19.* [Online] Available from: http://minrel.gov.cl/nacionales/minrel/2012-10-10/172919.html. [Accessed: 22nd November 2016].

Cornut, J. (2010). Why and When We Study the Arctic in Canada. *International Journal.* 65(4). pp. 943–953.

Corry, O. (2017). Governing the Arctic Climate: Geoengineering in a Global Polity. In Keil, K. & Knecht, S. (eds.) *Governing Arctic Change-Global Perspectives.* Basingstoke, Hampshire: Palgrave Macmillan. pp. 59–78.

Costa, S. (2011). Researching Entangled Inequalities in Latin America: The Role of Historical, Social and Transregional Interdependencies. *desiguALdades.net Working Paper Series.* Vol. 9. Berlin: desiguALdades.net Research Network on Interdependent Inequalities in Latin America.

Curia, W. (2008). Entrevista Jorge Taiana Canciller 'La presentación en la ONU es ejercitar nuestros derechos'. *Clarín.* [Print] 1st September.

Curry, B. (2010). An Apology Five Decades in the Making. *Globe and Mail.* [Print] 19th August 2007. Available from: ProQuest Database. [Accessed: 30th March 2014].

Dalby, S. (2003). Geopolitical Identities: Arctic Ecology and Global Consumption. *Geopolitics.* 8(1). pp. 181–202.

Dalby, S. (2012). *Environmental Geopolitics in the Twenty First Century.* Presentation to the Sussex Conference on Rethinking Climate change, Conflict and Security, 18th–19th October.

Dalby, S., Ó Tuathail, G. (1998). *Rethinking Geopolitics.* New York: Routledge.

Dalby, S., Routledge, P., Ó Tuathail, G. (2006). *The Geopolitics Reader.* Abingdon, New York: Routledge.

Degeorges, D., Ali, S. H. (2015). Chapter 9: Connecting China through Creative Diplomacy. Greenland, Australia, and Climate Cooperation in Polar Regions. In: Ali, S. H. & Pincus, R. (eds.) *Diplomacy on Ice: Energy and the Environment in the Arctic and Antarctic.* New Haven, London: Yale University Press. pp. 151–160.

Delacourt, S. (1991). Background: Inuit claim: The Long Journey to Our Land. *Globe and Mail.* [Print] 17th December. Available from: ProQuest Database. [Accessed: 30th March 2014].

De La Torre, V. (2005). Alaska Natives Offer a Herd of Reasons to Block Oil Drilling. *The Washington Post.* [Print] 20th September. Available from: ProQuest Database. [Accessed: 30th March 2014].

Delgado, J. P. (2000) A myth begins to melt. *Globe and Mail.* [Print] 26th August 2000. Available from: ProQuest Database. [Accessed: 30th March 2014].

Dexter, B. (1989). Canada Ignoring Cold War Thaw, Greenpeace Says. *Toronto Star.* [Print] 24th January. Available from: ProQuest Database. [Accessed: 31st March 2014].

Dinatale, M. (2012). El escenario. Una jugada política para ser tenidos en cuenta. *La Nación.* [Print] 13th June. Available from: Biblioteca Nacional Mariano Moreno.

Dirección Nacional del Antártico. (2004). *Decisión Administrativa 509/2004.* [Online] Available from: http://servicios.infoleg.gob.ar/infolegInternet/anexos/100000-104999/100093/norma.htm. [Accessed: 24th November 2016].

Dirección Nacional del Antártico. (2006). *Plan anual antártico 2006 científico técnico y de servicios.* [Online] Available from: http://iri.edu.ar/publicaciones_iri/anuario/CD%20Anuario%202006/Malvinas/MALV7.pdf. [Accessed: 24th November 2016].

Dirección Nacional del Antártico. (2010). *Plan anual antártico 2010. Científico técnico y de servicios.* [Online] Available from: http://marambio.aq/pdf/ PAN2011-12.pdf. [Accessed: 24th November 2016].

Dirección Nacional del Antártico. (2014). *Plan anual antártico 2014 científico técnico y de servicios.* [Online] Available from: http://iri.edu.ar/publicaciones_iri/anuario/CD%20Anuario%202006/Malvinas/MALV7.pdf. [Accessed: 24th November 2016].

Dirección Nacional del Antártico. (2015). *Plan annual antártico 2015. Científico técnico y de servicios.* [Online] Available from: http://www.dna.gov.ar/DIVULGAC/PAA1415. PDF. [Accessed: 24th November 2016].

Dodds, K. (1997). *Geopolitics of Antarctica: Views from the Southern Oceanic Rim.* Chichester, New York, Weinheim, Brisbane, Singapore, Toronto: Wiley John + Sons.

Dodds, K. (2001). Political Geography III: Critical Geopolitics after Ten Years. *Progress in Human Geography.* 25(3). pp. 469–484.

Dodds, K. (2010). The Geopolitics of Regionalism: The Valdivia Group and Southern Hemispheric Environmental Co-Operation. *Third World Quarterly.* 19(4). pp. 725–743.

Dodds, K. (2011). Sovereignty Watch: Claimant States, Resources, and Territory in Contemporary Antarctica. *Polar Record.* 47(3). pp. 231–243.

Dodds, K. (2014). *Geopolitics: A Very Short Introduction.* Oxford: Oxford University Press.

Dodds, K. (2015). Hope, Fear and Dread: Imagining the Arctic and Antarctic in 100 Years' Time. In Kennedy, D. (ed.) *The Arctic and Antarctica: Differing Currents of Change.* Wellington: New Zealand Institute of International Affairs. pp. 73–82.

Dodds, K., Hemmings, A. D. (2009). Frontier Vigilantism: Australia and Contemporary Representations of Australian and Antarctic Territory. *Australian Journal of Politics and History.* 55(4). pp. 513–529.

Dodds, K., Kuus, M., Sharp, J. (2013). *The Ashgate Research Companion to Critical Geopolitics.* Farnham, Surrey, England, Burlington, VT: Ashgate Publishing Limited.

Dodds, K., Nuttall, M. (2016). *The Scramble for the Poles.* Cambridge, Malden: Polity Press.

Doel, R. E., Wrakberg, U., Zeller, S. (2014). Science, Environment, and the New Arctic. *Journal of Historical Geography.* 44. pp. 2–14.

Duyck, S. (2011). Drawing Lessons for Arctic Governance from the Antarctic Treaty System. *The Yearbook of Polar Law.* 3. pp. 683–713.

Duyck, S. (2012). Participation of Non-State Actors in Arctic Environmental Governance. *Nordia Geographical Publications.* 40(4). pp. 99–110.

Duyck, S. (2015). Chapter 1: Polar Environmental Governance and Nonstate Actors. In: Ali, S. H. & Pincus, R. (eds.) *Diplomacy on Ice: Energy and the Environment in the Arctic and Antarctic.* New Haven, London: Yale University Press. pp. 13–40.

The Economist. (2015). Antarctica: Core Values. [Online] 31st January. Available from: http://economist.com/news/international/21641239-southern-contine. [Accessed: 6th March 2018].

Editor and Publisher. (2013). *Newspaper Data Book: The Encyclopedia of the Newspaper Industry: Book 1.* Irvine: Duncan McIntosh Co., Inc.

Egan, T. (1991). The Great Alaska Debate. *New York Times.* [Print] 4th August. Available from: ProQuest Database. [Accessed: 30th March 2014].

Eilperin, J. (2010). Climate Change Experts Adjust Strategy. *The Washington Post.* [Print] 30th November. Available from: ProQuest Database. [Accessed: 30th March 2014].

Eilperin, J. (2012). In a Cold Place, Warming Is Most evident. *The Washington Post.* [Print] 6th August. Available from: ProQuest Database. [Accessed: 30th March 2014].

Eilperin, J., Mufson, S. (2013b). Shell Suspends Exploratory Drilling Off Coast of Alaska. *The Washington Post.* [Print] 28th February. Available from: ProQuest Database. [Accessed: 30th March 2014].

Einhorn, B. (2014). *As China Goes Exploring, Antarctica Becomes Another Frontier.* [Online] Available from: http://bloomberg.com/news/articles/2014-01-03/as-china-goes-exploring-antarctica-becomes-another-frontier. [Accessed: 1st April 2016].

Elliot-Meisel, E. (2009). Politics, Pride, and Precedent: The United States and Canada in the Northwest Passage. *Ocean Development & International Law.* 40(2). pp. 204–232.

El Mercurio. (1990). *Conservacionistas: Piden Declarar Parque Mundial a la Antártida.* [Print] 14th November. Available from: Biblioteca Nacional de Chile.

El Mercurio. (1991a). *En Encuentro Juvenil: Aylwin Anunció Pronta Política del Ambiente.* [Print] 28th September. Available from: Biblioteca Nacional de Chile.

El Mercurio. (1991b). *Por 50 años: Prohibirán la Exploración Minera en la Antártida.* [Print] 1st October. Available from: Biblioteca Nacional de Chile.

El Mercurio. (1997). *En Kioto, potencias difi en sobre nivel de reducción de gases de efecto invernadero. Pocas Esperanzas para Conferencia Climática.* [Print] 1st December. Available from: Biblioteca Nacional de Chile.

El Mercurio. (2008). *Los recursos del Polo Sur, en la mira de Rusia.* [Print] 19th February. Available from: Biblioteca Nacional de Chile.

El Mercurio. (2010). *La política mundial necesita de predicciones científi as más contundentes.* [Print] 2nd December. Available from: Biblioteca Nacional de Chile.

El Mercurio. (2012). *Proyección antártica.* [Print] 19th January. Available from: Biblioteca Nacional de Chile.

El president de la Nación Argentina. (1990). *Política Nacional Antártica.* [Online] Available from: https://www.dipublico.org/doc/legislacion/Decreto2316-1990.pdf. [Accessed: 24th November 2016].

Emirbayer, M. (1997). Manifesto for a Relational Sociology. *American Journal of Sociology.* 103(2). pp. 281–317.

Emmerson, C. (2010). *The Future History of the Arctic.* New York: Public Affairs.

Emmerson, C. (2011). *The Future History of the Arctic: How Climate, Resources and Geopolitics Are Reshaping the North, and Why It Matters to the World.* London: Vintage Books.

English, J. (2013). *Ice and Water: Politics, Peoples, and the Arctic Council.* Toronto, Ontario: Penguin Canada Books Inc.

Entman, R. M. (1993). Framing: Toward Clarification of a Fractured Paradigm. *Journal of Communication.* 43(4). pp. 51–58.

Epple, A. (2014). *Geographical Imaginaries: Global Microhistory as an Answer to the Shortcomings of Global Syntheses.* Presentation, Workshop Conceptualizing (Geo) Political Imaginaries, 13th January, Centrum for Interdisciplinary Research, Bielefeld.

Ette, O. (2012). Globalisierung IV. Im Netz transarchipelischer Beziehungen: Von der Fülle des Polyperspektivischen und der Falle einsprachiger Globalisierung. In Ette, O. (ed.) *TransArea.* Berlin: De Gruyter. pp. 221–314.

Ezcurra, E. (2013). Opinión. Lo peligroso es el petróleo. *Página 12.* [Print] 17th October. Available from: Biblioteca Nacional Mariano Moreno.

Fagan, D. (1989). Development Proves a Mixed Blessing for North. *Globe and Mail.* [Print] 24th April. Available from: ProQuest Database. [Accessed: 30th March 2014].

Fagan, D. (1990). VALDEZ A Year Later Alaskan Push toward Development under Fire. *Globe and Mail.* [Print] 22nd March. Available from: ProQuest Database. [Accessed: 30th March 2014].

Fairclough, N. (1992). *Discourse and Social Change.* Cambridge: Polity Press.

Faist, T., Özveren, E. (2004). *Transnational Social Spaces: Agents, Networks and Institutions.* Aldershot: Ashgate.

Farnsworth, C. H. (1992). Canada to Divide Its Northern Land. *New York Times.* [Print] 6th May. Available from: ProQuest Database. [Accessed: 30th March 2014].

Ferguson, D. (1994). Compensation for Inuit Urged 1950s Relocation Was Inhumane, Commission Says. *Toronto Star.* [Print] 14th July. Available from: ProQuest Database. [Accessed: 30th March 2014].

Ferguson, Y. H., Jones, R. J. B. (2002). Political Space and Global Politics. In Ferguson, Y. H. & Barry Jones, R. J. (eds.) *Political Space: Frontiers of Change and Governance in a Globalizing World*. New York: State University of New York Press. pp. 1–20.

Ferguson, Y. H., Mansbach, R. W. (2004). *Remapping Global Politics. History's Revenge and Future Shock*. Cambridge: Cambridge University Press.

Firestone, D. (2003). Drilling in Alaska: A Priority for Bush: Fails in the Senate. *New York Times*. [Print] 20th March. Available from: ProQuest Database. [Accessed: 30th March 2014].

Fitch, C. (2016). *What Does Argentina's Territory Expansion Mean for the Falklands?* [Online] 7th April. Available from: http://geographical.co.uk/geopolitics/geopolitics/item/1625-argentina-s-atlantic-territoryexpan. [Accessed: 1st May 2016].

Foucault, M. (2015). *The Archeology of Knowledge*. 1st Ed. 1972. [Online] Available from: https://rosswolfe.files.wordpress.com/2015/01/michel-foucault-the-archaeology-of-knowledge.pdf. [Accessed: 6th February 2018].

Fountain, H. (2013). Runaway Oil Rig Off Alaska under Control, Shell Says. *New York Times*. [Print] 1st January. Available from: ProQuest Database. [Accessed: 30th March 2014].

Fraga, J. A. (1992). *La Antartida Reserva Ecologica. Al cumplir 30 años su tratado*. Buenos Aires: Instituto de Publicaciones Navales.

Fraga, R. (2012). Malvinas se enmarca en otras disputas por recursos naturales. Reclamos estratégicos sobre el mar. *La Nación*. [Print] 30th January. Available from: Biblioteca Nacional Mariano Moreno.

Friedman, T. L. (2001). Drilling in the Cathedral. *New York Times*. [Print] 2nd March. Available from: ProQuest Database. [Accessed: 30th March 2014].

Gaffoglio, L. (2012). Un viaje a un paraíso amenazado. *La Nación*. [Print] 13th October. Available from: Biblioteca Nacional Mariano Moreno.

Galloway, G. (2010). Offshore Drilling Given Thumbs-Up by Senators. *Globe and Mail*. [Print] 19th August. Available from: ProQuest Database. [Accessed: 30th March 2014].

Gamba, V. (2012). Opinión. Dos bases británicas con mucho valor estratégico. *La Nación*. [Print] 4th April. Available from: Biblioteca Nacional Mariano Moreno.

García, R. (2013). Queremos mostrar que a Chile le importa la Antártica. *El Mercurio*. [Print] 23th May. Available from: Biblioteca Nacional de Chile.

Genest, E. (2004). *Política Antártica Argentina*. [Online] Available from: http://dna.gov.ar/DIVULGAC/SINAPA04.DOC. [Accessed: 4th January 2016].

Gerth, J. (2005). Big Oil Steps Aside in Battle Over Arctic. *New York Times*. [Print] 21st February. Available from: ProQuest Database. [Accessed: 30th March 2014].

Gilbert, R. (2005). Raising Awareness of Our Role in Arctic. *Toronto Star*. [Print] 23rd September. Available from: ProQuest Database. [Accessed: 30th March 2014].

Gillies, R. (2006). Giant Ice Shelf Breaks Free in Arctic; Climate Change Cited as Major Factor. *The Washington Post*. [Print] 30th December 2006. Available from: ProQuest Database. [Accessed: 30th March 2014].

Global Affairs Canada. (2016). *Canada's Extended Continental Shelf*. [Online] Available from: http://international.gc.ca/arctic-arctique/continental/index.aspx?lang=eng. [Accessed: 1st April 2016].

Globe and Mail. (1990). *No Apology for Inuit*. [Print] 20th November. Available from: ProQuest Database. [Accessed: 30th March 2014].

Globe and Mail. (1995a). *The World Decision Expected on Wildlife Refuge*. [Print] 18th September. Available from: ProQuest Database. [Accessed: 30th March 2014].

Globe and Mail. (1995b). *Ottawa May Deny Inuit Compensation*. [Print] 10th October. Available from: ProQuest Database. [Accessed: 30th March 2014].

Globe and Mail. (1997). *Up-Close Look at Global Warming.* [Print] 12th December. Available from: ProQuest Database. [Accessed: 30th March 2014].

Globe and Mail. (2004). *Report on Arctic Ice Warns 'The Big Melt Has Begun'.* [Print] 3rd November. Available from: ProQuest Database. [Accessed: 30th March 2014].

Globe and Mail. (2006). *Reclaiming Canada's Role as a World Player.* [Print] 22nd September 2006. Available from: ProQuest Database. [Accessed: 30th March 2014].

Globe and Mail. (2007a). *Canada's Polar Stake.* [Print] 3rd March. Available from: ProQuest Database. [Accessed: 30th March 2014].

Globe and Mail. (2007b). *The U.S. Interest in Canada's Claim.* [Print] 22nd August. Available from: ProQuest Database. [Accessed: 30th March 2014].

Globe and Mail. (2008a). *China Eyes Arctic Regions for Resources, Expert Says.* [Print] 26th February. Available from: ProQuest Database. [Accessed: 30th March 2014].

Globe and Mail. (2008b). *Quote of the Day.* [Print] 27th August. Available from: ProQuest Database. [Accessed: 30th March 2014].

Globe and Mail. (2008c). *Arctic Sea Ice Melt Reaches Second-Lowest Dimension in Generation.* [Print] 17th September. Available from: ProQuest Database. [Accessed: 30th March 2014].

Globe and Mail. (2009). *Real Priority, But Unsolved Misery.* [Print] 15th August. Available from: ProQuest Database. [Accessed: 30th March 2014].

Goar, C. (1995). Arctic Reserve Will be Saved, Clinton Vows. *Toronto Star.* [Print] 28th October. Available from: ProQuest Database. [Accessed: 30th March 2014].

Gorostegui, J., Waghorn, R. (2012). *Chile en la Antártica. Nuevos desafíos y perspectivas.* Santiago: Lom.

Gorrie, P. (2007). Landmark UN Study Backs Climate Theory; 2,000 Scientists All But End the Debate Human Activity Causes Global Warming. *Toronto Star.* [Print] 19th January. Available from: ProQuest Database. [Accessed: 30th March 2014].

Government of Alaska. (2015a). *Arctic Policy and Climate Change.* [Online] Available from: https://gov.alaska.gov/Walker_media/transition_page/arctic-policy-and-climate-change.final.pdf. [Accessed: 17th September 2016].

Government of Alaska. (2015b). *Final Report of the Alaska Arctic Policy Commission.* [Online] Available from: http://akarctic.com/wp-content/uploads/2015/01/AAPC_final_report_lowres.pdf. [Accessed: 17th September 2016].

Government of Canada. (2009). *Canada's Northern Strategy: Our North, Our Heritage, Our Future.* [Online] Available from: http://northernstrategy.gc.ca/cns/cns-eng.asp. [Accessed: 23rd November 2016].

Government of Canada. (2010). *Statement on Canada's Arctic Foreign Policy: Exercising Sovereignty and Promoting Canada's Northern Strategy Abroad.* [Online] Available from: http://international.gc.ca/arctic-arctique/assets/pdfs/canada_arctic_foreign_policy-eng.pdf. [Accessed: 17th September 2016].

Goycoolea, L. U., Luco, N. R. (2007). En la Antártica, lo peor es el turismo. *El Mercurio.* [Print] 30th November. Available from: Biblioteca Nacional de Chile.

Graczyk, P., Koivurova, T. (2014). A New Era in the Arctic Council's External Relations? Broader Consequences of the Nuuk Observer Rules for Arctic Governance. *Polar Record.* 50(3). pp. 225–236.

Graczyk, P., Koivurova, T. (2015). The Arctic Council. In Jensen, L. C. & Hønneland, G. (eds.) *Handbook of the Politics of the Arctic.* Cheltenham, Northampton: Edward Elgar Publishing. pp. 298–327.

Graham, W. (2002). *Future Perspectives on the Arctic: Intervention by the Canadian Minister of Foreign Affairs.* [Speech] Arctic Council 3rd Ministerial Meeting in Inari,

Finland, 9th–10th October. [Online] Available from: https://oaarchive.arctic-council. org/handle/11374/1584. [Accessed: 1st April 2016].

Grant, S. (2002). A Warning from a Chunk of Antarctic Ice. *New York Times.* [Print] 24th March. Available from: ProQuest Database. [Accessed: 30th March 2014].

Grant, S. (2010). Troubled Arctic Waters. *Globe and Mail.* [Print] 30th June. Available from: ProQuest Database. [Accessed: 30th March 2014].

Grid-Arendal. (2013). Boundaries of the Arctic Council Working Groups. [Online] Available from: http://grida.no/graphicslib/detail/boundaries-of-the-arctic-counc8385. [Accessed: 9th December 2015].

Griffiths, F. (2006). Breaking the Ice on Canada-U.S. Arctic Co-Operation. *Globe and Mail.* [Print] 22nd February. Available from: ProQuest Database. [Accessed: 30th March 2014].

Grondona, M. (2012). Sin las Malvinas, ¿es la nuestra una nación inconclusa? *La Nación.* [Print] 1st April. Available from: Biblioteca Nacional Mariano Moreno.

Gugliotta, G. (2003). Ice Shelf Break in Arctic Attributed to Climate Warming. *The Washington Post.* [Print] 23rd September. Available from: ProQuest Database. [Accessed: 30th March 2014].

Gundersen, A. (2011). *Statement by Arlene Gundersen: On Behalf of the Aleut International Association.* Seventh Ministerial Meeting Nuuk, Greenland, 12th May. [Speech] Arctic Council 7th Ministerial Meeting in Nuuk, Greenland, 12th May. [Online] Available from: https://oaarchive.arctic-council.org/handle/11374/530. [Accessed: 1st April 2016].

Gundersen, A. (2013). *Written Statement by AIA President Arlene Gundersen.* Eight Ministerial Meeting in Kiruna, Sweden, 15th May. [Online] Available from: https://oaarchive. arcticcouncil.org/bitstream/handle/11374/1569/ACMM08_Kiruna_2013_Statement_ AIA_Arlene_Gundersen.pdf?sequence=6&isAllowed=y. [Accessed: 1st April 2016].

Guzman, L. H. (2008). América se une en la Antártica. *El Mercurio.* [Print] 6th September. Available from: Biblioteca Nacional de Chile.

Haas, P. (1990). *Knowledge, Power, and International Policy Coordination.* New York: Columbia University Press.

Haase Ligget, D. (2009). *Tourism in the Antarctic: Modi Operandi and Regulatory Effectiveness.* Saarbrücken: VDM.

Hall, H. R. (1986). *Antarctica and World Politics: The Significance of political factors in Antarctic Affairs during the Twentieth Century.* Master thesis, University of Tasmania. [Online] Available from: http://eprints.utas.edu.au/15759/. [Accessed: 2nd February 2016].

Hänsch, L, Riekenberg, M. (2008). Das Politische System Argentiniens. In Stüwe, K. & Rinke, S. (eds.) *Die politischen Systeme in Nord und Lateinamerika.* Wiesbaden: Verlag für Sozialwissenschaften. pp. 59–84.

Hansom, J. D., Gordon, J. E. (2014). *Antarctic Environments and Resources: A Geographical Perspective.* Addison Wesley Longman: Harlow.

Harden-Donahue, A. (2010). Moratorium Needed on Arctic Drilling. *Toronto Star.* [Print] 12th July. Available from: ProQuest Database. [Accessed: 30th March 2014].

Harper, T. (2005). U.S. Closer to Arctic Oil Drilling. *Toronto Star.* [Print] 30th September. Available from: ProQuest Database. [Accessed: 30th March 2014].

Haycox, S. W., Mangusso, M. C. (1996). *An Alaskan Anthology: Interpreting the Past.* Seattle, London: University of Washington Press.

Hecht, G. (2011). *Entangled Geographies: Empire and Technopolitics in the Global Cold War.* Cambridge, London: The MIT Press.

Heininen, L. (2004). Circumpolar International Relations and Geopolitics. *Arctic Human Development Report.* pp. 207–225.

Heininen, L. (2013). *The Arctic of Regions vs. The Globalized Arctic*. [Presentation] UArctic TNs' Meeting, Arkhangelsk, January.

Heininen, L., Exner-Pirot, H., Plouffe, J. (2014). *The Arctic Yearbook 2014*. Akureyri: Northern Research Forum.

Heininen, L., Southcott, C. (2010). *Globalization and the Circumpolar North*. Fairbanks: University of Alaska Press.

Hemmings, A. D. (2009). From the New Geopolitics of Resources to Nanotechnology: Emerging Challenges of Globalism in Antarctica. *Yearbook of Polar Law I*. pp. 55–72.

Hemmings, A. D. (2015). Common Challenge: International Equity in the Arctic and Antarctica. In Kennedy, D. (ed.) *The Arctic and Antarctica: Different Currents of Change*. Wellington: New Zealand Institute of International Affairs. pp. 66–72.

Herr, R. A. (1997). The Changing Roles of Non-Governmental Organisations in the Antarctic Treaty System. In Schram Stokke, O., & Vidas, D. (eds.) *Governing the Antarctic: The Effectiveness and Legitimacy of the Antarctic Treaty System*. Cambridge: Cambridge University Press. pp. 91–112.

Higginbotham, J. (2013a). *A North American Alternative to a Russian Monopoly: Centre for International Governance Innovation*. [Online] Available from: https://cigionline.org/articles/north-american-alternativerussian-monopoly. [Accessed: 16th December 2016].

Higginbotham, J. (2013b). Nunavut and the New Arctic. *CIGI Policy Brief*. 27. pp. 1–10.

Higginbotham, J., Charron, A., Manicom, J. (2012). Canada-US Arctic Marine Corridors and Resource Development. *CIGI Policy Brief*. 24. pp. 1–10.

Higgins, M. (2009). Is Antarctica Getting Too Popular? *New York Times*. [Print] 22nd March. Available from: ProQuest Database. [Accessed: 30th March 2014].

Hinz, K. (2011). Wem gehört die zentrale Arktis? Geologie, Bathymetrie und das Seerecht. In Sapper, M., et al. (eds.) *Logbuch Arktis. Der Raum, die Interessen und das Recht*. Osteuropa: Zeitschrift für Gegenwartsfragen des Ostens. 61(2–3). pp. 87–92.

Hirst, D. (2014). *Negotiating Climates: The Politics of Climate Change and the Formation of the Intergovernmental Panel on Climate Change (IPCC), 1979–1992*. Master thesis, Manchester, UK: The University of Manchester.

Hitchins, D. R. M. (2011). An Alaskan Perspective: The Relationship between the US and Canada in the Arctic. *International Journal*. 66(4). pp. 971–977.

Hogan, M. P., Pursell, T. (2007). The Real Alaskan: Nostalgia and Rural Masculinity in the Last Frontier. *Men and Masculinities*. 11(1). pp. 63–85.

Hønneland, G. (2013). *The Politics of the Arctic*. Cheltenham, Northampton: Edward Elgar.

Hooghe, L., Marks, G. (2001). *Multi-Level Governance and European Integration*. New York, Oxford: Rowman & Littlefield Publishers, Inc.

Hough, P. (2013). *International Politics of the Arctic: Coming in from the Cold*. London, New York: Routledge.

Howard, R. (2009). *The Arctic Gold Rush: The New Race for Tomorrows Natural Resources*. Auckland: Pindar NZ.

Howkins, A. (2008). Reluctant Collaborators: Argentina and Chile in Antarctica during the International Geophysical Year, 1957–58. *Journal of Historical Geography*. 34(4). pp. 596–617.

Howkins, A. (2016). *The Polar Regions: An Environmental History*. Cambridge: Polity Press.

Huebert, R. (2007a). The Battle for the Arctic Is Heating Up. *Globe and Mail*. [Print] 30th July. Available from: ProQuest Database. [Accessed: 30th March 2014].

Huebert, R. (2007b). Canadian Arctic Maritime Security: The Return to Canada's Third Ocean. *Canadian Military Journal*. [Online] Available from: http://journal.dnd.ca/vo8/no2/huebert-eng.asp. [Accessed: 12th December 2016].

Huebert, R. (2009a). United States Arctic Policy: The Reluctant Arctic Power. *SPP Briefing Papers*. University of Calgary, The School of Public Policy.

Huebert, R. (2009b). Foreword by Rob Huebert: Inuit Endurance and the Arctic Transformation. In Zellen, B. S. (ed.) *On Thin Ice: The State, and the Challenge Sovereignty*. Maryland, Plymouth: Lexington Books. pp. vii–x.

Huebert, R. (2011a). Submarines, Oil Tankers, and Icebreakers: Trying to Understand Canadian Arctic Sovereignty and Security. *International Journal*. 66(4). pp. 809–824.

Huebert, R. (2011b). Canadian Arctic Sovereignty and Security in a Transforming Circumpolar World. In Griffiths, F., Huebert, R. & Lackenbauer, P. (eds.) *Canada and the Changing Arctic: Sovereignty, Security and Stewardship*. Waterloo: Wilfrid Laurier University Press. pp. 13–68.

Huebert, R. (2013). Cooperation or Conflict in the New Arctic? Too Simple of a Dichotomy. In Berkman, P. A. & Vylegzhanin, A. N. (eds.) *Environmental Security in the Arctic Ocean*. Dordrecht: Springer. pp. 195–204.

Huesseini de Araújo, S. (2012). Die zweite Heimat ‚des Orients'. In Dzudzek, I., Reuber, P. & Strüver, A. (eds.) *Die Politik räumlicher Repräsentationen – Beispiele aus der empirischen Forschung*. Berlin: LIT Verlag. pp. 132–151.

Hughes, J. (2007). Límites antárticos. *La Nación*. [Print] 26th December. Available from: Biblioteca Nacional Mariano Moreno.

Humrich, C. (2016). Sustainable Development in Arctic International Environmental Cooperation and the Governance of Hydrocarbon Related Activities, Chapter 2. In Pelaudeix, C. & Basse, E. M. (eds.) *The Governance of Arctic Offshore Oil and Gas*. Farnham: Ashgate, Gower.

Humrich, C. (2017). Coping with Institutional Challenges for Arctic Environmental Governance, Chapter 5. In Keil, K. & Knecht, S. (eds.) *Governing Arctic Change: Global Perspectives*. Basingstoke: Palgrave MacMillan. pp. 81–100.

Humrich, C., Wolf, K. D. (2012). *From Meltdown to Showdown? Challenges and Options for Governance in the Arctic*. Report No. 113. Frankfurt: Peace Research Institute Frankfurt (PRIF).

Ilulissat Declaration. (2008). [Online] Available from: https://regjeringen.no/global assets/upload/ud/080525_arctic_ocean_conference-_outcome.pdf. [Accessed: 1st April 2016].

Indigenous and Northern Affairs Canada. (2016). *Shared Arctic Leadership Model Engagement 2016*. [Online] Available from: https://aadnc-aandc.gc.ca/eng/1469120834151/14 6912090542. [Accessed: 21st November 2016].

Infante Caffi, M. T. (2006). La Política Antártica Chilena: Nuevas Realidades. *Revista de Estudios Internacionales*. 39(155). pp. 37–51.

Instituto Antártico Chileno. (2000). *Aspectos Institucionales y Jurídicos de la Actividad Antártica*. [Online] Available from: http://www.inach.cl/inach/wp-content/uploads/2009/10/ Politica_Antartica.pdf. [Accessed: 24th November 2016].

Inter-American Treaty of Reciprocal Assistance (1947). [Online] Available from: http:// www.oas.org/juridico/english/treaties/b-29.html. [Accessed: 30th April 2016].

Intergovernmental Panel on Climate Change. (2014). *Climate Change 2014: Impacts, Adaptation, and Vulnerability*. [Online] Available from: http://ipcc.ch/report/ar5/wg2/. [Accessed: 6th February 2018].

Irigoin Barrenne, J. (2011). *Chile y la Antártica. Academia Nacional de Estudios Políticos y Estratégicos*. [Online] Available from: http://anepe.cl/2011/04/chile-y-la-antartica/. [Accessed: 7th December 2013].

Jabour, J. (2014). Strategic Management and Regulation of Antarctic Tourism. In Tin, T., Liggett, D., Maher, P. T. & Lamers, M. (eds.) *Antarctic Futures: Human Engagement with the Antarctic Environment.* Dordrecht, Heidelberg, London, New York: Springer. pp. 273–286.

Jackson, D. (2008). Bush Calls for End to Ban on Offshore Oil Drilling: Such Action Would Help Companies More Than Customers, Dems Say. *USA Today.* [Print] 19th June. Available from: ProQuest Database. [Accessed: 30th March 2014].

Jacobson, H. K., Reisinger, W. M., Mathers, T. (1986). National Entanglements in International Governmental Organizations. *The American Political Science Review.* 80(1). pp. 141–159.

Javo, A. (2013). *Statement by Ms Áile Javo, President of the Saami Council, on the occasion of the Eight Ministerial Meeting of the Arctic Council Giron 15. May 2013.* [Speech] Arctic Council 8th Ministerial Meeting in Kiruna, Sweden, 15th May. [Online] Available from: https://oaarchive.arctic-council.org/handle/11374/528. [Accessed: 1st April 2016].

Javo, A. (2015). *Statement by Ms Áile Javo, President of the Saami Council, on the occasion of the Ninth Ministerial Meeting of the Arctic Council, Iqaluit 24. April 2015.* [Speech] Arctic Council 9th Ministerial Meeting in Iqaluit, Canada, 24th–25th April. [Online] Available from: https://oaarchive.arcticcouncil.org/handle/11374/376. [Accessed: 1st April 2016].

Jenisch, U. (2011). Arktis und Seerecht. Seegrenzen, Festlandsockelansprüche und Verkehrsrechte. In Sapper, M., et al. (eds.) *Logbuch Arktis. Der Raum, die Interessen und das Recht.* 61(2–3). Osteuropa: Zeitschrift für Gegenwartsfragen des Ostens. pp. 57–76.

Joyner, C. C. (2011). United States Foreign Policy Interests in the Antarctic. *The Polar Journal.* 1(1). pp. 17–35.

Jull, P. (1990). Inuit Concerns and Environmental Assessment. In Vanderzwaag, D. L. & Lamson, C. (eds.) *The Challenge of Arctic Shipping: Science, Environmental Assessment, and Human Values.* London, Buffalo: McGill-Queen's University Press. pp. 139–153.

Kaelble, H. et al. (2002). *Transnationale Öffentlichkeiten und Identitäten im 20. Jahrhundert.* Frankfurt a. M.: Campus.

Kahn, J. (2001). A New Role for Greens: Public Enemy. *New York Times.* [Print] 25th March. Available from: ProQuest Database. [Accessed: 30th March 2014].

Kaiser, K. (1969). Transnationale Politik. In Czempiel, E.-O. (ed.) *Die anachronistische Souveränität.* Cologne-Opladen: Westdeutscher Verlag. pp. 80–109.

Kaiser, W., Leucht, B., Gehler, M. (2010). Transnational Networks in European Integration Governance: Historical Perspective on an Elusive Phenomenon. In Kaiser, W., et al. (eds.) *Transnational Networks in Regional Integration: Governing Europe 1945–83.* Hampshire: Palgrave. pp. 1–17.

Kaltmeier, O. (2014). Inter-American Perspectives for the Rethinking of Area Studies. *Forum for Inter-American Research.* 7(3). pp. 171–182.

Kanamine, L. (1991). Con: Don't Disturb 'National Treasure'. *USA Today.* [Print] 13th March. Available from: ProQuest Database. [Accessed: 30th March 2014].

Kankaanpää, P., Smieszek, M. (2014). *Assessments in Policy-Making: Case-Studies from the Arctic Council: Preparatory Action, Strategic Environmental Impact Assessment of Development of the Arctic.* Rovaniemi: Arctic Centre, University of Lapland.

Kankaanpää, P., Young, O. R. (2012). The Effectiveness of the Arctic Council. *Polar Research.* 31. pp. 1–14.

Kanter, J., Revkin, A. C. (2007). World Scientists Near Consensus on Warming. *New York Times.* [Print] 30th January. Available from: ProQuest Database. [Accessed: 30th March 2014].

Kao, S. M., Pearre, N. S., Firestone, J. (2012). Adoption of the Arctic Search and Rescue Agreement: A Shift of the Arctic Regime toward a Hard Law Basis? *Marine Policy.* 36. pp. 832–838.

Katcoff, D. (2001). Tap Arctic Oil Reserves. *New York Times.* [Print] 21st January. Available from: ProQuest Database. [Accessed: 30th March 2014].

Keck, M., Sikkink, K. (1999). Transnational Advocacy Networks in International and Regional Politics. *International Social Science Journal.* 51(159). pp. 89–101.

Keil, K., Knecht, S. (2017). Introduction. In Keil, K. & Knecht, S. (eds.) *Governing Arctic Change: Global Perspectives.* Basingstoke, Hampshire: Palgrave Macmillan. pp. 1–18.

Kelly, P., Child, J. (1988). *Geopolitics of the Southern Cone and Antarctica.* Boulder: Rienner.

Kenna, K. (1990). Environmental Protection a Top Priority, Candidates Say. *Toronto Star.* [Print] 5th March. Available from: ProQuest Database. [Accessed: 30th March 2014].

Kennworthy, T. (1995). Study Condemns Arctic Oil Drilling. *The Washington Post.* [Print] 27th August. Available from: ProQuest Database. [Accessed: 30th March 2014].

Kerry, J. (2013). *Remarks at the Arctic Council Ministerial Session: John Kerry Secretary of State: Kiruna City Hall Kiruna, Sweden, May 15, 2013.* [Speech] Arctic Council 8th Ministerial Meeting in Kiruna, Sweden, 15th May. [Online] Available from: https:// oaarchive.arctic-council.org/handle/11374/528. [Accessed: 1st April 2016].

Keskitalo, E. C. H. (2004). The Arctic as an International Region: But for Whom? In Koivurova, T., Joona, T., & Shnoro, R. (eds.) *Arctic Governance.* Rovaniemi: Oy Sevenprint. pp. 2–26.

Keskitalo, E. C. H. (2012). Setting the Agenda on the Arctic: Whose Policy Frames the Region. *Brown Journal of World Affairs.* 19(1). pp. 155–164.

Keskitalo, E. C. H. (2015). The Role of Discourse Analysis in Understanding Spatial Systems. In Jensen, L. C. & Hønneland, G. (eds.) *Handbook of the Politics of the Arctic.* Cheltenham, Northampton: Edward Elgar Publishing. pp. 421–433.

Keskitalo, E. C. H., Nuttall, M. (2015). Globalization of the 'Arctic'. In Evengard, B., Nymand Larsen, J. & Paasche, O. (eds.) *The New Arctic.* Heidelberg, New York, Dordrecht, London: Springer. pp. 175–188.

Kharlampyeva, N. K. (2013). The Transnational Arctic and Russia. In Peimani, H. (ed.) *Energy Security and Geopolitics in the Arctic: Challenges and Opportunities in the 21st Century.* Singapore: World Scientific Publishing Co. pp. 95–126.

Kinossian, N. (2016). Re-Colonising the Arctic: The Preparation of Spatial Planning Policy in Murmansk Oblast, Russia. *Environment and Planning C: Government and Policy.* May. [Online] Available from: http://epc.sagepub.com/content/early/2016/05/10/02637 74X16648331.abstract. [Accessed: 23rd May 2016].

Klotz, F. G. (1990). *America on the ice: Antarctic Policy Issues.* Washington, DC: National Defense University Press.

Knecht, S. (2016a). Procedural Reform at the Arctic Council: The Amended 2015 Observer Manual. *Polar Record.* April. [Online] Available from: http://journals.cambridge.org/ action/displayAbstract?fromPage=online&aid=10268636&fulltextType=RC&fileI d=S0032247416000. [Accessed: 11th April 2016].

Knecht, S. (2016b). Stakeholder Participation in Arctic Council Meetings (STAPAC) Dataset. *Harvard Dataverse.* [Online] Available from: http://dx.doi.org/10.7910/DVN/ OMQAEW. [Accessed: 1st June 2016].

Kobelev, A. (2006). *Statement by Alexander Kobelev, President of the Sámi Council.* [Speech] Arctic Council 5th Ministerial Meeting in Salekhard, 25th–26th October. [Online] Available from: https://oaarchive.arcticcouncil.org/handle/11374/1564. [Accessed: 1st April 2016].

Koch, M. (2011). Non-State and State Actors in Global Governance. In Reinalada, B. (ed.) *The Ashgate Research Companion to Non-State Actors*. Farnham: Ashgate. pp. 197–208.

Koch, W. (2014). Demystifying Climate Change: In Report, Scientists Answer Some of the Top Questions. *USA Today*. [Print] 28th February. Available from: ProQuest Database. [Accessed: 30th March 2014].

Koivurova, T. (2008). Alternatives for an Arctic Treaty: Evaluation and a New Proposal. *Review of European Community & International Environmental Law*. 17(1). pp. 14–26.

Koivurova, T. (2009). Limits and Possibilities of the Arctic Council in a Rapidly Changing Scene of Arctic Governance. *Polar Record*. 46(2). pp. 1–11.

Koivurova, T. (2010). Sovereign States and Self-Determining Peoples: Carving Out a Place for Transnational Indigenous Peoples in a World of Sovereign States. *International Community Law Review*. 12(2). pp. 191–212.

Koivurova, T. (2013). Multipolar and Multilevel Governance in the Arctic and the Antarctic. *American Society of International Law: Proceedings of the Annual Meeting*. 107. pp. 443–446.

Koivurova, T., Keskitalo, E. C. H., Bankes, N. (2009). *Climate Governance in the Arctic*. Luxembourg: Springer Science+Business Media.

Koivurova, T., Vanderzwaag, D. L. (2007). The Arctic Council at 10 Years: Retrospect and Prospects. *University of British Columbia Law Review*. 40(1). pp. 121–194.

Kolton, A. (2000). Drill and Spoil. *The Washington Post*. [Print] 16th December. Available from: ProQuest Database. [Accessed: 30th March 2014].

Koring, P. (2001). Bush Foes Confident Refuge Is Safe from Drilling: Canada among Group Expecting U.S. Senate Will Stop Plan to Drill Oil in Alaskan Arctic. *Globe and Mail*. [Print] 3rd August. Available from: ProQuest Database. [Accessed: 30th March 2014].

Kornbluh, P. (2000). *CIA Acknowledges Ties to Pinochet's Repression*. [Online] Available from: http://nsarchive.gwu.edu/news/20000919/. [Accessed: 11th October 2016].

Kothari, U. (2013). Political Discourses of Climate Change and Migration: Resettlement Policies in the Maldives. *The Geographical Journal*. 180(2). pp. 130–140.

Kramer, A. E. (2013). Vast Amnesty by Russians Include Case of Greenpeace. *New York Times*. [Print] 25th December. Available from: ProQuest Database. [Accessed: 31st March 2014].

Krauss, C. (2010). Shell Presses for Drilling in Arctic. *New York Times*. [Print] 6th November. Available from: ProQuest Database. [Accessed: 30th March 2014].

Krauss, C. (2014). ConocoPhilipps Suspends Its Arctic Drilling Plans. *New York Times*. [Print] 11th April 2013. Available from: ProQuest Database. [Accessed: 30th March 2014].

Kristof, N. D. (2003). Casting a Cold Eye on Arctic Oil. *New York Times*. [Print] 10th September. Available from: ProQuest Database. [Accessed: 30th March 2014].

Kristof, N. D. (2006). Warm, Warmer, Warmest. *New York Times*. [Print] 5th March. Available from: ProQuest Database. [Accessed: 30th March 2014].

Kristof, N. D. (2007). The Big Melt. *New York Times*. [Print] 16th August. Available from: ProQuest Database. [Accessed: 31st March 2014].

Krugman, P. (2010). Who Cooked the Planet? *New York Times*. [Print] 26th July. Available from: ProQuest Database. [Accessed: 30th March 2014].

Kuckartz, U. (2014). *Qualitative Inhaltsanalyse. Methoden, Praxis, Computerunterstützung*. Weinheim, Basel: Beltz Juventa.

Kyoto Protocol (1997). [Online] Available from: https://unfccc.int/process/the-kyoto-protocol. [Accessed: 30th April 2016].

La Nación. (1989). *Clima notablemente seco en el último semestre de 1988.* [Print] 11th January. Available from: Biblioteca Nacional Mariano Moreno.

La Nación. (2001a). *Una década de confl diplomático.* [Print] 6th September. Available from: Biblioteca Nacional Mariano Moreno.

La Nación. (2001b). Día Nacional del Antártico. *La Antártida, un gran laboratorio a cielo abierto.* [Print] 22nd February. Available from: Biblioteca Nacional Mariano Moreno.

La Nación. (2002a). *Antártida: se enfría, a pesar del clima global.* [Print] 14th February. Available from: Biblioteca Nacional Mariano Moreno.

La Nación. (2002b). *Se celebra hoy el Día de la Antártida.* [Print] 22nd February. Available from: Biblioteca Nacional Mariano Moreno.

La Nación. (2003a). *Científicos prevén graves efectos climáticos por la actividad del hombre.* [Print] 3rd December. Available from: Biblioteca Nacional Mariano Moreno.

La Nación. (2003b). *Editorial II, La Argentina en la Antártida.* [Print] 2nd March. Available from: Biblioteca Nacional Mariano Moreno.

La Nación. (2004). *Nuestra política antártica.* [Print] 16th September. Available from: Biblioteca Nacional Mariano Moreno.

La Nación. (2005). Los hielos antárticos se derriten a ritmo récord. *Lo atribuyen al cambio climático.* [Print] 4th August. Available from: Biblioteca Nacional Mariano Moreno.

La Nación. (2006a). *Los hielos del Ártico.* [Print] 7th October. Available from: Biblioteca Nacional Mariano Moreno.

La Nación. (2006b). *Rumbo al Año Internacional 2007: científicos internacionales advierten sobre un problema que puede dambiar el mundo.* [Print] 11th June. Available from: Biblioteca Nacional Mariano Moreno.

La Nación. (2007). Luego del naufragio del crucero Explorer. *Estudian restringir los cruceros turisticos a la zona antártica.* [Print] 26th November. Available from: Biblioteca Nacional Mariano Moreno.

La Nación. (2009). *El mundo, convertido en un infi sin ozono.* [Print] 20th March. Available from: Biblioteca Nacional Mariano Moreno.

La Nación. (2010). *Malvinas, con más presión.* [Print] 14th August. Available from: Biblioteca Nacional Mariano Moreno.

La Nación. (2011a). *Registran cambios en los lagos antárticos.* [Print] 26th July. Available from: Biblioteca Nacional Mariano Moreno.

La Nación. (2011b). *Editorial II, Presencia chilena en la Antártida.* [Print] 6th April. Available from: Biblioteca Nacional Mariano Moreno.

La Nación. (2011c). *Opinion: Política Antártica.* [Print] 26th April. Available from: Biblioteca Nacional Mariano Moreno.

La Nación. (2012a). *Revelan el misterio de la expansión de la Antártida y el derretimiento del Ártico.* [Print] 12th November. Available from: Biblioteca Nacional Mariano Moreno.

La Nación. (2012b). *La OEA respalda a la Argentina en su denuncia por Malvinas.* [Print] 10th February. Available from: Biblioteca Nacional Mariano Moreno.

La Nación. (2013a). Editorial II. *El Ártico, en peligro.* [Print] 9th October. Available from: Biblioteca Nacional Mariano Moreno.

La Nación. (2013c). *Natalia Oreiro visitó a los activistas de Greenpeace que habían sido detenidos en Rusia.* [Print] 12th December. Available from: Biblioteca Nacional Mariano Moreno.

La Tercera. (1989). *Chile propone reunión de ecólogos sobre Antártica.* [Print] 24th February. Available from: Biblioteca Nacional de Chile.

La Tercera. (1990a). Ecólogos están en contra de esta posición. *Chile apoya moratoria para explotación de la Antártica.* [Print] 16th November. Available from: Biblioteca Nacional de Chile.

La Tercera. (1990b). *Piden declarar 'Parque Mundial' a la Antártica*. [Print] 16th November. Available from: Biblioteca Nacional de Chile.

La Tercera. (1990c). *Greenpeace postula protección permanente para la Antártica*. [Print] 22nd November. Available from: Biblioteca Nacional de Chile.

La Tercera. (1997a). *Industriales temen un frenazo del crecimiento*. [Print] 14th December. Available from: Biblioteca Nacional de Chile.

La Tercera. (1997b). *Países reducirán emisión de contaminantes*. [Print] 12th December. Available from: Biblioteca Nacional de Chile.

La Tercera. (1997c). En Cumbre Climática. *Pesimismo en Kyoto*. [Print] 7th December. Available from: Biblioteca Nacional de Chile.

La Tercera. (2004). Tras años de debate el Presidente Vladimir Puten determinó ayer enviar su decisión al Parlamento. *Rusia ratifica protocolo de Kioto y facilita su vigencia desde 2005*. [Print] 1st October 2004. Available from: Biblioteca Nacional de Chile.

La Tercera. (2007a). Se estima que en la región se encuentra un 25% de las reservas petroleras del planeta, además de yacimientos de gas. *EE.UU. compite con Rusia en carrera por riqueza del Ártico*. [Print] 15th August. Available from: Biblioteca Nacional de Chile.

La Tercera. (2007b). *Ola polar en Chile y cambio climático*. [Print] 11th August. Available from: Biblioteca Nacional de Chile.

Lackenauer, P. W. (2010). High Arctic Theatre for All Audiences. *Globe and Mail*. [Print] 17th August. Available from: ProQuest Database. [Accessed: 30th March 2014].

Lackenauer, P. W., Manicom, J. (2013). East Asia-Arctic Relations: Boundary, Security and International Politics: Canada's Northern Strategy and East Asian Interests in the Arctic. *CIGI Policy brief*. 5. pp. 1–22.

Lackenauer, P. W., Manicom, J. (2015). Asian States and the Arctic: National Perspectives on Regional Governance. In Jensen, L. C. & Hønneland, G. (eds.) *Handbook of the Politics of the Arctic*. Cheltenham, Northampton: Edward Elgar Publishing. pp. 517–532.

Laclau, E., Mouffe, C. (2001). *Hegemony and Socialist Strategy: Towards a Radical Democratic Politics*. 2nd Ed. London, New York: Verso.

Lajeunesse, A. (2008). The Northwest Passage in Canadian Policy: An Approach for the 21st Century. *International Journal*. 63(4). pp. 1037–1052.

Leal, J. (1994). Para este pionero del polo, el territorio debe ser un espacio integración regional. Antártida Sudamericana. *Clarín*. [Print] 7th December. Available from: Biblioteca Nacional Mariano Moreno.

Lean, G. (2007). Acuerdo de países ricos y pobres para combatir el cambio climatico. *Página 12*. [Print] 23rd September. Available from: Biblioteca Nacional Mariano Moreno.

Leeder, J. (2008). Huge Chunk Snaps Off Storied Arctic Ice Shelf. *Globe and Mail*. [Print] 29th July. Available from: ProQuest Database. [Accessed: 30th March 2014].

Lee Myers, S. (2013). Arctic Council Adds 6 Nations as Observer States, Including China. *New York Times*. [Print] 16th May. Available from: ProQuest Database. [Accessed: 30th March 2014].

Le Mière, C., Mazo, J. (2013). *Arctic Opening: Insecurity and Opportunity*. London, New York: Routledge.

Lewis, P. (1989). Canada about to Sign Major Land Agreement with Eskimos. *New York Times*. [Print] 21st August. Available from: ProQuest Database. [Accessed: 30th March 2014].

Liggett, D. (2009). *Tourism in the Antarctic: Modi Operandi and Regulatory Effectiveness*. Saarbrücken: VDM Verlag.

Liggett, D. (2015). Chapter 3: An Erosion of Confidence? The Antarctic Treaty System in the Twenty-First Century. In Ali, S. H. & Pincus, R. (eds.) *Diplomacy on Ice: Energy and the Environment in the Arctic and Antarctic*. New Haven, London: Yale University Press. pp. 61–71.

Lillebuen, S. (2006). Giant Ice Shelf Snaps Off from Arctic Island; Unusually Warm Temperatures Definitely Played a Major Role. *Toronto Star*. [Print] 29th December. Available from: ProQuest Database. [Accessed: 30th March 2014].

Lipcovich, P. (2009). Continente blanco y menos frío. *Página 12*. [Print] 22nd January. Available from: Biblioteca Nacional Mariano Moreno.

Lowman, R. (1989). Canada Too Reliant on U.S. for Security, Report Says. *Toronto Star*. [Print] 15th September. Available from: ProQuest Database. [Accessed: 30th March 2014].

Lynge, A. (2011). *Statement by Inuit Circumpolar Council Aqqaluk Lynge, Chair*. [Speech] Arctic Council 7th Ministerial Meeting in Nuuk, Greenland, 12th May. [Online] Available from: https://oaarchive.arcticcouncil.org/handle/11374/530. [Accessed: 1st April 2016].

Lynge, A. (2013). *Aqqaluk Lynge, Chair, Inuit Circumpolar Council-Eighth Ministerial Meeting of the Arctic Council 14–15 May 2013, Kiruna, Sweden*. [Speech] Arctic Council 8th Ministerial Meeting in Kiruna, Sweden, 15th May. [Online] Available from: https://oaarchive.arctic-council.org/handle/11374/528. [Accessed: 1st April 2016].

Mahoney, J. (1999). Column One: The Birth of Nunavut: A Long-Held Inuit Dream Is Realized at Last: Nunavut becomes Canada's Newest Territory April 1. *Globe and Mail*. [Print] 29th March. Available from: ProQuest Database. [Accessed: 30th March 2014].

Manicom, J. (2013). Identity Politics and the Russia-Canada Continental Shelf Dispute: An Impediment to Cooperation. *Geopolitics*. 18(1). pp. 60–76.

Mann, M. (2006). Globalization, Macro-Regions and Nation-States. In Budde, G., et al. (eds.) *Transnationale Geschichte. Themen, Tendenzen und Theorien*. Göttingen: Vandenhoeck & Ruprecht. pp. 32–55.

Mansilla, L. A. (1994). Ecologistas piden integrar el Mercosur. *La Nación*. [Print] 6th December. Available from: Biblioteca Nacional Mariano Moreno.

Martello, M. L. (2008). Arctic Indigenous Peoples as Representations and Representatives of Climate Change. *Social Studies of Science*. 38(3). pp. 351–376.

Mason, A. (2015). Chapter 8: Growth Imperative: Intermediaries, Discourse Frameworks, and the Arctic. In Ali, S. H., & Pincus, R. (eds.) *Diplomacy on Ice: Energy and the Environment in the Arctic and Antarctic*. New Haven, London: Yale University Press. pp. 141–150.

Mason, G. (2008). When It Comes to the North, Canada Lacks a Sense of Direction. *Globe and Mail*. [Print] 12th December. Available from: ProQuest Database. [Accessed: 30th March 2014].

Mason, G. (2009). The Flame Arrives, But Inuit Still Await an Apology. *Globe and Mail*. [Print] 10th November. Available from: ProQuest Database. [Accessed: 30th March 2014].

Mastio, D. (1999). Global Warming Propaganda Trumps Science. *USA Today*. [Print] 16th December. Available from: ProQuest Database. [Accessed: 30th March 2014].

May, P. J., Jones, B. D., Beem, B. E., Neff-Sharum, E. A., Poague, M. K. (2005). Policy Coherence and Component-Driven Policymaking: Arctic Policy in Canada and the United States. *The Policy Studies Journal*. 33(1). pp. 37–63.

Mayring, P. (2007). *Qualitative Inhaltsanalyse: Grundlagen und Techniken*. Weinheim: Beltz.

Mayring, P. (2014). *Qualitative Content Analysis. Theoretical Foundation, Basic Procedures and Software Solution.* Klagenfurt, Austria. [Online] Available from http://ssoar.info/ssoar/handle/document/39517 [Accessed: 17th June 2015].

McBeath, J. (2013). East Asia-Arctic Relations: Boundary, Security and International Politics. *CIGI Policy Brief.* 4. pp. 1–9.

McCallum, M., Sheiban, N., Stawicki, S. (2013). Implementing Canada's Arctic Council Priorities. *CIGI Junior Fellows Policy Brief.* 7. pp. 1–8.

McCarthy, S. (2009). Arctic Holds 13 Per Cent of World's Undiscovered Oil. *Globe and Mail.* [Print] 29th May. Available from: ProQuest Database. [Accessed: 30th March 2014].

McCarthy, S. (2010). BP's Spill, Obama's Burden. *Globe and Mail.* [Print] 28th May. Available from: ProQuest Database. [Accessed: 30th March 2014].

McCarthy, S., Vanderklippe, N. (2010). Spill Puts New Oil Frontiers at Risk. *Globe and Mail.* [Print] 11th June. Available from: ProQuest Database. [Accessed: 30th March 2014].

McDorman, T. L., Schofield, C. (2015). Maritime Limits and Boundaries in the Arctic Ocean: Agreements and Disputes. In Jensen, L. C. & Hønneland, G. (eds.) *Handbook of the Politics of the Arctic.* Cheltenham, Northampton: Edward Elgar Publishing. pp. 207–226.

McGhee, R. (2007). *The Last Imaginary Place: A Human History of the Arctic World.* Chicago: University of Chicago Press.

McInnes, C. (1989). Agreement Reached on Inuit land claim. *Globe and Mail.* [Print] 9th December. Available from: ProQuest Database. [Accessed: 30th March 2014].

McKenna, B. (2008). Record Prices Put Arctic Oil within Reach. *Globe and Mail.* [Print] 25th July. Available from: ProQuest Database. [Accessed: 30th March 2014].

McKibben, B. (2007). The Race against Warming. *The Washington Post.* [Print] 29th September. Available from: ProQuest Database. [Accessed: 30th March 2014].

McLarien, C. (1990). Greenpeace Crew May Face Charges Ship Seized Near Nuclear Test Site Violated Territory, Soviets Say. *Globe and Mail.* [Print] 13th October. Available from: ProQuest Database. [Accessed: 30th March 2014].

McRae, D. (2007). Arctic Sovereignty? What Is at Stake? *Behind the Headlines.* 64(1). [Online] Available from: https://questia.com/magazine/1G1-158959250/arctic-sovereign. [Accessed: 21st November 2016].

Medvedev, S. (2015). The Blank Space: Glenn Gould, Russia, Finland and the North. *International Politics.* 38(1). pp. 91–102.

Mendez Araya, J. (2004). Canadá: La otra tierra de las oportunidades. *El Mercurio.* [Print] 12th November. Available from: Biblioteca Nacional de Chile.

Merton, R., Kendall, P. (1946). The focused interview. *American Journal of Sociology.* 51(6). pp. 541–557.

Mignolo, W. (2002). The Geopolitics of Knowledge and the Colonial Difference. *South Atlantic Quarterly.* 101(1). pp. 56–96.

Ministerio de Defensa. (1995). *Decreto N° 1041/95. Apruébase su estructura organizativa.* [Online] Available from: http://servicios.infoleg.gob.ar/infolegInternet/anexos/25000-29999/25111/norma.htm. [Accessed: 24th November 2016].

Ministerio de Relaciones Exteriores de Chile. (2013). Antártica. *Dirección de Antártica.* [Online] Available from: http://minrel.gob.cl/minrel/site/edic/base/port/antartica.html. [Accessed: 21st November 2016].

Ministerio de Relaciones Exteriores de Chile. (2015). *Chile en la Antártica. Visión Estratégica al 2035.* [Online] Available from: http://minrel.gov.cl/minrel/site/artic/20121010/asocfile/20121010172919/vision_estrategica.pdf. [Accessed: 17th September 2016].

Ministerio de Relaciones Exteriores de Chile. (2016a). *Política Antártica Nacional del Año 2000*. [Online] Available from: http://minrel.gov.cl/nacionales/minrel/2012-10-10/172919.html. [Accessed: 17th September 2016].

Ministerio de Relaciones Exteriores de Chile. (2016b). *Dirección de Antártica*. [Online] Available from: http://minrel.gov.cl/minrel/site/edic/base/port/antartica.html. [Accessed: 21st November 2016].

Ministry of Foreign Affairs of Finland. (2017). *Finland's Chairmanship of the Arctic Council in 2017–2019*. [Online] Available from: http://formin.fi/public/default.aspx?nodeid=50020&contentlan=2&culture=en-US. [Accessed: 10th January 2017].

Mitchell, A. (2001). Scientists Raise Alarm of Climate Catastrophe: Rapid Warming Trend Could Be Reversed by Move to Alternate Power Sources. *Globe and Mail*. [Print] 22nd January. Available from: ProQuest Database. [Accessed: 31st March 2014].

Mitchell, A. (2002). Fate of Polar Bears Rests on Thinning Ice: The Marine Mammal Could Be Extinct within the Next 100 Years, Scientists Warn. *Globe and Mail*. [Print] 26th December. Available from: ProQuest Database. [Accessed: 30th March 2014].

Mittelstaedt, M. (2001). Industrial Activity Threatens Arctic, Group Says. *Globe and Mail*. [Print] 8th September. Available from: ProQuest Database. [Accessed: 31st March 2014].

Mittelstaedt, M. (2007). The Fallout of Global Warming: 1,000 Years: In Stark Terms, Scientists Confirm That Climate Change Is 'unequivocal'. *Globe and Mail*. [Print] 31st January. Available from: ProQuest Database. [Accessed: 30th March 2014].

Montera, J. P. (2001). Bush y el Cambio Climático. *El Mercurio*. [Print] 24th April. Available from: Biblioteca Nacional de Chile.

Moreno, M. A. (2007). Jornadas de calorón y con futuro derretido. *Clarín*. [Print] 7th February. Available from: Biblioteca Nacional Mariano Moreno.

Mose, J., Reuber, P. (2011). Zwischen Separatismus and Transnationalisierung – Nationale Identitäten in Spanien und Katalonien vor dem Hintergrund der europäischen Integration. In Dzudzek, I., Reuber, P. & Strüver, A. (eds.) *Die Politik räumlicher Repräsentation. Beispiele aus der empirischen Forschung*. Berlin, Münster, Wien, Zürich, London: LIT Verlag. pp. 171–196.

Mouawad, J. (2007). Tension at the Edge of Alaska. *New York Times*. [Print] 4th December. Available from: ProQuest Database. [Accessed: 30th March 2014].

Mufson, S. (2008). Offshore Drilling Backed as Remedy for Oil Prices. *The Washington Post*. [Print] 14th July. Available from: ProQuest Database. [Accessed: 30th March 2014].

Mufson, S. (2009). Shell Wins Federal Approval to Drill for Oil Off Alaska Coast. *The Washington Post*. [Print] 20th October. Available from: ProQuest Database. [Accessed: 30th March 2014].

Mufson, S. (2012). Shell Gets Go-Ahead to Prepare for Alaska Drilling. *The Washington Post*. [Print] 31st August. Available from: ProQuest Database. [Accessed: 30th March 2014].

Mufson, S. (2013). U.S. to Review Arctic Oil, Gas Drilling. *The Washington Post*. [Print] 9th January. Available from: ProQuest Database. [Accessed: 30th March 2014].

Mufson, S. (2014). Shell Delays Plans to Drill in Alaska. *The Washington Post*. [Print] 31st January. Available from: ProQuest Database. [Accessed: 30th March 2014].

Murkowski, F. H. (1991). America Needs Alaska's Oil. *New York Times*. [Print] 1st April. Available from: ProQuest Database. [Accessed: 30th March 2014].

Murray, R. W. (2012). Arctic Politics in the Emerging Multipolar System: Challenges and Consequences. *The Polar Journal*. 2(1). pp. 7–20.

Murray, R. W., Dey Nuttall, A. (2014). *International Relations and the Arctic: Understanding Policy and Governance*. Amherst: Cambria.

Nader, L. (2011). Ethnography as Theory. *Journal of Ethnographic Theory*. 1(1). pp. 211–219.

Naidu, S., Jr. (2008). Claiming the Last Global Frontier: Overlapping Geographical Claims of Antarctic Territory. *Transnational Law and Contemporary Problems*. 17(2). pp. 529–552.

National Ocean Council. (2011). *Changing Conditions in the Arctic*. [Online] Available from: http://nmfs.noaa.gov/ocs/mafac/reports/docs/arctic_fact_sheet.pdf. [Accessed: 23rd November 2016].

The National Science Foundation. (2011). *Antarctic Research*. [Online] Available from: https://nsf.gov/pubs/2011/nsf11532/nsf11532.htm. [Accessed: 21st November 2016].

New York Times. (2006). Drilling in Alaska: The Moral Divide. [Print] 1st January. Available from: ProQuest Database. [Accessed: 30th March 2014].

New York Times. (2008). Arctic in Retreat. [Print] 8th September. Available from: ProQuest Database. [Accessed: 30th March 2014].

New York Times. (2013). Further Protection for Antarctica. [Print] 15th July. Available from: ProQuest Database. [Accessed: 30th March 2014].

Nick, J. (2010). BP Tragedy Gives Us Pause Here in Alaska. *USA Today*. [Print] 8th June. Available from: ProQuest Database. [Accessed: 30th March 2014].

Nicol, H. N. (2015). Canada's Arctic Agenda: Failing to Make a Case for Economic Development as an International Strategy in the Circumpolar North? In Jensen, L. C. & Hønneland, G. (eds.) *Handbook of the Politics of the Arctic*. Cheltenham, Northampton: Edward Elgar Publishing. pp. 51–65.

Niebieskikwiat, N. (2007). Reclamo ante la ONU de un millon de kilometros cuadrados de lecho marino. Antártida: Argentina y Chile se unen frente a Gran Bretaña. *Clarín*. [Print] 27th October. Available from: Biblioteca Nacional Mariano Moreno.

Niebieskikwiat, N. (2009). Londres reclamó más lecho marino en Malvinas: protesta argentina. *Clarín*. [Print] 12th May. Available from: Biblioteca Nacional Mariano Moreno.

Nord, D. C. (2016). *The Arctic Council: Governance within the Far North*. Abingdon, New York: Routledge.

Northwest Territories, Yukon, Nunavut. (2007). *A Northern Vision: A Stronger North and a Better Canada*. [Online] Available from: http:// anorthernvision.ca/documents/newvision_english.pdf. [Accessed: 23rd November 2016].

Northwest Territories, Yukon, Nunavut. (2011). *Pan-Territorial Adaptation Strategy*. [Online] Available from: http://anorthernvision.ca/strategy/ index.html. [Accessed: 23rd November 2016].

Nuorgam, A. (2002). *Intervention by Anne Nuorgam, President of the Saami Council*. [Speech] Arctic Council 3rd Ministerial Meeting in Inari, 9th–10th October 2002. [Online] Available from: https://oaarchive.arcticcouncil.org/bitstream/handle/11374/1584/MM03_Saami_Council_Intervention%20%282%29.pdf?sequence=7&isAllowed=y. [Accessed: 19th May 2016].

Nuttall, M. (1998). *Protecting the Arctic: Indigenous Peoples and Cultural Survival*. Oxon: Routledge.

Nuttall, M., Callaghan, T. V. (2000). *The Arctic: Environment, People, Policy*. Amsterdam: Harwood Academic Publishers.

Nye, J. S. (1968). *International Regionalism: Regions*. Boston: Little, Brown and Company.

Nye, J. S., Keohane, R. O. (1971). Transnational Relations and World Politics: An Introduction. *International Organization*. 25(3). pp. 329–349.

OAS. (2016). *Environment*. [Online] Available from: http://oas.org/en/topics/environment. asp. [Accessed: 13th May 2016].

O'Bryan, R. (1997). Regulación Ambiental en Chile. *El Mercurio*. [Print] 14th December. Available from: Biblioteca Nacional de Chile.

Oliva. L. (2007). Política internacional. Disputa polar: las nuevas fronteras de la geopolítica. *La Nación*. [Print] 11th November. Available from: Biblioteca Nacional Mariano Moreno.

Olli, E. (2013). *Statement by Egil Olli, President of the Sámi Parliament in Norway*. [Speech] Arctic Council 8th Ministerial Meeting in Kiruna, Sweden, 15th May. [Online] Available from: https://oaarchive.arcticcouncil.org/ handle/11374/528. [Accessed: 1st April 2016].

O'Neill, K. (2008). Arctic Ice Shelf Now Split in Three, Mission Finds. *Globe and Mail*. [Print] 14th April. Available from: ProQuest Database. [Accessed: 30th March 2014].

O'Neill, K. (2009). No Thaw in Spat over Symbol of Arctic Sovereignty. *Globe and Mail*. [Print] 8th September. Available from: ProQuest Database. [Accessed: 30th March 2014].

On Think Tanks. (2018). *Open Think Tank Directory*. [Online] Available from: https://ottd. onthinktanks.org/. [Accessed: 13th June 2018].

Ortúzar, P. (2010). Antártida: ¿Cómo cuidamos el laboratorio del mundo? *Def Online*. [Online] Available from: http://defonline.com.ar/?p=34584. [Accessed: 5th February 2016].

Osherenko, G., Young, O. R. (1989). *The Age of the Arctic: Hot Conflicts and Cold Realities*. New York, Port Chester, Melbourne, Sydney: Cambridge University Press.

Ó Tuathail, G. (1992). 'Pearl Harbor without Bombs': A Critical Geopolitics of the US-Japan 'FSX' Debate. *Environment and Planning A*. 24. pp. 975–994.

Ó Tuathail, G. (1996). *Critical Geopolitics: The Politics of Writing Global Space*. Minneapolis: University Of Minnesota Press.

Ó Tuathail, G. (2006). General Introduction: Thinking Critically about Geopolitics. In Ó Tuathail, G., Dalby, S. & Routledge, P. (eds.) *The Geopolitics Reader*. Abingdon, New York: Routledge. pp. 1–32.

Ó Tuathail, G., Agnew, J. (2006). Geopolitics and Discourse: Practical Geopolitical Reasoning in American Foreign Policy. In Ó Tuathail, G., Dalby, S. & Routledge, P. (eds.) *The Geopolitics Reader*. 2nd Ed. Abingdon, New York: Routledge. pp. 94–102.

Página 12. (2013). Zuaín: Malvinas es uno de los territorios más militarizados del mundo. [Print] 28th January. Available from: Biblioteca Nacional Mariano Moreno.

Página 12. (2014). El Día de la Antártida. [Print] 26th February. Available from: Biblioteca Nacional Mariano Moreno.

Palosaari, T. (2012). The Amazing Race: On Resources, Conflict, and Cooperation in the Arctic. In Heininen, L. & Rouge-Oikarinen, R. (eds.) *Nordia Geographical Publications Yearbook 2011: Theme Issue on 'Sustainable Development in the Arctic through Peace and Stability'*. Oulu: Geographical Society of Northern Finland. pp. 13–30.

Palosaari, T. (2016). *Solving the Arctic Paradox: Climate Change, Natural Resources and Security in the Arctic*. [Online] Available from: http://uta.fi/yky/en/research/tapri/ Research/arcticparadox.html. [Accessed: 6th February 2018].

Palosaari, T., Tynkkynen, N. (2015). Arctic Securitization and Climate Change. In Jensen, L. C. & Hønneland, G. (eds.) *Handbook of the Politics of the Arctic*. Cheltenham, Northampton: Edward Elgar Publishing. pp. 87–104.

Paredes, J. U. (2009). El Tratado del Antártico, posición de Chile como país Puente. *UNISCI Discussion Papers*. 21. pp. 138–147.

Parodi, S. (2007). La Antártida. Un Raro Caso de Continuidad en La Línea de Política Exterior Argentina. *Jornadas del CENSUD.* pp. 1–16.

Pear, R. (2005). Arctic Drilling Push Is Seen as Threat to Budget Bill. *New York Times.* [Print] 3rd November. Available from: ProQuest Database. [Accessed: 30th March 2014].

Pedersen, T. (2012). Debates over the Role of the Arctic Council. *Ocean Development & International Law.* 43(2). pp. 146–156.

Pelletier, S., Lasserre, F. (2012). Arctic Shipping: Future Polar Express Seaways? Shipowners' Opinion. *Journal of Maritime Law & Commerce.* 43(4). pp. 553–564.

Petersen, M., Buitrago, C., Tyrell, P.-M., Wehrmann, D. (2016). Introduction: Researching (Geo-)Political Imaginaries in the Americas. *Forum for Inter-American Research.* 9(1). pp. 1–13.

Petersen, M., Wehrmann, D. (2015). Geopolitics. InterAmerican Wiki: Terms-Concepts-Critical Perspectives. [Online] Available from: www.uni-bielefeld.de/cias/wiki/g_Geopolitics.html. [Accessed: 8th February 2018].

Petrov, A. N., Cavon, P. A. (2013). Creative Alaska: Creative Capital and Economic Development Opportunities in Alaska. *Polar Record.* 49(251). pp. 348–361.

Pianin, E. (2001). Norton Argues for Arctic Drilling. *The Washington Post.* [Print] 20th January. Available from: ProQuest Database. [Accessed: 30th March 2014].

Piattoni, S. (2010). *The Theory of Multi-Level Governance.* Oxford: Oxford University Press.

Pincus, R. (2015). Chapter 10: Security in the Arctic: A Receding Wall. In Ali, S. H. & Pincus, R. (eds.) *Diplomacy on Ice: Energy and the Environment in the Arctic and Antarctic.* New Haven, London: Yale University Press. pp. 161–170.

Pizzi, N. (2013). Siguen las demoras inexplicables en la campaña antártica. *Clarín.* [Print] 13th April. Available from: Biblioteca Nacional Mariano Moreno.

Platiel, R. (1993). Inuit Accused of Making False Claims for Compensation Sovereignty not the Goal of Relocation, Commission Told. *Globe and Mail.* [Print] 30th June. Available from: ProQuest Database. [Accessed: 30th March 2014].

Plaut, S. (2012). 'Cooperation Is the Story': Best Practices of Transnational Indigenous Activism in the North. *The International Journal of Human Rights.* 16(1). pp. 193–215.

Polack, M. E. (2009). La Argentina pide ampliar la plataforma continental. *La Nación.* [Print] 22nd April. Available from: Biblioteca Nacional Mariano Moreno.

Postigo, J. C., Blanco Wells, G., Chacón Cancino, P. (2013). Social Sciences at the Crossroads: Global Environmental Change in Latin America and the Caribbean. In *World Social Science Report.* ISSC/UNESCO: OECD Publishing and UNESCO Publishing. [Online] Available from: http://oecdilibrary.org/social-issues-migration-health/world-social-science-report-2013/social-sciences-at-the-crossroads global-environmental-change-in-latin-ameri9789264203419-19-en. [Accessed: 16th December 2016].

Potapov, E. (2009). *The Scramble for the Arctic: Ownership, Exploitation and Conflict in the Far North.* London: Frances Lincoln.

Potts, T., Schofield, C. (2008). Current Legal Developments: The Arctic. *The International Journal of Marine and Coastal Law.* 23. pp. 151–176.

Potts, T., Schofield, C. (2013). Climate change and evolving regional ocean governance in the Arctic. In Scheiber, H. & Paik, J. (eds.) *Regions, Institutions and The Law of the Sea: Studies in Ocean Governance.* pp. 437–466.

Powell, R. C., Dodds, K. (2014). *Polar Geopolitics: Knowledges, Resources and Legal Regimes.* Aldershot, Brookfield: Elgar.

Press, T. (2014). Foreword: Polar Oceans Governance: An Antarctic Perspective. In Stephens, T. & Vanderzwaag, D. L. (eds.) *Polar Oceans Governance in an Era of Environmental Change*. Cheltenham, Northampton: Edward Elgar. pp. x–xii.

Pries, L. (2007). *Die Transnationalisierung der sozialen Welt*. Frankfurt a. M.: Suhrkamp.

Pries, L. (2010). *Transnationalisierung: Theorie und Empirie grenzüberschreitender Vergesellschaftung*. Wiesbaden: VS Verlag für Sozialwissenschaften.

Putnam, R. D. (1988). Diplomacy and Domestic Politics: The Logic of Two-Level Games. *International Organization*. 42(3). pp. 427–460.

Ramos, C. M. (2008). No tomamos en serio la protección del medio ambiente. *La Nación*. [Print] 26th November. Available from: Biblioteca Nacional Mariano Moreno.

Raussert, W. (2014). Mobilizing 'America/América': Toward Entangled Americas and a Blueprint for Inter-American 'Area Studies'. *Forum for Inter-American Research*. 7(3). pp. 59–97. [Online] Available from: http://.unibielefeld.de/cias/fiar/pdf/073/FIAR073-59-97-Raussert.pdf. [Accessed: 17th September 2015].

Reporter without Borders. (2013). *World Press Freedom Index*. [Online] Available from: http://fr.rsf.org/IMG/pdf/classement_2013_gb-bd.pdf. [Accessed: 18th June 2015].

Reuber, P. (2012). *Politische Geographie*. Stuttgart: UTB.

Reuber, P., Strüver, A., Wolkersdorfer, G. (2012). *Politische Geographien Europas – Annäherungen an ein umstrittenes Konstrukt*. Münster: LIT Verlag.

Reuber, P., Wolkersdorfer, G. (2001). Die neuen Geographien des Politischen und die neue politische Geographie – eine Einführung. In Reuber, P. & Wolkersdorfer, G. (eds.) *Politische Geographie: Handlungsorientierte Ansätze und Critical Geopolitics*. Heidelberg: Selbstverlag des Geographischen Instituts der Universität Heidelberg. pp. 1–17.

Revkin, A. C. (2000). When Will We Be Sure? *New York Times*. [Print] 10th September. Available from: ProQuest Database. [Accessed: 30th March 2014].

Revkin, A. C. (2002). Large Ice Shelf in Antarctica Disintegrates at Great Speed. *New York Times*. [Print] 20th March. Available from: ProQuest Database. [Accessed: 30th March 2014].

Revkin, A. C. (2003a). Experts Conclude Oil Drilling Has Hurt Alaska's North Slope. *New York Times*. [Print] 5th March. Available from: ProQuest Database. [Accessed: 30th March 2014].

Revkin, A. C. (2003b). Huge Ice Shelf Is Reported to Break Up in Canada. *New York Times*. [Print] 23rd September. Available from: ProQuest Database. [Accessed: 30th March 2014].

Revkin, A. C. (2004a). Alaska Thaws, Complicating the Hunt for Oil. *New York Times*. [Print] 13th January. Available from: ProQuest Database. [Accessed: 30th March 2014].

Revkin, A. C. (2004b). As the Arctic Warms. *New York Times*. [Print] 9th November. Available from: ProQuest Database. [Accessed: 30th March 2014].

Revkin, A. C. (2004c). Big Arctic Perils Seen in Warming. *New York Times*. [Print] 30th October. Available from: ProQuest Database. [Accessed: 30th March 2014].

Revkin, A. C. (2006a). By 2040, Greenhouse Gases Could Lead to an Open Arctic Sea in Summers. *New York Times*. [Print] 12th December. Available from: ProQuest Database. [Accessed: 31st March 2014].

Revkin, A. C. (2006b). World Briefing Americas: Canada: After 3,000 Years, Ice Shelf Broke Off. *New York Times*. [Print] 30th December. Available from: ProQuest Database. [Accessed: 31st March 2014].

Rinke, S. (2008). Das politische System Chiles. In Stüwe, K. & Rinke, S. (eds.) *Die politischen Systeme in Nord und Lateinamerika*. Wiesbaden: Verlag für Sozialwissenschaften. pp. 138–166.

Risse-Kappen, T. (1995). *Bringing Transnational Relations Back In: Non-State Actors, Domestic Structures and International Institutions*. Cambridge: Cambridge University Press.

Roberts, A. (2010). Cold as Ice. *Globe and Mail*. [Print] 27th August. Available from: ProQuest Database. [Accessed: 30th March 2014].

Roberts, D. (1998). Disputed Territory. *Globe and Mail*. [Print] 4th July. Available from: ProQuest Database. [Accessed: 30th March 2014].

Rojas, N. L. (2009). A prepararse para un mundo sin glaciares. *El Mercurio*. [Print] 29th July. Available from: Biblioteca Nacional de Chile.

Rojas-Kienzle, D. (2013). Duopol für den Neoliberalismus. *Lateinamerika Nachrichten*. [Online] Available from: http://lateinamerika-nachrichten.de/wp-content/uploads/2015/01/Dossier_Medien_Web.pdf. [Accessed: 18th June 2015].

Rosenau, J. (1969). Towards the Study of National-International Linkages. In Rosenau, J. (ed.) *Linkage Politics: Essays on the Convergence of National and International Systems*. New York: Free Press. pp. 44–63.

Roth, J. (2014). Decolonizing American Studies: Toward a Politics of Intersectional Entanglements. *Forum for Inter-American Research*. 7(3). pp. 135–170. [Online] Available from: http://interamericaonline.org/volume-7-3/roth/. [Accessed: 29th September 2015].

Rother, L. (2005). Antarctica, Warming, Looks Ever More Vulnerable. *New York Times*. [Print] 25th January. Available from: ProQuest Database. [Accessed: 30th March 2014].

Rothwell, D. (2015). The Polar Regions and the Law of the Sea. In Kennedy & Roy, A. C. (eds.) (1992). *La Antártida. Catedral del hielo*. Maidenhead: S.A. McGraw-Hill.

Sale, R., Potapov, E. (2010). *The Scramble for the Arctic: Ownership, Exploitation and Conflicts in the Far North*. London: Frances Lincoln Limited.

Salzman, A. (2006). A Plan for Heating Aid Runs into a Cold Front from Alaska. *New York Times*. [Print] 1st January. Available from: ProQuest Database. [Accessed: 30th March 2014].

Sanata, V. (1997a). Hay mayor conciencia ambiental. *La Nación*. [Print] 5th June. Available from: Biblioteca Nacional Mariano Moreno.

Sanata, V. (1997b). El recalentamiento haría colapsar la barrera de hielos. *La Nación*. [Print] 9th February. Available from: Biblioteca Nacional Mariano Moreno.

Santana, V. (1996). Greenpeace: 'No hay una política ambiental'. *La Nación*. [Print] 15th September. Available from: Biblioteca Nacional Mariano Moreno.

Santiago Declaration. (2016). *Santiago Declaration on the Twenty Fifth Anniversary of the Signing of the Protocol on Environmental Protection to the Antarctic Treaty*. XXXIX Antarctic Treaty Consultative Meeting. [Online] Available from: http://atcm39chile.gov.cl/2016/05/espanol-declaracion-de-santiago-en-ocasion. [Accessed: 6th June 2016].

SATCM. (1990). [Online] Available from: https://www.ats.aq/devAS/ats_meetings.aspx. [Accessed: 30th April 2016].

Scassola, A. (2013). All Is Well in the High North? Contemporary Sources of Tension in the Arctic. *New Global Studies*. 7(2). pp. 183–204.

Schiller, B. (2010). China Warming up to Be an Arctic Player. *Toronto Star*. [Print] 1st March. Available from: ProQuest Database. [Accessed: 31st March 2014].

Schmit, J. (2010). Deep-Water Drilling Hits Still Waters: Catastrophic BP Spill Brakes Industry's Steady Expansion. *USA Today*. [Print] 28th May. Available from: ProQuest Database. [Accessed: 30th March 2014].

Scott, N. (2007). Global Warming Will Hit Poor Countries Harder, Report Says. *Globe and Mail*. [Print] 2nd August. Available from: ProQuest Database. [Accessed: 30th March 2014].

Secretariat of the Antarctic Treaty. (2013). *The Protocol on Environ Mental Protection to the Antarctic Treaty*. [Online] Available from: http://ats.aq/e/ep.htm. [Accessed: 13th August 2015].

Seelye, K. Q. (2001). Facing Obstacles on Plan for Drilling for Arctic Oil, Bush Says He'll Look Elsewhere. *New York Times*. [Print] 30th March. Available from: ProQuest Database. [Accessed: 30th March 2014].

Seguin, R. (2008). Scientists Predict Seasonal Ice-Free Arctic by 2015. *Globe and Mail*. [Print] 12th December. Available from: ProQuest Database. [Accessed: 30th March 2014].

Senado de la Nación. (2014). *Proyecto de ley*. [Online] Available from: http://www.senado.gov.ar/parlamentario/parlamentaria/356138/downloadPdf. [Accessed: 24th November 2016].

Shadian, J. (2010). From States to Polities: Re-Conceptualizing Sovereignty through Inuit Governance. *European Journal of International Relations*. 16(3). pp. 485–510.

Shadian, J. (2014). The Arctic Gaze: Redefining the Boundaries of the Nordic Region. In Soerlin, S. (ed.) *Science, Geopolitics and Culture in the Polar Region: Norden beyond Borders*. Surrey: Ashgate. pp. 259–292.

Shadian, J., Olsen, I. H. (2016). Greenland & the Arctic Council: Subnational Regions in a Time of Arctic Westphalianisation. In Heininen, L., Exner-Pirot, H. & Plouffe, J. (eds.) *The Arctic Yearbook 2016*. Akureyri: Northern Research Forum. pp. 230–250.

Shadian, J., Wirpsa, L. (2006). *Transnational Indigenous and Community Responses to Hydrocarbon Hegemony in the Arctic and the Andes*. Conference Paper, International Studies Association.

Shapiro, M. (1989). Textualizing Global Politics. In Der Derian, D. & Shapiro, M. (eds.) *International/Intertextual Relations*. Lexington: Lexington Books. pp. 11–22.

Sharp, J. P., et al. (2000). Entanglements of Power: Geographies of Domination/Resistance. In Sharp, J. P., Routledge, P., Philo, C. & Paddison, R. (eds.) *Entanglements of Power: Geographies of Domination/Resistance (Critical Geographies)*. London, New York: Routledge. pp. 1–42.

Sheppard, E. (2012). Trade, Globalization and Uneven Development: Entanglements of Geographical Political Economy. *Progress in Human Geography*. 36(1). pp. 44–71.

Shields, R. (1991). *Places on the Margin: Alternative Geographies of Modernity*. London, New York: Routledge.

Sierra, G. (2008). LA ANTARTIDA SE DERRITE (SEGUNDA NOTA) Comenzó la conquista de la Antártida, última 'tierra mítica'. *Clarín*. [Print] 3rd March. Available from: Biblioteca Nacional Mariano Moreno.

Sierra, G. (2012). Argentina debe desarrollar una estrategia para el Atlántico Sur, Malvinas y la Antártida. *Clarín*. [Print] 6th May. Available from: Biblioteca Nacional Mariano Moreno.

Simpson, J. (2009a). It's an Arctic Policy Worth Building On. *Globe and Mail*. [Print] 31st July. Available from: ProQuest Database. [Accessed: 30th March 2014].

Simpson, J. (2009b). Harper Can Command, Obama Can Cajole: But Canada Has So Little to Say. *Globe and Mail*. [Print] 16th September. Available from: ProQuest Database. [Accessed: 30th March 2014].

Simpson, J. (2009c). It Gets Harder to Ignore the Signs of Climate Change. *Globe and Mail*. [Print] 2nd October. Available from: ProQuest Database. [Accessed: 30th March 2014].

Skliarevsky, F. (1997). Antártida, el misterio de hielo. *La Nación*. [Print] 11th May. Available from: Biblioteca Nacional Mariano Moreno.

Smieszek, M., Kankaanpää, P. (2015). Role of the Arctic Council Chairmanship. In Heininen, L. (ed.) *Arctic Yearbook*. Akureyri, Iceland: Northern Research Forum. pp. 1–16.

Smith, H. A. (2010). Choosing Not to See: Canada, Climate Change, and the Arctic. *International Journal*. 65(4). pp. 931–942.

Solanas, P. (2012). Opinión. Una estrategia soberana para Malvinas. *La Nación*. [Print] 6th April. Available from: Biblioteca Nacional Mariano Moreno.

Solberg, H.-M. (2006). *Statement by Ms Hill-Marton Solberg, Chair of the Standing Committee of Parliamentarians of the Arctic Region (SCPAR)*. [Speech] Arctic Council 5th Ministerial Meeting in Salekhard, 25th–26th October. [Online] Available from: https://oaarchive.arctic-council.org/handle/11374/1564. [Accessed: 1st April 2016].

Sollie, F. (1983). The Development of the Antarctic Treaty System. In Wolfrum, R. (ed.) *Antarctic Challenge: Conflicting Interests, Cooperation Environmental Protection, Economic Development*. Berlin: Duncker & Humblot. pp. 17–37.

Spence, J. (2015). The Arctic Council Leadership Merry-Go-Round: Words of Advice as the United States Assumes the Arctic Council Chairmanship. *CIGI Policy Brief*. 55. pp. 1–5.

Speth, J. G. (2006). The Globe Is Warming: Why Aren't We Marching? *New York Times*. [Print] 24th February. Available from: ProQuest Database. [Accessed: 30th March 2014].

Spinelli, O. (2003). Punto de vista. Hora de exigir. *Clarín*. [Print] 17th October. Available from: Biblioteca Nacional Mariano Moreno.

Stavenhagen, R. (2011). Making the Declaration on the Rights of Indigenous Peoples Work: The Challenge Ahead. In Allen, S. & Xanthaki, A. (eds.) *Reflections on the UN Declaration on the Rights of Indigenous Peoples*. Oxford: Hart Publishing Ltd. pp. 147–170.

Stein, R. (1998). Chile y el Cambio Climático. *El Mercurio*. [Print] 7th January. Available from: Biblioteca Nacional de Chile.

Steinberg, P. E., Tasch, J., Gerhardt, H. (2015). *Contesting the Arctic: Politics and Imaginaries in the Circumpolar North*. London, New York: Tauris.

Stepien, A., Koivurova, T., Kankaanpää, P. (2015). The Region of Uncertainty: Arctic Change and Possible Pathways for the EU. In Stepien, A., Koivurova, T. & Kankaanpää, P. (eds.) *The Changing Arctic and the European Union*. Leiden, Boston: Brill Nijhoff. pp. 317–330.

Stevens, W. K. (1997). If Climate Changes, Who Is Vulnerable? *New York Times*. [Print] 30th September. Available from: ProQuest Database. [Accessed: 30th March 2014].

Stickman, M. (2011). *AAC – Remarks by Chief Michael Stickman. Arctic Council Ministerial, May 12th, 2011 Nuuk, Greenland*. [speech] Arctic Council 7th Ministerial Meeting in Nuuk, Greenland, May 12 2011. [Online] Available from: https://oaarchive.arctic-council.org/handle/11374/530. [Accessed: 1st April 2016].

Stickman, M. (2013). *Michael Stickman, International Chair, Arctic Athabascan Council Kiruna, Sweden, May 15, 2013*. [Speech] Arctic Council 8th Ministerial Meeting in Kiruna, Sweden, 15th May. [Online] Available from: https://oaarchive.arctic-council.org/handle/11374/528. [Accessed: 1st April 2016].

Stokke, O. S. (1991). Conference Report: Preservation or Exploitation? *Marine Policy*. March. pp. 137–141.

Stokke, O. S. (2011). Environmental Security in the Arctic: The Case for Multilevel Governance. *Canada's Journal of Global Policy Analysis*. 66(4). pp. 835–848.

Stokke, O. S. (2015). Institutional Complexity in Arctic Governance: Curse or Blessing? In Jensen, L. C. & Hønneland, G. (eds.) *Handbook of the Politics of the Arctic*. Cheltenham, Northampton: Edward Elgar Publishing. pp. 328–351.

Stokke, O. S., Hønneland, G. (2006). *International Cooperation and Arctic Governance: Regime Effectiveness and Northern Region Building*. New York: Routledge.

Strauss, S. (2003). Largest Arctic Ice Shelf Breaks Up, Wiping Out Unique Ecosystem. *Globe and Mail*. [Print] 23rd September. Available from: ProQuest Database. [Accessed: 31st March 2014].

Stüwe, K. (2008). Das politische System der USA. In Stüwe, K. & Rinke, S. (eds.) *Die politischen Systeme in Nord und Lateinamerika*. Wiesbaden: Verlag für Sozialwisenschaften. pp. 540–580.

Subrahmanyam, S. (1997). Connected Histories: Notes towards a Reconfiguration of Early Modern Eurasia. *Modern Asian Studies*. 31(3). pp.735–762.

Sullivan, A. (1995). Big Oil Divides Alaskan Native Groups Plan to Develop Wildlife Refuge Pits Caribou dependent Gwich'in against Neighbours. *Globe and Mail*. [Print] 2nd November. Available from: ProQuest Database. [Accessed: 30th March 2014].

Sulyandziga, R. (2013). *Raipon Statement. Rodion Sulyandziga. 15 May 2013, Kiruna, Sweden, Arctic Council, Ministerial Meeting*. [Speech] Arctic Council 8th Ministerial Meeting in Kiruna, Sweden, 15th May. [Online] Available from: https://oaarchive. arctic-council.org/handle/11374/528. [Accessed: 1st April 2016].

Suzuki, D. (1989). More 'Valdez' Disasters May Await. *Globe and Mail*. [Print] 27th May. Available from: ProQuest Database. [Accessed: 31st March 2014].

Suzuki, D. (2006). Ignoring the Canaries; in the Year 2020, the World Is Paying the Price for Its Indifference to the Failing Health of the Environment. *Toronto Star*. [Print] 30th December. Available from: ProQuest Database. [Accessed: 30th March 2014].

Suzuki, D. (2010). UN Climate Talks: Who Gives a Damn? *Toronto Star*. [Print] 12th December. Available from: ProQuest Database. [Accessed: 30th March 2014].

Sylvester, S. (2009). We're Not Waiting for Our Politicians to Think Globally. *Globe and Mail*. [Print] 2nd March. Available from: ProQuest Database. [Accessed: 30th March 2014].

Tadjdeh, Y. (2016). Coast Guard: We Are Increasingly 'Concerned' about China's Arctic Ambitions. *Business Insider*. [Online] 16 June 2016. http://businessinsider.com. au/coast-guard-we-are-increasingly-concerned-aboutchinas-arctic-. [Accessed: 15th August 2016].

Tanaka, Y. (2014). Reflections on Transboundary Air Pollution in the Arctic: Limits of Shared Responsibility. *Nordic Journal of International Law*. 83(3). pp. 213–250.

Taylor, C. (2004). *Modern Social Imaginaries*. Durham: Duke University Press.

Tèlam, A. (2013). El Gobierno denunció que Gran Bretaña quiere controlar la Antártida desde Malvinas. *La Nación*. [Print] 28th January. Available from: Biblioteca Nacional Mariano Moreno.

Tennberg, M. (2000). *Arctic Environmental Cooperation: A Study in Governmentality*. Surrey: Ashgate.

Tennberg, M. (2015). Arctic Change through a Political Reading. In Jensen, L. C. & Hønneöamd, G. (eds.) *Handbook of the Politics of the Arctic*. Northampton: Edward Elgar Publishing. pp. 408–420.

The Guardian. (2013). China Eyes Antarctica's Resource Bounty. *The Guardian*. [Online] 8th November. Available from: http://theguardian.com/environment/2013/nov/08/ china-antarctica-trip-icebreaker-snow-dragon. [Accessed: 2nd April 2016].

Tinati, R., Halford, L., Carr, C. P. (2014). Big Data: Methodological Challenges and Approaches for Sociological Analysis. *Sociology*. 48(4). pp. 663–681.

Tittor, A. (2014). *Transnational Entanglements and the Transformation of Gender Relations in the Americas*. Presentation. Congress of the Latin American Studies Association. Chicago, IL, 21st–24th May.

Tokatlian, J. G. (2013). Argentina y Chile, unidos en la Antártida. *Clarín*. [Print] 10th September. Available from: Biblioteca Nacional Mariano Moreno.

Toronto Star. (1989). Native Groups Oppose Export of Arctic Gas. [Print] 11th April. Available from: ProQuest Database. [Accessed: 30th March 2014].

Toronto Star. (1990a). Our Fading Claim to Arctic Waters. [Print] 9th September. Available from: ProQuest Database. [Accessed: 30th March 2014].

Toronto Star. (1990b). MP Attack Ottawa's Refusal to Apologize to Relocated Inuit. [Print] 22nd November. Available from: ProQuest Database. [Accessed: 30th March 2014].

Toronto Star. (1990c). Inuit Seeking Apology for Forced Move to Far North. [Print] 2nd April. Available from: ProQuest Database. [Accessed: 30th March 2014].

Toronto Star. (1991a). Animal-Rights Movement Hurting Inuit, Activist Says. [Print] 19th June. Available from: ProQuest Database. [Accessed: 30th March 2014].

Toronto Star. (1991b). Global Warming 'Real,' Ottawa Says. [Print] 19th December. Available from: ProQuest Database. [Accessed: 30th March 2014].

Toronto Star. (1993a). Better Protection for Our Arctic. [Print] 18th March. Available from: ProQuest Database. [Accessed: 30th March 2014].

Toronto Star. (1993b). Apologize to Inuit. [Print] 21st March. Available from: ProQuest Database. [Accessed: 30th March 2014].

Toronto Star. (1998). Canada Leads Way in Policy for North: Arctic Countries Meet to Solve Area's Woes, Plan Development. [Print] 18th September. Available from: ProQuest Database. [Accessed: 30th March 2014].

Toronto Star. (1999). Birth of the Inuit Homeland. [Print] 28th March. Available from: ProQuest Database. [Accessed: 30th March 2014].

Toronto Star. (2004). Enjoy the Arctic While It's Still Here. [Print] 18th November. Available from: ProQuest Database. [Accessed: 30th March 2014].

Toronto Star. (2008). Who Owns Resource-Rich Arctic? Map Blending Geological Data, International Law Off a New Look at Overlapping Claims. [Print] 7th August. Available from: ProQuest Database. [Accessed: 30th March 2014].

Transport Canada. (2015). *Tanker Safety and Spill Prevention*. [Online] Available from: http://tc.gc.ca/eng/marinesafety/menu-4100.htm. [Accessed: 6th February 2018].

Travers, J. (2010). Conservatives Push the Politics of Fear. *Toronto Star*. [Print] 14th December. Available from: ProQuest Database. [Accessed: 30th March 2014].

Triggs, G. D., Riddell, A. (eds.). (2007). *Antarctica: Legal and Environmental Challenges for the Future*. London: British Institute for International & Compara.

Truedsson, C. G. R. (2013). *Russia: The 'Other' in the Arctic? A Critical Geopolitical Analysis of Western Discursive Representations of the Russian Federation's Actions towards the Arctic*. Master thesis. Edinburgh: The University of Edinburgh.

Turner, J. (2005). Killing the Drilling Is a Must. *Globe and Mail*. [Print] 16th August. Available from: ProQuest Database. [Accessed: 30th March 2014].

UNCLOS. (2016). Submissions, through the Secretary-General of the United Nations, to the Commission on the Continental Shelf, Pursuant to Article 76, Paragraph 8, of the United Nations Convention on the Law of the Sea of 10 December 1982. [Online] Available from: http://un.org/depts/los/clcs_new/commission_ submissions.htm. [Accessed: 5th February 2016].

United Nations. (2013). *United Nations Convention on the Law of the Sea*. [Online] Available from: http://un.org/depts/los/conventionagreements/texts/unclos/unclose.pdf. [Accessed: 5th February 2016].

UN Declaration on the Rights of Indigenous Peoples. (2007). [Online] Available from: https://www.un.org/development/desa/indigenouspeoples/declaration-on-the-rights-of-indigenous-peoples.html. [Accessed: 30th April 2016].

United States Coast Guard. (2013). *Arctic Strategy*. [Online] Available from: https://uscg.mil/seniorleadership/docs/cgarctic_strategy.pdf. [Accessed: 23rd November 2016].

Urquidi Fell, J. C. (1998). Compensaciones Ambientales. *El Mercurio*. [Print] 27th January. Available from: Biblioteca Nacional de Chile.

USA Today. (1990). Nation Needs the Oil: Drill Where the Oil Is. [Print] 23rd October. Available from: ProQuest Database. [Accessed: 30th March 2014].

USA Today. (2005). Arctic Drilling Makes Sense, But Don't Expect Miracles. [Print] 23rd March. Available from: ProQuest Database. [Accessed: 30th March 2014].

U.S. Arctic Chairmanship Program. (2001). Submitted by Julia L. Gourley, U.S. Senior Arctic Official. Personal Communication June 2, 2016.

United States Coast Guard. (2013). *Arctic Strategy*. [Online] Available from: https://www. uscg.mil/Portals/0/Strategy/cg_arctic_strategy.pdf. [Accessed: 24th November 2016].

U.S. Department of Defense. (1996). *PDD/NSC-26: US Antarctica Policy*. [Online] Available from: https://fas.org/irp/offdocs/pdd26.htm. [Accessed: 17th September].

U.S. Department of Defense. (2013). *Arctic Strategy*. [Online] Available from: http://archive.defense.gov/pubs/2013_Arctic_Strategy.pdf. [Accessed: 30th April 2016].

U.S. Department of State. (2015). *One Arctic: U.S. Chairmanship Arctic Council 2015–2017: Chairmanship Brochure*. [Online] Available from: http://arcticcouncil.org/images/PDF_attachments/US_Chairmanship/Chairmanship_Brochure_2_page_public. pdf. [Accessed: 5th February 2016].

U.S. Department of State. (2016). *Antarctic Treaty*. [Online] Available from: http://state. gov/t/avc/trty/193967.htm. [Accessed: 5th February 2016].

U.S. Geological Survey. (2008). *Circum-Arctic Resource Appraisal: Estimates of Undiscovered Oil and Gas North of the Arctic Circle*. [Online] Available from: http:// pubs. usgs.gov/fs/2008/3049/fs2008-3049.pdf. [Accessed: 6th February 2018].

U.S. Geological Survey (2016). *GIS Representation of Coal-Bearing Areas in Antarctica*. [Online] Available from: http://energy.usgs.gov/Coal/AssessmentsandData/Coal Assessments.aspx. [Accessed: 19th March 2016].

U.S. Government Publishing Office. (2011). *15 U.S.C. 4111: Arctic Defined*. [Online] Available from: http://gpo.gov/fdsys/granule/USCODE-2010-title15USCODE-2010-title15-chap67-sec4111. [Accessed: 6th February 2018].

U.S. National Security Strategy. (2010). [Online] Available from: http://nssarchive.us/ national-security-strategy-2010/. [Accessed: 30th April 2016].

Valle, M. (1990). En Reunión Internacional: Chile Propondrá un Protocolo de Defensa Ecológica Antártica. *El Mercurio*. [Print] 10th November. Available from: Biblioteca Nacional de Chile.

Verhovek, S. H. (2001). Drill, Say Alaskans, Who Know Their Pockets Are Lined with Oil. *New York Times*. [Print] 18th March. Available from: ProQuest Database. [Accessed: 30th March 2014].

Voeten, E. (2014). Does Participation in International Organizations Increase Cooperation? *The Review of International Organizations*. 9(3). pp. 285–308.

Waghorn Gallegos, R. (2007). *Análisis y Proyecciones de la Política Antártica Nacional. Universidad de Santiago de Chile*. [Online] Available from: http://hemisfericosypolares.cl/ tesis-premiomorla/Tesis%20Rodrigo%20Waghorn.pdf. [Accessed: 9th December 2013].

Wald, M. L. (1989). Oil Means Comfort to Alaska Natives But Peril to Their Culture. *New York Times*. [Print] 1st May. Available from: ProQuest Database. [Accessed: 30th March 2014].

Walkom, T. (1999). Our Land: Nunavut, the Inuit Homeland Carved Out of the Eastern Arctic. *Toronto Star*. [Print] 28th March. Available from: ProQuest Database. [Accessed: 30th March 2014].

252 *Bibliography*

Warrick, J. (1997). Arctic Implicates Humans in Warming. *The Washington Post.* [Print] 14th November. Available from: ProQuest Database. [Accessed: 30th March 2014].

Watt-Cloutier, S. (2006). Don't Abandon the Arctic to Climate Change. *Globe and Mail.* [Print] 24th May. Available from: ProQuest Database. [Accessed: 30th March 2014].

Weber, B. (2008). Cracks in Arctic Ice Shelf Signals Its Demise: 'Map of Canada Has Changed,' Scientist Say after Ranger See Collapse of Polar Landmark First-Hand. *Toronto Star.* [Print] 12th April. Available from: ProQuest Database. [Accessed: 30th March 2014].

Weber, B. (2010). Canada's Arctic Meltdown Grows at Alarming Pace. *Toronto Star.* [Print] 7th January. Available from: ProQuest Database. [Accessed: 30th March 2014].

Wehrmann, D. (2016). Shaping Changing Circumpolar Agendas: The Identification and Significance of 'Emerging Issues' Addressed in the Arctic Council. In Heininen, L. (ed.) *The Arctic Yearbook 2016.* Akureyri, Iceland: Northern Research Forum. pp. 90–103.

Wehrmann, D. (2017). Non-State Actors in Arctic Council Governance. In Keil, K. & Knecht, S. (eds.) *Governing Arctic Change: Global Perspectives.* Basingstoke: Palgrave Macmillan. pp. 187–206.

Weitzman, C. B. (2008). La Antártica y su influenca en el cambio climático. *Revistamar.* 4. pp. 313–316.

Welch, D. A. (2013). East Asia-Arctic Relations: Boundary, Security and International Politics. *CIGI Policy Brief.* 6. pp. 1–8.

Werner, M., Zimmermann, B. (2002). Vergleich, Transfer, Verflechtung. Der Ansatz der Histoire croisée und die Herausforderung des Transnationalen. *Geschichte und Gesellschaft.* 28(4). pp. 607–636.

Westermeyer, W. E., Shusterich, K. M. (1984). *United States Arctic Interests: The 1980s and 1990s.* New York: Springer Verlag.

The Arctic Institute. (2016). *A Quick Start Guide to the Law of the Seas in the Arctic.* [Online] Available from: https://drive.google.com/file/d/0B_FFtdNAwiw AOE5kQTZEQkJTN0k/view?pref=2&pli=1. [Accessed: 2nd April 2016].

The White House. (1994). *Presidential Decision Directive/NSC-26: United States Policy on the Arctic and Antarctic Regions.* [Online] Available from: https://fas.org/irp/offdocs/pdd/pdd-26.pdf. [Accessed: 17th September 2016].

The White House. (2009). *Arctic Region Policy: NSPD 66/HSPD 25.* [Online] Available from: http://nsf.gov/geo/plr/opp_advisory/briefings/may2009/nspd66_hspd25.pdf. [Accessed: 17th September 2016].

The White House. (2013a). *Working Together to Understand and Predict Arctic Change.* [Online] Available from: https://whitehouse.gov/blog/2013/02/19/working-together-understand-and-predict-arctic-change. [Accessed: 22nd January 2016].

The White House. (2013b). *National Strategy for the Arctic Region.* [Online] Available from: https://whitehouse.gov/sites/default/files/docs/ nat_arctic_strategy.pdf. [Accessed: 17th September 2016].

The White House. (2014). *Implementation Plan for the National Strategy for the Arctic Region.* [Online] Available from: https://whitehouse.gov/sites/default/files/docs/implementation_plan_for_the_national_strategy_ for_the_arctic_region_-_fi. . . . pdf. [Accessed: 23rd November 2016].

The White House. (2015). *Executive Order: Enhancing Coordination of National Efforts in the Arctic.* [Online] Available from: https://whitehouse.gov/the-press-office/2015/01/21/executive-order-enhancingcoordination-nat. [Accessed: 23rd November 2016].

Warrick, J. (1998). Higher Oil Estimate May Boost Pressure to Tap Alaska Refuge. *The Washington Post.* [Print] 16th May 1998. Available from: ProQuest Database. [Accessed: 30th March 2014].

Wilson, D. (1993). Russia Confirms Nuclear Dumping Greenpeace Says Radioactivity High. *Globe and Mail*. [Print] 24th March. Available from: ProQuest Database. [Accessed: 30th March 2014].

Winson, T. (2010). Drilling Is Desirable: To Whom? *Globe and Mail*. [Print] 6th May. Available from: ProQuest Database. [Accessed: 30th March 2014].

Wodak, R., Meyer, M. (2009). Critical Discourse Analysis: History, Agenda, Theory, and Methodology. In Wodak, R. & Meyer, M. (eds.) *Methods for Critical Discourse Analysis*. 2nd Ed. London: Sage. pp. 1–33.

Wolfrum, R. (1983). The Principle of the Common Heritage of Mankind. *Zeitschrift für ausländisches öffentliches Recht und Völkerrecht*. 43. pp. 312–337.

Woods, A. (2011). Staking Claim to an 'Arctic Renaissance'. *Toronto Star*. [Print] 20th August. Available from: ProQuest Database. [Accessed: 30th March 2014].

Wright, J. R. (2013). Keynote Speech: 'Canada's Arctic Foreign Policy'. In Berkman, P. A. & Vylegzhanin, A. N. (eds.) *Environmental Security in the Arctic Ocean*. Cambridge: Springer. pp. 103–108.

WWF. (2011). *WWF Statement to the Nuuk Ministerial Meeting of the Arctic Council May 12, 2011*. [Speech] Arctic Council 7th Ministerial Meeting in Nuuk, Greenland, May 12 2011. [Online] Available from: https://oaarchive.arctic-council.org/handle/11374/530. [Accessed: 1st April 2016].

Young, O.R. (1989). *International Cooperation. Building Regimes for Natural Resources and the Environment*. Ithaca and London: Cornell University Press.

Young, O. R. (1998). *Creating Regimes: Arctic Accords and International Governance*. Ithaca, New York: Cornell University Press.

Young, O. R. (2005). Governing the Arctic: From Cold War Theater to Mosaic of Cooperation. *Global Governance*. 11(1). pp. 9–15.

Young, O. R. (2010). Arctic Governance: Pathways to the Future. *Arctic Review on Law and Politics*. 2(1). pp. 164–185.

Young, O. R. (2011). The Future of the Arctic: Cauldron of Conflict or Zone of Peace. *International Affairs*. 87(1). pp. 185–193.

Young, O. R. (2012a). Arctic Tipping Points: Governance in Turbulent Times. *AMBIO*. 41. pp. 75–84.

Young, O. R. (2012b). Arctic Politics in an Era of Global Change. *Brown Journal of World Affairs*. 19(1). pp. 165–178.

Young, O. R., Osherenko, G. (1993). *Polar Politics: Creating International Environmental Regimes*. Ithaca: Cornell University Press.

Zebich-Knos, M. (2015). Chapter 5: Managing Polar Policy through Public and Private Regulatory Standards: The Case of Tourism in the Antarctic. In Ali, S. H. & Pincus, R. (eds.) *Diplomacy on Ice: Energy and the Environment in the Arctic and Antarctic*. New Haven, London: Yale University Press. pp. 94–110.

Zegras, C. (1998). Negociaciones de Kioto. *El Mercurio*. [Print] 9th January. Available from: Biblioteca Nacional Mariano Moreno.

Zellen, B. S. (1993). Risks and Promises in the New North Nunavut. *Globe and Mail*. [Print] 28th May. Available from: ProQuest Database. [Accessed: 30th March 2014].

Zellen, B. S. (2009a). *Arctic Doom, Arctic Boom: The Geopolitics of Climate Change in the Arctic*. Santa Barbara: Praeger.

Zellen, B. S. (2009b). *On Thin Ice: The State, and the Challenge Sovereignty*. Maryland, Plymouth: Lexington Books.

Index

Made in the USA
Las Vegas, NV
28 January 2022